The Surgeon's Guide to Antimicrobial Chemotherapy

By

John M. B. Smith MSc PhD
Associate Professor, Department of Microbiology, University of Otago, Dunedin, New Zealand

John E. Payne MS FRCS(Ed) FRACS FACS
Professorial Surgical Unit, Repatriation General Hospital, Concord, NSW, Australia

Thomas V. Berne MD
Professor of Surgery, University of Southern California Medical School, Los Angeles, CA, USA

A member of the Hodder Headline Group
LONDON
Co-published in the United States of America by Oxford University Press Inc., New York

First published in Great Britain in 2000 by
Arnold, a member of the Hodder Headline Group,
338 Euston Road, London NW1 3BH

http://www.arnoldpublishers.com

Distributed in the United States of America by
Oxford University Press Inc.,
198 Madison Avenue, New York, NY10016
Oxford is a registered trademark of Oxford University Press

© 2000 John M. B. Smith, John E. Payne and Thomas V. Berne

All rights reserved. No part of this publication may be reproduced or transmitted in any form or by any means, electronically or mechanically, including photocopying, recording or any information storage or retrieval system, without either prior permission in writing from the publisher or a licence permitting restricted copying. In the United Kingdom such licences are issued by the Copyright Licensing Agency: 90 Tottenham Court Road, London W1P 9HE.

Whilst the advice and information in this book are believed to be true and accurate at the date of going to press, neither the authors nor the publisher can accept any legal responsibility or liability for any errors or omissions that may be made. In particular (but without limiting the generality of the preceding disclaimer) every effort has been made to check drug dosages; however it is still possible that errors have been missed. Furthermore, dosage schedules are constantly being revised and new side-effects recognized. For these reasons the reader is strongly urged to consult the drug companies' printed instructions before administering any of the drugs recommended in this book.

British Library Cataloguing in Publication Data
A catalogue record for this book is available from the British Library

Library of Congress Cataloging-in-Publication Data
A catalog record for this book is available from the Libary of Congress

ISBN 0 340 74196 1

1 2 3 4 5 6 7 8 9 10

Commissioning Editor: Nick Dunton
Production Editor: Julie Delf
Production Controller: Sarah Kett

Typeset in 10/13pt Sabon by Saxon Graphics, Derby.
Printed and bound in Great Britain by MPG Books Ltd, Bodmin, Cornwall

The Surgeon's Guide to

Antimicrobial Chemotherapy

CONTENTS

Preface ix
Part One   General Considerations
1. Surgical infection 3
    History 3
    Principal factors in surgical site infection 4
    Host response/abscesses 11
2. Biology of bacteria and related microbes 14
    General properties of bacteria 14
    Bacterial cell wall 15
    Bacterial growth kinetics 16
    Genetic change/gene transfer 18
    Normal flora 20
    Changes in normal flora 20
Part Two   Antimicrobial Agents
3. Antibacterial agents – general information 25
    History 25
    Mechanisms of action 27
    Bacterial resistance to antibiotics 31
    Treatment strategies and principles 49
4. Selected antibacterial agents 67
    β-Lactams 68
    Aminoglycosides 88
    Quinolones 94
    Anti-anaerobe agents 98
    Glycopeptides 101
    Fusidic acid 104
    Rifamycins 104
    Trimethoprim and cotrimoxazole 105
    Macrolides 106
    Tetracyclines 107
    Investigational antimicrobial agents 107
    Topical antibacterial agents 108
5. Antibacterial prophylaxis in surgery 111
    General principles 111
    Appendectomy 114
    Biliary surgery 114
    Clean procedures 115
    Invasive investigational procedures 115

		Route and timing of antibiotic administration	115
		Possible prophylactic antibiotic choices and regimens	117
		Conclusions	121
		Selective decontamination of the digestive tract	122
		Organ transplant patients	123
	6.	Antifungal agents	124
		Mode of action	124
		Range of antifungal agents	126
		Treatment of selected fungal diseases	136
		Resistance to antifungal agents	146
	7.	Other antimicrobial agents	148
		Antiviral agents	148
		Antiparasitic agents	156
	8.	Laboratory aspects of antimicrobial use	160
		Antimicrobial sensitivity testing	160
		Serum assays on patients during antibiotic therapy	164
Part Three	**Surgical Infections**		
	9.	Neurosurgical infections	169
		Brain abscess	169
		Epidural abscess	172
		Intramedullary abscess of the spinal cord	173
		Surgical site infection following spinal infusion	174
	10.	Infections of the head and neck	175
		Sinusitis	175
		Face	175
		Oral cavity	176
	11.	Cardiovascular infections	179
		Vascular graft infection	179
		Intravenous and central catheter-related infections	181
	12.	Chest infections	183
		Pulmonary cavity (abscess)	183
	13.	Intra-abdominal infections	186
		Peritonitis	186
		Liver abscess/biliary tract infections	203
		Antibiotic-associated colitis	206
		Renal abscess	208
		Gynaecological infections	209
	14.	Skin and soft-tissue infections of surgical importance	211
		Necrotizing soft-tissue infections	211
		Diabetic foot infections	218
		Bite and puncture wounds	221
	15.	Locomotor/orthopaedic infections	225
		Osteomyelitis	225
		Prosthetic infections	233
		Microbiological aspects of large-joint sepsis	236

16.	Infection in the immunocompromised and transplant patient	**238**
	Host barriers to infection	238
	Bacterial infections	240
	Viral infections	242
	Fungal infections	244
	Toxoplasmosis	245
References		**246**
Index		**275**

PREFACE

The development of antimicrobial drugs has been one of the most important achievements of medicine in the twentieth century. Antimicrobial agents have enabled the provision of a scope of surgical therapy which is only limited by technology, and which is attended by morbidity and mortality no longer mainly related to sepsis or septic complications. There has been a remarkable evolution in the treatment of surgical infection for the betterment of patient outcome, which has only recently been tempered by the emerging problems of antimicrobial resistance.

For optimal treatment to be initiated, and for informed evaluation of the current and future numbers of anti-infective drugs, the surgeon should fully understand the microbiological and pharmacological principles from a clinical perspective. Surgeons are major users of antimicrobial agents, and infections represent a significant component of surgery and surgical critical care.

A recent survey of the Surgical Infectious Disease Society (North America) membership suggested that special credentialling of surgeons in the area of infectious disease was unnecessary (Cohn and Fisher, 1996). We consider that the management of sepsis has such an essential role in surgical critical care that knowledge of fundamental mechanisms and therapy is obligatory.

During the preparation of examination material for the Basic Clinical Sciences Examination of the Royal Australasian College of Surgeons (RACS) in the mid-1990s, it became obvious to two of the authors (J.M.B.S. and J.E.P.) that it was impossible readily to source general texts and/or other collated reading material pertaining to antimicrobial agents and their use in the practice of surgery. At around the same time, it also became obvious that many colleges were now initiating training and education programmes, in addition to their more historical examining roles. The need for a single text containing information on the fundamentals of antimicrobial agents and their use in the prophylaxis and treatment of surgery-related microbial disease thus became obvious.

We believe that the following short text will provide both trainee and practising surgeons with the rationale for some of the critical decisions to be made when choosing an antimicrobial agent for the treatment or prevention of infection in surgical patients.

JOHN M. B. SMITH
JOHN E. PAYNE
THOMAS V. BERNE

PART ONE

GENERAL CONSIDERATIONS

chapter 1

SURGICAL INFECTION

What is surgical infection? The most important type is a deep infection which usually results from perforation of a hollow abdominal viscus containing multiple pathogenic microbes often acting in synergy. This results in proliferation and possible systemic invasion of virulent microbes with release of growth products from a very large mesothelial cavity. Surgical infection may also follow a blunt or penetrating injury, to which the trauma of a surgical operation is added. Furthermore, the above type of infection may be accompanied by such an overwhelming inflammatory response that these multiple insults result in multi-organ system failure. Alternatively, the infections are localized and amenable to surgical drainage, or occur post-operatively.

HISTORY

The Edwin Smith Surgical Papyrus (C. 1500 BC) contains case histories of 48 patients who had wounds or fractures of varying severity (see Breasted, 1930). Clean wounds were managed by bandaging with fresh meat, followed by daily dressings with grease and honey. These dressings might have restricted major exogenous contamination or, in the case of honey, possessed antimicrobial properties. One of the patients had tetanus, another had spreading (possibly streptococcal) cellulitis, and yet another had multiple pustular swellings over the breast. Various leaves, salt, green pigment (possibly copper sulphate) and dung were used for treatment, while the pustular wounds were opened by cautery and abscesses were lanced with sharpened reeds. The general recommendation about abscesses was that they were to point before drainage.

Circumventing the natural method for cure of appendiceal abscesses led Abraham Groves to perform the first appendicectomy on 10 May 1883. Prior to this (in May 1847), Semmelweiss (a Hungarian surgeon/gynaecologist) first realized that surgical infections were transmissible after noting a significantly elevated rate of puerperal sepsis among women under the care of surgeons who had earlier examined women at necropsy. By altering the daily routine of rounds and initiating hand disinfection with 'chlorina liquida' (a solution of sodium hypochlorite) or the cheaper chlorinated lime, he was able to reduce the puerperal sepsis rate dramatically (see Rotter, 1998). However, the benefit of handwashing was earlier apparent to Fling (AD 23–79), who claimed it was a remedy for 'auger' (redness) in wounds, although it is clear that, in the clinical situation, handwashing may not always be sufficiently effective in reducing the risk of infection-generating microbial transfer.

In the 1860s, Joseph Lister, who realized the implications of Pasteur's germ theory, initiated antiseptic practice by showing that disinfection of the surgical environment with carbolic acid reduced subsequent sepsis rates. Although his ideas were only slowly accepted, the implementation of aseptic technology has probably reduced the mortality associated with surgery more than any other single development.

PRINCIPAL FACTORS IN SURGICAL SITE INFECTION

While complete elimination of wound infections is an unattainable objective, clearly it is possible to reduce infection to a level which has minimal impact on patient mortality and morbidity and institutional resources. In 1992, the Surgical Wound Infection Task Force (including representatives from the Society for Hospital Epidemiology of America, the Association for Practitioners in Infection Control, the Centers for Disease Control and the Surgical Infection Society) suggested that wound infections should be renamed 'surgical site infections' (see Sawyer and Pruett, 1994; Malangoni, 1998). They considered the term 'wound infection' to be confusing in that it could encompass superficial as well as deep infections, and they favoured the use of 'surgical site infection', which could be either superficial or deep, or involve an organ or space underlying the incision. While it seems that acceptance of the term 'surgical site infection' is gathering momentum, much debate still occurs regarding what in fact constitutes significant infection, and what level of post-operative infection actually does occur. The latter is complicated by the fact that infections occur at variable times after the operation depending on a variety of factors, including the microbe involved, the presence or absence of a foreign implant, and the host's response to the invading microbe.

ASEPTIC TECHNIQUE – AEROSOLS

The importance of sterile and no-touch techniques in minimizing post-operative surgical site infection is long established and accepted. However, it seems that a number of other operating-room rituals, that have been religiously considered to be significant, may in fact have little influence on subsequent infection rates. The fact that a procedure is a ritual does not by itself imply that it is 'good' or 'bad', nor does it imply that it should persist or be eliminated (Quebbeman, 1996). Protection of the patient is only one aspect of infection control, and clearly at least as significant in the present-day era of HIV and other blood-borne pathogens is protection of the health-care worker from microbes carried by the patient. Items such as masks, gowns and overshoes should be used, and should be designed with this in mind. In addition, it seems that most surgeons underestimate the risks both of eye-splash injury in surgery (Marasco and Woods, 1998) and of blood-borne pathogens, and do not routinely use double gloves (Patterson et al., 1998).

PATIENT FACTORS

What therefore are the significant factors that independently influence post-operative surgical site infection rates (Table 1.1). Three proven and accepted influences are the degree of microbial contamination during the operation, the duration of the operation, and patient physiological status as measured by parameters such as the Acute Physiology and Chronic Health Evaluation-II (APACHE-II) score.

MICROBIAL CONTAMINATION

In 1964, The National Research Council *ad hoc* Committee on Trauma introduced their now legendary wound classification system, which has repeatedly been shown to have a strong correlation with wound infection rates. Peter Cruse and Rosemary Foord at the Foothills Hospital in Calgary, Canada, were responsible for one of the earliest and largest prospective studies of wound infections. Their studies reported in 1973 involved almost 24 000 wounds over a 5-year period, and in 1980 almost 63 000 wounds over a 10-year period (Sawyer and Pruett, 1994).

Table 1.1 Factors associated with an increased risk of surgical site infection

Major Proven Risk Factors
   Degree of microbial contamination (surgical wound class)
   Length of operation
   Pre-existing illness (comorbidity)[a] – infection rate increases with increasing APACHE-II or ASA pre-operative score

Other Proven Risk Factors
1 Patient-related factors
      Increasing age
      Obesity

2 Surgery-related factors
      Razor shaving of operation site
      Poor tissue perfusion/circulation
      Airborne contamination
      Emergency procedure (possible or probable contamination)

Other Suggested But Largely Unproven Risk Factors
   Length of pre-operative hospitalization ⎫
   Re-operation   ⎬ possibly related to pre-existing illness
   Glove punctures, powder
   Time of day (highest rate is in operations performed between midnight and 8.00 a.m.)
   Month of year (infection rates are highest in the summer)
   Use of adhesive drapes
   Use of wound irrigation

[a] More significant examples include diabetes mellitus, chronic obstructive pulmonary disease (especially sternal wounds in CABG patients), advanced malignancy, concurrent infection at a remote site (e.g. bladder) and malnutrition.

They were able to confirm the importance of the extent of wound contamination as a predictor of subsequent sepsis. Gross contamination produces appreciable post-operative infection.

The Study on the Efficacy of Nosocomial Infection Control (SENIC) risk index for wound infections was largely based on the survey of Haley *et al.* (1985). This index was subsequently modified by Culver *et al.* (1991) and validated by a study of 84 691 surgical procedures in 44 hospitals from 1987 to 1990. The modified index utilized the American Society of Anesthesiologists (ASA) pre-operative score, rather than the earlier category of the number of medical diagnoses at the time of release from hospital.

WOUND SIZE

While many other patient-, surgeon- or environment-related factors have been suggested, including age, pre-existing illness (e.g. malignancy, diabetes mellitus, remote infections), malnutrition, obesity, length of pre-operative hospitalization, body site involved, previous cigarette smoking, glove punctures, emergency or elective procedure, time of day or year, use of adhesive drapes, peritoneal or wound irrigation/lavage, these invariably do not withstand rigorous statistical evaluation as independent influencing factors (see Sawyer and Pruett, 1994). In a study of wound infections after saphenous vein harvest for coronary revascularization, the only variable that was found to be significantly associated with subsequent leg wound infection was the length of the wound (Wong *et al.*, 1997).

SKIN SHEDDING

In cases where hair removal is necessary, the use of a depilatory or clipping device immediately prior to operation is clearly better than early clipping or any form of razor shaving. Staff with overt infections should not be allowed access to operating areas. It is also apparent that staff may carry microbes into the operating-theatre from some external source (e.g. pet animals), and that these microbes may contaminate the operating-theatre environment and result in subsequent sepsis. Surgeons should also not shower in the change room prior to operating, as this may result in large numbers of bacteria being shed from the skin shortly afterwards. Surgeons should use a brush only for cleaning their fingernails when preparing their hands for surgery, and brushes should not be used on the skin, where they may create injury and predispose to increased microbial colonization. The availability of DNA (molecular) typing techniques has greatly facilitated investigations into potential reservoirs of post-operative surgical infections.

BLOOD SUPPLY

Adequate perfusion of wounds is essential for healing, with the adequate delivery of oxygen (which has been described as an unappreciated 'antibiotic') and neutrophils being major requirements. It seems that the activity and attraction of neutrophils is likely to be impaired in the normal wound environment where oxygen levels are low and the pH is acidic. The exacerbation of microbial growth in the presence of foreign bodies (e.g. sutures) may be partly explained by a decreased oxygen tension. Studies that support the use of hyperoxia to aid wound healing and reduce infection are gradually appearing in the literature (Heimbach, 1993; Thom, 1993; Morgan et al., 1995).

POST-OPERATIVE DRESSINGS

While it is generally accepted that the surgical site is only at risk of microbial contamination and subsequent infection during the surgical procedure, it is apparent that even when the wound is closed, microbes may gain access to the underlying tissues. It has been demonstrated that post-operative bathing at the bedside can lead to contamination of tissues despite the presence of apparently occlusive dressings. Wherever possible, wounds should be left alone for at least 48 h post-operatively.

LATENT INTERVAL

Another interesting fact that has emerged in recent years is that most wound infections occur after the patient has been discharged from hospital, and may take from weeks to months to manifest themselves (Santos et al., 1997). In addition, the classical risk factors associated with surgical site infections do not appear to apply to post-discharge wound infections (Lecuona et al., 1998). Clearly we still have much to learn about microbial attachment and invasion, host-defence mechanisms, and the effect of prophylactic antibiotics (see p. 111) and microbial virulence with regard to surgery-related infections. Proper evaluation of these issues mandates appropriate patient review (Poulsen and Meyer, 1996).

EXOGENOUS SOURCES

As already stated, surgical infections result from microbial contamination following an invasive procedure, trauma or perforation of a viscus. The infection is usually by endogenous microbes; occasionally causal micro-organisms come from the environment, as demonstrated in an outbreak of multi-resistant *Pseudomonas aeruginosa* infection associated with contaminated sinks/water in a neurosurgery intensive-care unit (Bert et al., 1998). Numerous other examples of contaminated water being a source of *Pseudomonas aeruginosa* have recently appeared in the literature (e.g. Ferroni et al., 1998). In addition, microbes may contaminate parenterally administered fluids (e.g. blood, lipid-containing substances) and be a significant source of sepsis. The lipid-based intravenous anaesthetic agent propofol can support bacterial growth and has resulted in post-operative infections (Seeberger et al., 1998). Whether or not contamination of stethoscopes by skin microbes such as *Staphylococcus epidermidis*, constitutes a significant risk factor regarding the spread of microbes is largely unknown.

ENDOGENOUS INFLUENCES

It seems that there is a certain amount of endogenous infection with bacteria such as *Staphylococcus aureus* in nearly every surgical setting. More recent studies (see

Kluytmans et al., 1997) have suggested that the proportion of endogenous *Staphylococcus aureus* infection as opposed to cross-infection may be higher today than it was in the past. Before 1970 a series of studies revealed 467 infections in 5 030 carriers and 348 infections in 6 844 non-carriers. This results in a relative risk of 1.8. In comparison, results from a relatively small number of studies in the 1990s found 50 *Staphylococcus aureus* infections in 628 carriers and 33 such infections in 2 962 non-carriers, giving a relative risk of 7.1. Further studies concerning the importance of nasal carriage of *Staphylococcus aureus* as a risk factor in various populations of surgical patients are clearly warranted, especially in view of newer cross-infection policies and our ability to type microbes at the molecular level. However, it does seem that nasal carriage of *Staphylococcus aureus* is a significant risk factor for surgical site infection following some forms of surgery, including cardiac surgery (Ahmed et al., 1998; Kluytmans, 1998).

Clearly it is not always possible to make a distinction between endogenous and environmentally acquired infections. In addition, ongoing infection may suggest apparent failure of antimicrobials which have been shown in laboratory tests to be active against the infecting microbes. This is because of complex factors that reduce the efficacy of the drug or prevent its access to the relevant tissue, or other host factors. Therefore the control of the source of contamination, and the removal by surgical debridement and drainage of foci that are inaccessible to antimicrobial treatment, should always be an integral part of the therapy of surgical infections. Removal of infected tissue by surgical debridement is well exemplified by the management of chronic osteomyelitis (see p. 227).

CHANGING MICROBIAL CAUSATION

Taking into account the considerable changes in surgical technique, availability and use of antibiotics and patient care in the last 35 years or so, it is not surprising that there have been changes in the types of microbes associated with nosocomial infections. While it appeared that *Staphylococcus aureus*, and to a lesser extent *Escherichia coli*, were the dominant microbes recovered from surgical site infections in the early 1960s, recent data from the USA and the UK strongly suggest that while the incidence of *Staphylococcus aureus* has remained static overall, infections attributable to *E. coli* have decreased. Emerging pathogens have included enterococci, coagulase-negative staphylococci (usually line associated), methicillin-resistant *Staphylococcus aureus* (MRSA), yeasts such as *Candida albicans*, and a variety of more resistant Gram-negative bacilli such as *Pseudomonas aeruginosa* and enterobacters. Coagulase-negative staphylococci are now the most common cause of nosocomial bacteraemia in critically ill patients in many hospitals (Vallés et al., 1997). In some hospitals the incidence of MRSA has risen from almost zero in the mid-1970s to around 30% at the present time.

In a large study covering the period 1984–1992, 7.26% of 281 797 blood cultures taken at the Mayo Clinic were considered to be positive for microbes (Cockerill et al., 1997). Over this 8-year period the value of anaerobic techniques was endorsed, as was the growing significance of staphylococci (both coagulase-positive and coagulase-negative), *Enterococcus faecalis*, and non-*albicans Candida* species. Overall, the occurrence of Gram-negative bacilli such as *E. coli* and *Pseudomonas aeruginosa* appeared to be declining. The most common isolates were *Staphylococcus aureus, E. coli, Candida albicans*, coagulase-negative staphylococci, *Pseudomonas aeruginosa, Enterococcus faecalis*, obligate anaerobes (mainly of the *Bacteroides fragilis* group) and *Klebsiella pneumoniae*. Isolation rates increased overall in the second half of the 8-year period.

A major advance related to general surgery has been the recognition of the significant role of obligate anaerobes in many infections, and the availability of the anti-anaerobe drug, metronidazole. The importance of anaerobes became apparent to Sydney Finegold during his Naval service in World War Two and his subsequent studies in the USA during the 1950s. Originally used for the treatment of protozoal infections, metronidazole has subsequently been shown to possess potent antibacterial activity, mostly confined to obligate anaerobes. Trials in the early 1980s endorsed its clinical application in controlling intestinal-associated sepsis, (e.g. peritonitis associated with a perforated appendix) (Brennan *et al.*, 1982). However, despite the availability of metronidazole and other novel antimicrobial agents, anaerobic infections continue to pose significant and often unresolvable problems in surgical patients in whom debridement and drainage are further complicated by factors such as raised intra-abdominal pressure and organ failure.

WOUND CLOSURE

With intra-abdominal sepsis, the surgical and chemotherapeutic management implemented around 1970 resulted in the infection rate falling to 25–40%. Delayed wound closure was introduced around 1975 in an attempt to further reduce complications and mortality. Staged abdominal repair (STAR) involving planned re-exploration every day or on alternate days, regardless of clinical signs or symptoms, also enabled improved outcome by permitting evaluation of gut viability, control of the source of contamination and closure of the fascia without undue tension. Use of this Wittmann STAR approach has been shown to result in a significant drop in mortality in patients with APACHE-II scores of 25 or less (Condon, 1996a), e.g. a halving of the predicted mortality with scores between 11 and 20 (from 28% down to 14% in the group that scored 11–15).

PROTECTED MICROBIAL ENVIRONMENTS

With abscesses, drainage of pus rather than antibiotic therapy is the clinician's primary aim, while removal of dead tissue or other foreign bodies (e.g. sequestrae, renal or biliary calculi) is required before recurrent or chronic infections can be eradicated. Acidity and/or binding of antibiotics to protein within pus and large haematomas may result in a reduction in their antimicrobial activity. Anaerobic conditions and the intracellular location and/or non-dividing nature of bacteria within abscesses may also impair antimicrobial activity, while penetration of most antimicrobial agents will be affected where the blood supply is poor.

There are many instances where abscesses can be sampled or drained (Lerner, 1996) or, in selected cases, treated only with antimicrobials (Bamberger, 1996). Such management (i.e. antimicrobials only) carries the risk of selection of resistant microbes, and must take into account factors such as abscess size, the microbes involved and the antimicrobial agent to be employed.

ANTIMICROBIAL RESISTANCE

Resistance to antimicrobial agents has been recognized since the beginning of the chemotherapeutic era. When penicillin was first introduced in the 1940s, most staphylococci were known to be susceptible. However, soon after its introduction, increasing numbers of penicillinase-producing staphylococci were reported, so that today most staphylococci recovered from humans are penicillin resistant. Areas with high rates of antibiotic resistance are typically those in which antibiotic use is high and conditions favour the easy spread of bacteria (e.g. intensive-care units) (see Murray, 1997). Prolonged use of antibiotics confers an enormous selective pressure on microbes that may readily spread under situations of poor hygiene. Undoubtedly the 1990s have seen the world-wide emergence of antimicrobial resistance, and the spectra of untreatable infectious disease. Of major concern to the surgeon are methicillin (flucloxacillin, nafcillin)-resistant *Staphylococcus aureus* (MRSA), vancomycin-resistant enterococci (VRE), Gram-negative bacilli that are resistant to many of the extended-spectrum β-lactams (e.g. ceftazidime), the increasing resistance of a number of anaerobes to clindamycin and the anti-anaerobe cephalosporins, and fluconazole resistance in *Candida albicans* and related yeasts (Murray, 1997). Other microbes, such as penicillin-resistant pneumococci and multi-resistant mycobacteria, perhaps present more alarming prospects but are of less significance to surgeons in general.

NEW TECHNOLOGY

The application of new technologies developed largely by surgeons has been one of the main reasons for the growth in numbers and sizes of hospitals since the 1960s. Safe anaesthesia has allowed major procedures and longer operations. The development of cardiopulmonary bypass for open cardiac surgery led to the use of invasive haemodynamic monitoring, ventilatory support, recovery rooms and intensive-care units (Wilson, 1993). Soon after World War Two, surgery was revolutionized by the introduction of the pump oxygenator, total parenteral nutrition, arteriography, aortic aneurysm repair, haemodialysis and organ transplantation. Beginning in the early 1960s, the development of bioprosthetic implants and joint replacement led to increased hope for those with lifelong disabilities. Hospitals were being viewed as the source of definitive cures for previously fatal or incurable debilitating diseases.

Unfortunately, these surgical advances were not without problems, many of which were microbiologically orientated. In the intensive-care unit patient, naso-tracheal and nasogastric tubes and intravascular lines serve as sources of infection. Microbes adhering to the surface of these devices may result in overt disease and/or facilitate the introduction of microbes, many of which are inherently less susceptible to antimicrobial agents. Total parenteral nutrition also has a disconcertingly high rate of complications, especially infections, while prosthetic implants (e.g. hip, heart, blood vessels) appear to be particularly prone to microbial adherence and colonization following significant bacteraemias or fungaemias. In a recent Australia-wide survey covering 15 hospitals, over 800 cases of bacteraemia or fungaemia were identified, and it was apparent that intravenous catheters were not only the most common primary focus for hospital-acquired bacteraemia, but also the most common overall cause of positive blood cultures (Collignon, 1995). When present as biofilms on the surface of prostheses, microbes tend to exhibit phenotypically reduced susceptibility to antimicrobial agents. This has complicated the treatment of infections associated with implants and prostheses, which often requires surgery and/or removal of the device in addition to the use of antimicrobial agents.

HOST RESPONSE/ABSCESSES

The unique feature of all surgical infections is tissue necrosis. This may be induced by mechanical or other physical trauma, or it may occur spontaneously by the normal pathophysiological process. Inflammation, with its surface features – first described by Celsus and then refined by Galen – of rubor (redness), tumor (swelling), calor (heat), dolor (pain) and functio laesa (loss of function), is the response to tissue necrosis. When controlled and properly regulated, these responses result in elimination of necrotic material and invading microbes, and initiate tissue repair.

During inflammation, increased blood flow and vascular permeability result in the accumulation of various cellular and humoral antibacterial factors at the site of microbial growth and tissue damage. Included in the cellular array are initially polymorphonuclear phagocytic cells, followed by mononuclear cells which, as well as scavenging any remaining debris, initiate an immune response characterized by the secretion of appropriate cytokines. Also present in the plasma may be specific immunoglobulins (antibodies) and a variety of other potentially antibacterial proteins (e.g. complement). In cases where the microbial burden is overwhelming or highly virulent, and/or where the host's ability to mount a satisfactory inflammatory response is impaired, continued microbial growth may lead to progressive necrosis, abscess formation, bacteraemia and multisystem malfunction.

Where tissue injury and the number of bacteria exceed the capability of the host to terminate an infection locally, an abscess may form. Plasma components such as fibrin and phagocytic cells attempt to wall off the centrifugally expanding bacterial population through the formation of a pyogenic membrane. Inside this primitive 'abscess', degenerating phagocytes and bacteria release enzymes and toxins that liquefy the contents. The resulting high osmolarity attracts water, thereby increasing the pressure within the abscess capsule. Poor oxygen and nutrient diffusion through the capsule promotes anaerobic glycolysis. Abscesses are thus characterized by high pressure, an acidic pH and low oxygen tension, and represent an ideal environment for the growth of anaerobic bacteria. Many abscesses are polymicrobial, containing a mixture of facultative anaerobes (e.g. *E. coli*, enterococci) which utilize the available oxygen and create strict oxygen-free anaerobiosis suitable for the growth of oxygen-sensitive obligate anaerobes such as *Bacteroides fragilis*. Antibiotics penetrate poorly, if at all, into abscesses. The treatment of choice for abscesses is therefore drainage, which should be undertaken with appropriate peri-operative antibiotic cover.

As already stated, in cases where local containment or elimination of bacteria is not achieved, invasion of the bloodstream and dissemination to distant organs may occur. Where the bacteraemia is associated with persistence and/or multiplication of the bacteria in the blood, septicaemia occurs. This uncontrolled microbial growth can be associated with liberation of endotoxin (from Gram-negative bacteria) or a variety of exotoxins (mainly from Gram-positive bacteria) capable of stimulating massive release of cytokines such as tumour necrosis factor (TNFα). Failure to down-regulate the liberation of these inflammatory substances leads to blood vessel endothelial damage and leakage of blood into tissue spaces and cavities. Death may occur either immediately due to septic shock, or later as a consequence of multiple organ failure (MOF), the outcome being determined by the host response and the number and severity of hits or injury episodes.

Recently, MOF has been renamed multiple organ dysfunction syndrome (MODS) or systemic inflammatory response syndrome (SIRS). Although the pathophysiology of MODS and SIRS has yet to be clarified, it is apparent that inflammation rather than bacterial overgrowth may be the real problem (Wittmann *et al.*, 1996). MODS and SIRS occur in the absence of infection. The clinical signs and symptoms and the morphological alterations and changes in the oxygen transport system possibly result from phagocyte activation (Sauaia *et al.*, 1996). An important clinical inference that can be drawn from this is that a conservative approach to the management of infection is more likely to minimize the adverse consequences of an excessive inflammatory response.

Bacteraemia is accompanied by fever and the release of acute-phase reactants (e.g. cytokines). The high, spiking fever and accompanying chills may result from the release of bacterial toxins, and may be the prodrome of abscess formation. While the total leucocyte count may not be particularly abnormal, and may even be low because of consumption of polymorphonuclear cells, there is a shift to the left, with toxic granulation. Petechial lesions on the skin or conjunctivae occur in septicaemia and are attributable to specific bacteria (e.g. streptococci, meningococci). Anaemia secondary to haemolysis (associated with abnormal liver function tests) may occur with septicaemia due to exotoxin-producing microbes such as staphylococci and clostridia. Metastatic abscesses in organs such as the bone, brain, spleen or kidneys are not infrequent after a septicaemic episode, and any injured tissue is easily infected during septicaemia. Involvement of heart valves or prostheses necessitates awareness of the early signs of infection, and mandates the early initiation of empirical therapy.

Provided that any contaminating source is controlled, intra-abdominal abscesses may be drained by the least invasive means (e.g. computed tomography-guided percutaneous catheter drainage). The drains are left in place until the cavity is obliterated. It seems unlikely that this type of abscess can be treated adequately with antibiotics alone, although there are reports of appendiceal abscesses being successfully treated without surgery or drainage. The critical surgical factor is the eradication of all sources of hollow viscus contamination or other major contributing factors such as necrotic tissue and foreign bodies. An ultraconservative approach may be an effective alternative to immediate surgery or some form of imagery-guided drainage (e.g. in patients who have a coagulopathy; abscesses with difficult surgical access) (Hoffmann et al., 1991).

chapter 2

BIOLOGY OF BACTERIA AND RELATED MICROBES

Microbiology covers a diverse range of micro-organisms, from the non-cellular viruses through prokaryotes such as bacteria, chlamydiae, rickettsiae and mycoplasmas, to the eukaryotic fungi, algae and protozoa. Some basic differences between bacteria and related microbes are shown in Table 2.1.

GENERAL PROPERTIES OF BACTERIA

In general, prokaryotic cells have no discrete nucleus, possess both DNA and RNA as well as ribosomes, have no mitochondria or lysosomes, and have flagellae which (when present) differ in structure to the cilia of eukaryotic cells. Mycoplasmas are really bacteria without a cell wall, and hence they are pleomorphic and osmotically fragile. Rickettsiae can be broadly considered as small obligate intracellular parasites resembling Gram-negative bacteria in structure. Chlamydiae are obligate intracellular bacteria-like microbes that possess a rigid cell wall devoid of peptidoglycan (murein), and which form virus-like inclusion bodies in the host cell cytoplasm. The growth of all of these prokaryotes can be inhibited by appropriate antibiotics.

It is not the intention of this text to discuss the structure and characteristics of bacteria in any detail, and this type of information can be found in any basic microbiology text. However, a few features are worthy of brief mention.

Table 2.1 Some properties of bacteria and related microbes

Microbe	Prokaryotic	Obligate intracellular parasites	DNA + RNA	Ribosomes	Cell wall murein	Multiply by binary fission
Bacteria	+	−	+	+	+	+
Mycoplasmas	+	−	+	+	−	+
Rickettsiae	+	+	+	+	+	+
Chlamydiae	+	+	+	+	−	+
Viruses	Non-cellular	+	−	−	−	−

As there are no fossil bacteria, their origins are only speculative. It has been suggested that they represent the mitochondria of eukaryotic cells which have become adapted for independent growth and survival. However, it is important to remember that (prokaryotic) bacteria do differ in many respects from mammalian (eukaryotic) cells (see Figure 2.1). These differences are exploited in the selective toxicity of most antibiotics. Particularly important in this respect are the cell wall and protein-building structures (ribosomes) of bacteria (see pp. 27–31).

The bacterial cell wall determines the cell shape, and as well as being a physical barrier it confers protective rigidity on the cell – without it, bacteria become osmotically fragile although they are still capable of survival. Its basic component is a mixed polymer called peptidoglycan, which is a bag-shaped macromolecule surrounding the cytoplasmic membrane. Peptidoglycan contains two components, namely a linear polymer (glycan strands) composed of two unique sugars (N-acetylglucosamine and N-acetylmuramic acid) and short identical side-chains of four amino acids (tetrapeptides) attached to the glycan strands at the N-acetylmuramic acid residues. These tetrapeptides are in turn linked to one another by short peptides forming cross-bridges between the glycans. Peptidoglycan is not found in mycoplasmas (or in certain bacteria that occur in high-salt environments), and in different bacteria the amino acid composition of the tetrapeptide and cross-linking peptides differs.

BACTERIAL CELL WALL

Two types of bacteria are recognized on the basis of cell wall structure (see Figure 2.2). These are termed either Gram-positive or Gram-negative depending on their staining reaction with the Gram stain. However, it must be remembered that this stains structures within the cell (probably cytoplasmic membrane) and not the cell

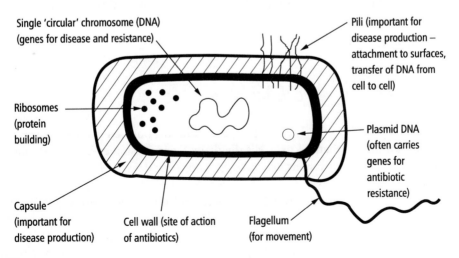

Figure 2.1 Schematic drawing of a bacterium, with brief explanation of important features.

wall itself. As Gram-positive (purple) bacteria die and start to lose their cell wall integrity, they may stain Gram-negative (pink) – an important point to remember when examining Gram stains of clinical material (e.g. pus).

The outer membrane of the cell wall of Gram-negative bacteria contains three regions of importance, namely lipid A (responsible for toxicity), a polysaccharide core and terminal repeating units of 'O' (somatic) antigenic side-chains, both of which are important taxonomically. In addition to these features, the outer membrane prevents leakage of the periplasmic proteins and protects the cell (in enteric bacteria) from bile salts and hydrolytic enzymes of the host environment. Proteinaceous pores (porins) in the outer membrane allow passage of low-molecular-weight solutes. Large antibiotic molecules penetrate the outer membrane relatively slowly, and this accounts for the relatively high natural (intrinsic) antibiotic resistance of Gram-negative bacteria. The permeability of this barrier varies widely between species (e.g. it is poor in *Pseudomonas aeruginosa*) and is an important antibiotic resistance mechanism.

The space between the cytoplasmic membrane and the outer membrane, which is known as the periplasmic space, is filled with a gel containing the hydrated peptidoglycan. Included in the periplasmic proteins of Gram-negative bacteria are a number of enzymes, including β-lactamases.

Although it is thicker, the cell wall of Gram-positive bacteria is less complex than that found in Gram-negative organisms. It consists simply of peptidoglycan-containing strands of teichoic acid, proteins and carbohydrates, depending on the species. It is not protected by an outer membrane, and is susceptible to degradation by lysozyme.

Capsules may be found external to the cell wall in both Gram-positive and Gram-negative bacteria. These are important virulence factors, and in non-immune patients they inhibit the early phase of phagocytosis.

Cell walls of mycobacteria differ from those of both Gram-positive and Gram-negative bacteria. They consist of a linked three-layered structure (peptidoglycan, arabinogalactan and lipid), with lipid contributing 35–60% of the wall (compared to 0.5% and 3% in Gram-positive and Gram-negative bacteria, respectively). These lipids are responsible for many of the important characteristics of the mycobacteria (e.g. intracellular survival, adjuvant effect, survival in the presence of acid).

BACTERIAL GROWTH KINETICS

Bacterial growth is rapid. Under optimal conditions, one cell produces over 10^8 progeny within less than 24 hs. Cells divide by binary fission every 10–20 min *in vitro*, although these divisions may take hours or even days *in vivo*.

Bacteria which grow only in the presence of air (requiring oxygen as the hydrogen acceptor) are described as aerobic, and those which grow only in the absence of oxygen are said to be anaerobic. In fact, many anaerobes are extremely sensitive to oxygen, being killed after exposure for only a few seconds. These are often termed obligate or oxygen-sensitive anaerobes, and the toxicity of oxygen results from its reduction by enzymes (e.g. flavoproteins) in the cell to hydrogen

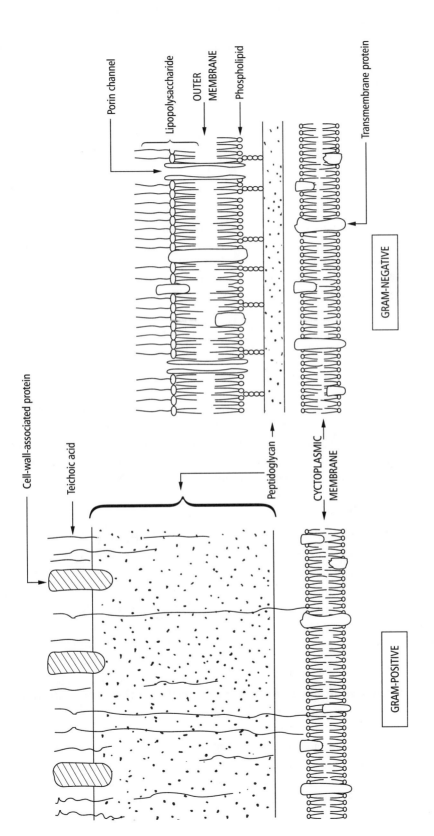

Figure 2.2 Schematic drawing of the cell wall structure of Gram-positive and Gram-negative bacteria.

peroxide, and by ferrous ions to the even more toxic free radical, superoxide (O_2^-). Aerobes and aerotolerant anaerobes are protected from these products by the presence of the enzymes catalase and superoxide dismutase. Facultative anaerobes are able to live aerobically or anaerobically, while micro-aerophilic microbes grow best at slightly reduced oxygen levels.

GENETIC CHANGE/GENE TRANSFER

MUTATION

An important feature of bacteria is their dynamic state and their ability to undergo continual and rapid genetic change. This may occur by spontaneous mutation or by gene transfer between cells. It must be remembered that mutations occur in the absence of any selection pressures (e.g. antibiotic). Environmental factors, such as the presence of an antibiotic, simply serve to select for further growth any spontaneously occurring resistant mutants. In any population of 10^8 or more bacterial cells (as occurs in any infectious process), at least one cell that is resistant to almost any antibiotic will be naturally present as a result of spontaneous mutation.

TRANSFORMATION

Genes may be transferred between cells by a variety of mechanisms, of which transformation, transduction and conjugation are the best understood. In the case of transformation, the recipient cell takes up soluble DNA released from the donor cell. It commonly occurs in pneumococci and gonococci – bacteria that are apparently readily capable of taking up high-molecular-weight DNA.

TRANSDUCTION

This is the transfer of genetic information (DNA) from a donor cell to a recipient cell via a virus (bacteriophage) vector. For the most part, transduction is carried out by temperate bacteriophages (i.e. viruses capable of lysogenizing host cells on infection). In the lysogenic response, the genes of the virus responsible for initiating the normal lytic event are repressed, and the viral DNA is integrated into the host bacterial DNA as prophage. The virus genes then replicate with the host cell genes, and all of the daughter cells will have the viral genes incorporated into their chromosome. Bacterial cells which have a viral genome incorporated into their chromosome are described as lysogenic (note that phages which always cause lysis of the cell that they infect are termed virulent or lytic bacteriophages).

When a lysogenic cell is induced to enter a lytic cycle, a small part of the bacterial genome adjacent to the prophage may become incorporated into the bacteriophage DNA and transferred to recipient cells, or during maturation of the viral particles an occasional mistake is made. Instead of packaging viral DNA, host DNA only of the same size is put into the viral protein coat (transducing particles) and transferred to a new host.

Although the DNA of the transducing particle usually becomes incorporated into the recipient chromosome, occasionally it does not become integrated but remains in the cytoplasm, where it is capable of normal function but not replication. This is termed abortive transduction. Transduction has been observed in many bacterial genera, both Gram-negative and Gram-positive.

CONJUGATION

This is gene transfer that occurs between sexually differentiated bacteria, and it requires cell-to-cell contact between viable donor and recipient cells. Genetic material is always transferred from the donor to the recipient cell, and never the other way round. Donor cells differ from recipients in that they possess the fertility or F factor. This is a typical plasmid – a small extrachromosal, self-replicating circular DNA fragment about 2% of the size of the bacterial chromosome. Cells carrying the F factor express protein appendages, termed pili, that are essential for the conjugation process, and are termed F+ cells. The F+ cells are capable of transferring (via the pili) the F factor to recipient or F– cells upon cell-to-cell contact. Within a short time, many F– cells thus become F+. However, the F+ cells remain F+, suggesting that the F factor is replicated before transfer. Like other plasmids, the F factor can be spontaneously lost – hence the finding that in time not all cells become F+.

In addition to the F factor, many Gram-negative bacteria carry conjugative plasmids that code for drug resistance. These R factors consist of two or more subunits – an element called resistant transfer factor (RTF) that codes for the intercellular transfer, and one or more elements that carry the resistance genes (R determinants). Many if not all R determinants are transposons (i.e. movable segments of DNA which can combine with other plasmids or with the bacterial host chromosome). The transfer of R factors appears to be an important means by which many bacteria, especially Gram-negative enterics and enterococci, acquire antibiotic resistance. Conjugative transposons can be exchanged between apparently diverse and unrelated genera. Unlike plasmids, they are not self-replicating in recipients, and must be integrated into plasmid or chromosomal DNA for replication to occur.

Some conjugative plasmids are capable of integration with the bacterial chromosome. For instance, the F factor can merge with the bacterial chromosome to form a single replicon, and thus chromosomal DNA, as well as F-DNA, is passed to the recipient during conjugation. However, it must also be remembered that not all plasmids are conjugative (e.g. the penicillinase-determining plasmid of staphylococci can only be transferred by transformation or phage-mediated transduction).

Apart from conferring drug resistance on recipients, plasmids may be associated with a number of other clinically significant properties. Some plasmids carry genes that code for toxins (e.g. enterotoxins of enteropathogenic *E. coli*, the neurotoxin of *Clostridium tetani*), while others carry genes whose products contribute in other ways to the virulence of the bacterium (e.g. adhesive pili of some bacteria). In addition, a number of antibiotics are encoded by plasmid genes, including many produced by *Streptomyces* species.

The ease with which plasmids can transfer from cell to cell, and the strong selection which chemotherapy has exerted for antibiotic resistance, have combined to produce striking effects in the 1990s. It is worth noting that while approximately 20% of the enteric Gram-negative bacilli in the 'pre-antibiotic era' (i.e. prior to 1950) carried conjugative plasmids, only rarely were genes for drug resistance (R determinants) also present. This situation has certainly changed, with drug resistance now being common.

NORMAL FLORA

Most infections that occur following surgery involve microbes that are usually found as part of the body's normal flora or indigenous microbiota. An understanding of the more common potential pathogens present on and in the body is therefore essential (see Table 2.2). It must be remembered that the human body is host to around 10^{14} bacteria, most of which are obligate anaerobes living in the large bowel (Hentges, 1993). In recent years, taxonomic aspects of obligate anaerobes have undergone considerable change, as has been highlighted in reviews of the Gram-positive anaerobic cocci (*Peptostreptococcus* species; see Murdoch, 1998) and of the clinically significant pigmented *Prevotella* and *Porphyromonas* species (which were originally members of other genera, e.g. *Bacteroides*; see Brook, 1995; Finegold, 1995a,b; Brook *et al.*, 1998). Some more unusual and recently recognized anaerobic pathogens include *Fusobacterium ulcerans* (skin ulcers), *Prevotella heparinolytica* and *Bacteroides tectum* (animal bite wounds), *Bilophila wadsworthia* (peritonitis), *Porphyromonas* (*Bacteroides*) *forsythus* (periodontal disease) and toxigenic strains of *Bacteroides fragilis* and *Anaerobiospirillum succiniciproducens* associated with diarrhoea (Goldstein and Ueno, 1996). It is only since the mid-1970s that the significance of anaerobes in surgery-related infections has been fully appreciated (Nichols and Smith, 1994).

As a general rule, normal flora microbes are found on the skin and in the mouth, alimentary tract and vagina, and in the respiratory tract as far down as the trachea. Airways distal to the trachea should be sterile, although microbes may transiently enter the lower respiratory areas. Similarly, the urethra and bladder may transiently be colonized by bacteria – these should be removed by the normal flushing process.

CHANGES IN NORMAL FLORA

It is important to remember that the numbers and composition of the normal flora can be dramatically influenced by various procedures. For example, keeping the skin excessively moist can result in colonization by 'unusual' microbes such as Gram-negative bacilli and yeasts. Hospitalization invariably results in replacement of indigenous microbes by similar microbes with increased antibiotic resistance and/or unusual species – e.g. *E. coli*, which is uncommon (< 2% of the population) in the oral cavity of non-hospitalized patients, may be recovered from the mouth of around 80% of hospitalized patients. Alkaline oral

Table 2.2 Examples of important microbes that are 'normally' found at defined body sites

Habitat	Type of microbe	Microbe	Comment
Skin			
	Obligate anaerobes	*Propionibacterium acnes*	Commensal in hair follicles
		Peptostreptococcus magnus	Common in soft-tissue infections
	Gram-positive cocci	*Staphylococcus aureus*	Transient; colonize damaged skin
		Staphylococcus epidermidis	
	Gram-negative bacilli	*Acinetobacter* species	Foot commensal
		Pseudomonas aeruginosa	Transient on moist/damaged skin
	Yeasts	*Candida albicans*	Transient on moist skin
		Malassezia furfur	Lipophilic commensal
Oral cavity			
	Obligate anaerobes	*Actinomyces israelii*	Actinomycete
		Bacteroides species	Related to *Prevotella*; some reclassified into other genera
		Fusobacterium species (e.g. *F. nucleatum*)	Gram-negative fusoforms
		Peptostreptococcus species (e.g. *P. micros*)	Gram-positive cocci
		Porphyromonas species (e.g. *P. gingivalis*)	Some previously pigmented *Bacteroides*
		Prevotella species (e.g. *P. melaninogenica*)	Related to *Bacteroides*; often produce β-lactamases
	Gram-positive cocci	*Staphylococcus aureus*	Commensal nasopharynx
		Streptococcus anginosus	Previously '*Streptococcus milleri*'
		Streptococcus pneumoniae	Pneumococcus
		Streptococcus pyogenes	Group A, β-haemolytic
		Viridans streptococci	α-Haemolytic
	Gram-negative bacilli	*Eikenella corrodens*	Bite infections
		Escherichia coli/coliforms[a]	Transients; numbers increase with hospitalization
	Chlamydiae	*Chlamydia pneumoniae*	
	Mycoplasmas	*Mycoplasma pneumoniae*	
	Yeasts	*Candida albicans*	
Stomach			
	Curved Gram-negative rods	*Helicobacter pylori*	
Small intestine			
	Gram-positive cocci	*Enterococcus faecalis*	Enterococcus
		Oral streptococci including *Streptococcus anginosus*	Possibly transients
	Gram-negative bacilli	*Escherichia coli*/coliforms[a]	
	Yeasts	*Candida albicans*	

Table 2.2 Continued

Habitat	Type of microbe	Microbe	Comment
Large intestine			
	Obligate anaerobes	Bacteroides fragilis	Most significant anaerobe; β-lactamase positive
		Bacteroides thetaiotaomicron	
		Other Bacteroides species	
		Bilophila wadsworthia	Abscess forming
		Clostridium difficile	Antibiotic-associated colitis
		Clostridium perfringens	Penicillin sensitive
		Other clostrida	
		Fusobacterium species	
		Peptostreptococcus species (e.g. P. anaerobius)	
	Gram-positive cocci	Enterococcus faecalis	
		Enterococcus faecium	
		Streptococci (e.g. Streptococcus anginosus)	
	Gram-negative bacilli	Escherichia coli/coliforms[a]	
		Pseudomonas aeruginosa	Transient; appendix region
	Yeasts	Candida albicans	
Vagina/female genital tract			
	Obligate anaerobes	Actinomyces israelii	Intra-uterine contraceptive device-related infections
		Bacteroides species (e.g. B. fragilis group)	
		Clostridium perfringens	
		Fusobacterium species	
		Peptostreptococcus species (e.g. P. magnus, P. asaccharolyticus)	
		Prevotella species (e.g. P. bivia)	
	Gram-positive cocci	Streptococcus agalactiae	Group B streptococcus
	Gram-negative bacilli	Escherichia coli/coliforms[a]	
	Gram-negative cocci	Neisseria gonorrhoeae	
	Chlamydiae	Chlamydia trachomatis	
	Mycoplasmas/ ureaplasmas	Mycoplasma hominis	
		Ureaplasma urealyticum	
	Yeasts	Candida albicans	

[a]Includes Enterobacter, Klebsiella, Proteus and Serratia species.

medications and H_2-antagonists may allow colonization of the stomach by increased numbers and types of bacteria, as may gastric malignancy. Blockage or stasis of food in the intestines often results in increased bacterial numbers, especially of anaerobes. All of these factors must be taken into account when possible antibiotic prophylaxis or treatment is being considered.

PART II

ANTIMICROBIAL AGENTS

chapter 3

ANTIBACTERIAL AGENTS – GENERAL INFORMATION

HISTORY

Empirical antimicrobial therapy was probably practised many centuries before Paul Ehrlich initiated his work on the 'magic bullet' in the early 1900s. For instance, the Egyptians used mould ('the film of dampness which is found on the wood of ships'), fermenting bread and several forms of yeast, which may have had antibacterial properties, for treating various abdominal and gynaecological ailments.

Although such an idea is possibly repugnant to us, animal dung was also used for wound dressings. Undoubtedly such material contained many potential pathogens (e.g. clostridia), but did it also cause debridement of slough, or stimulate leucocyte migration and activity, or even possess antimicrobial properties?

In 1932, Gerhard Domagk, working in the Bayer Wing of the German I.G. Farbenindustrie consortium, found that mice treated with a dye, prontosil red, survived otherwise fatal infection by streptococci. Surprisingly, this dye appeared to be ineffective against streptococci in laboratory tests. This paradox was subsequently explained when it was shown that in animals the dye was split into two components, one of which was the colourless compound sulphanilamide (the first sulphonamide).

Howard Florey and his team at the Sir William Dunn School of Pathology in Oxford first took an interest in penicillin in the late 1930s. However, the concept of antibiosis and its therapeutic potential was not new. The observation that microbes, including fungi, could produce substances that were capable of inhibiting the growth of other microbes had been known for years. One such substance, namely pyocyanase, produced by the bacterium *Pseudomonas aeruginosa*, had been used therapeutically by instillation into wounds at the turn of the century.

Alexander Fleming's observations in early September 1928 concerning lysis of staphylococci on a *Pencillium*-contaminated culture plate left on the bench are generally regarded as marking the beginning of the penicillin era, although it seems that similar observations were recorded by Frederick Dennis, a New York surgeon, in 1885! Early disconnected attempts to exploit penicillin (e.g. by instilling it into wounds) foundered on a failure to purify and concentrate the substance. It was the German biochemist, Ernst Chain, working with Florey, who first obtained a crude, and relatively impure, but stable extract of penicillin suitable for experimentation in animals and humans. The original preparations of penicillin were apparently a mixture of four closely related compounds – penicillins F, G, K and X. Benzylpenicillin (penicillin G), often simply referred to as penicillin, was the one chosen for further development, as it exhibited the most desirable properties, and could be produced in almost pure form from the mould *Penicillium chrysogenum*. The penicillin era had thus begun, although it was not until the recognition of 6-aminopenicillanic acid as the biologically active moiety, and the demonstration of its ability to be manipulated in the laboratory, that the true value of Fleming's original observation was realized.

A fungus of the genus *Cephalosporium* (*Cephalosporium acremonium*) was obtained in 1948 from sewage in Sardinia by the biochemist Guiseppe Brotzu, and when cultured was found to secrete a substance which inhibited both Gram-positive and Gram-negative bacteria. As local interest was minimal, the fungus was eventually sent to Sir Howard Florey's group at Oxford. They found two antibiotics among its metabolites, but ironically the one subsequently termed cephalosporin N (because of its activity on Gram-negative bacteria), which appeared to possess most of the effects noted by Brotzu, was in fact a penicillin (adicillin).

In 1953, the Oxford group detected a third antibiotic, cephalosporin C, as a minor component of cephalosporin N. In addition to its broad antibacterial activity, cephalosporin C was active against penicillinase (β-lactamase)-producing *Staphylococcus aureus*. A few years later, a mutant strain of this fungus was isolated which produced cephalosporin C in large amounts. Recognition of the biologically active moiety, 7-aminocephalosporanic acid, opened the way for molecular manipulation, and the cephalosporin era had begun.

The development of penicillin G and cephalosporin C and, subsequently, their numerous derivatives, is only one part of the antibiotic story. The other main route to present-day antimicrobial agents was via investigations into the ability of soil microbes to produce substances that were inhibitory to other micro-organisms, including bacteria and fungi. Selman Waksman, beginning in 1940, was one of the principal investigators. The first real breakthrough in these studies came in 1943 with the discovery of streptomycin, the first aminoglycoside. By the mid-1950s, representatives of most of the major families of antibiotics had been discovered, including aminoglycosides, chloramphenicol, tetracyclines, macrolides, rifamycins and fusidic acid.

Since that time, very few truly novel antibiotics have been discovered, although chemists and microbiologists have fortunately come up with a variety of synthetic antimicrobial drugs, including nitroimidazoles, quinolones and most antituberculous drugs. Undoubtedly the hunt for new antimicrobial agents, whether these be biological or synthetic, will continue, especially in the face of potentially untreatable microbes. The lanthionine-containing antibiotics (lantibiotics) are an example of a promising group of naturally occurring biological substances now under investigation.

The surgical patient who presents with infection or develops such infection while in hospital has always presented unique challenges to the clinician (Nichols, 1996). To optimize outcome, the therapy of very sick patients requires careful planning, including the selection and use of a variety of drugs. Unfortunately, the inappropriate use or indeed overuse of the myriad of antimicrobial agents which became available in the late 1980s and early 1990s has contributed significantly to the emergence of resistant microbes. These now pose a distinct threat in many areas of the world. Apart from a number of common bacteria (e.g. staphylococci, pneumococci and enterococci), some viruses (e.g. human immunodeficiency virus), fungi (e.g. *Candida* species) and parasites (e.g. malarial parasite) are now becoming increasingly resistant to many of the common antimicrobial agents which previously proved so efficient in their control.

The emergence of microbes that are resistant to antimicrobial agents is inevitable, albeit at varying rates. Careful consideration must therefore be given to the selection and use of antimicrobials both for prevention and for treatment of disease. Strict adherence to established guidelines is important, especially in prophylaxis. In therapy, the choice of antimicrobial agent must take into account the patient's physical and immune status, the type and location of the pathology, and whether or not the microbe is intracellular or in some other way protected from antimicrobial activity (e.g. within biofilms or abscesses). It is important that the antimicrobial agent selected reaches effective and sustained concentrations at the site where the microbe is growing.

MECHANISMS OF ACTION

Infection results when the equilibrium between the invading microbe (which is usually part of the normal flora) and the host defence mechanisms is pushed in favour of the microbe. At a minimum, the mere deceleration of microbial growth and multiplication may allow the host defences to eradicate or at least control the causal microbes.

Bacteria are prokaryotes, and as such they differ in several respects from mammalian (eukaryotic) cells. More important differences include the presence in bacteria of a rigid cell wall which is essential for the osmotic stability of the cell, and protein-synthesizing structures (ribosomes) composed of 30S and 50S subunits that are joined to form 70S particles (ribosomes from mammalian cells are 80S particles, composed of 40S and 60S subunits). The cytoplasmic membranes of bacterial cells are in many ways similar to those of human cells, and function in the selective transport of metabolites into and out of the cytoplasm. Individual bacteria show minor variations in the chemical composition of this membrane. Prokaryotic cells have no nuclear membranes, and the nuclear material consists of one or more copies of a large supercoiled circular strand of DNA. In addition, certain metabolic pathways are essential for the growth and survival of bacteria (e.g. those which lead to the formation of folic acid, which is an essential cofactor in nucleotide synthesis).

Antibiotic therapy is based on the use of drugs that are active against these unusual bacterial features. However, very few agents exert a toxic effect on the parasite without causing some injury to the host cells. The classical mechanisms of antibiotic action are shown in Figure 3.1, together with the more important examples of each group.

INHIBITION OF CELL WALL SYNTHESIS

The strength and rigidity of the bacterial cell wall depends on peptidoglycan, a polymer of N-acetylmuramic acid and N-acetylglucosamine (glycan), cross-linked by peptide bridges. While the Gram-positive cell wall consists of a thick peptidoglycan layer containing trapped proteins, polysaccharides and teichoic acids, the Gram-negative wall is thinner but more complex (see Figure 2.2). It contains a thin peptidoglycan layer and an outer membrane containing proteins and lipopolysaccharide in addition to the lipid bilayer. Proteins span the outer membrane and in some places form discrete water-filled channels or pores (porins). Antimicrobial agents must pass through the outer membrane, either through the porin channels

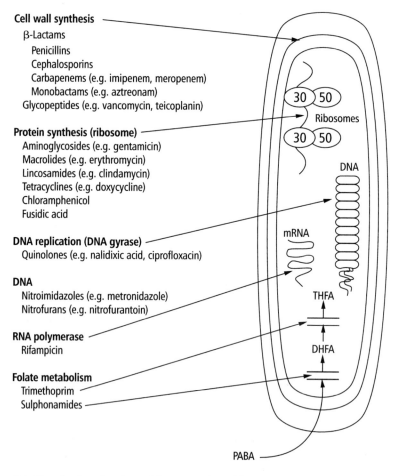

Figure 3.1 Site of action of antibacterial agents. Adapted from *Guide to pathogens and antibiotic treatment 1998* with kind permission of *New Ethicals,* Adis International and Dr Selwyn Lang, Medlab Auckland, New Zealand.

or the lipid bilayer, in order to reach the internal target sites. Many of the features of Gram-negative bacteria are attributable to this outer membrane.

β-Lactams (e.g. penicillins and cephalosporins) and glycopeptides (e.g. vancomycin) are the classical cell-wall-active antimicrobial agents. The β-lactams inhibit bacterial cell wall synthesis by inhibiting cross-linking between peptidoglycan strands, and are thus ineffective against bacteria that lack cell walls (e.g. mycoplasmas). Death of the cells tends to occur by osmotic rupture in Gram-negative bacteria, although in Gram-positive bacteria the process appears to be more complex, with release of lipoteichoic acid from the cell wall triggering a generalized autolytic dismantling of the peptidoglycan.

As the newly formed linear peptidoglycan strand is extruded from the cell membrane, pores are formed in the existing cell wall into which the new strand is to be inserted. Failure of the penicillin-binding proteins to link the new strand to the existing strands apparently initiates a signal that results in autolysis of the cell, although some cells may lyse because of osmotic forces and the poor structural integrity of the 'damaged' wall. This accounts for the high cidal activity of β-lactam drugs, something that is apparently not so evident with glycopeptides (e.g. vancomycin), which have their site of activity at an earlier stage in cell wall synthesis.

The target site of β-lactams consists of the enzymes associated with the cross-linking of the newly formed linear glycan strands (see Figure 3.2). These enzymes, which are referred to as the penicillin-binding proteins (PBPs), are responsible for a variety of enzymatic processes (e.g. transpeptidation). Up to five or more PBPs with varying functions are found in each bacterial species. The enzymes are trapped as inactive complexes by the β-lactam, and are unable to participate in peptidoglycan assembly. Of the seven PBPs that are found in *E. coli*, almost half seem to be unconnected with the antibacterial activity of β-lactams, while binding to one or more of the remainder can result in the formation of long filaments (inhibition of the cell division process), spherical cells or osmotically fragile cell-wall-deficient forms.

Glycopeptides (e.g. vancomycin and teicoplanin) act by binding to the D-alanyl-D-alanine portion of the pentapeptide peptidoglycan precursor, thereby preventing the addition of new building blocks to the growing cell wall. Cell wall synthesis is thus halted at an earlier stage than transpeptidation (β-lactams). The glycopeptides are too bulky to penetrate the outer membrane of Gram-negative bacteria, so their spectrum of activity is generally restricted to Gram-positive microbes.

INHIBITION OF PROTEIN SYNTHESIS

Although the mechanism of protein synthesis is essentially similar in bacterial and mammalian cells, there are several important differences in the structure of the ribosomes and accessory factors which form the targets for selective inhibition by antimicrobial agents. As already stated, bacterial ribosomes are smaller (70S) than those found in human cells (80S), and in many cases selective inhibition of bacterial growth is associated with binding of the antibiotic to either the bacterial 50S or 30S subunit.

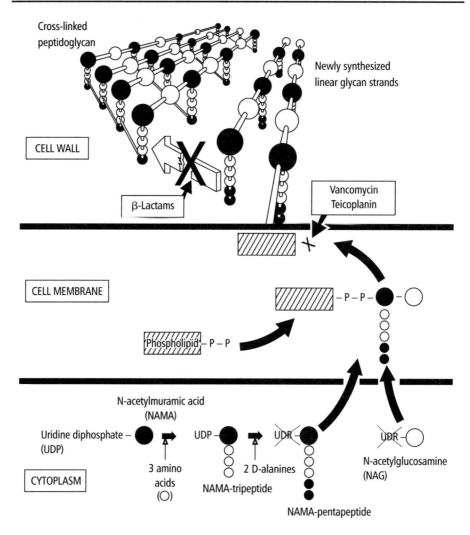

Figure 3.2 Biosynthesis of peptidoglycan. Adapted from Lambert, P.A. (1992) Mechanisms of action of antibiotics. In Hugo, W.B. and Russell, A.D. (eds), *Pharmaceutical microbiology*, 5th edn. Oxford: Blackwell Scientific Publications, 189–207.

Aminoglycosides (e.g. gentamicin) and tetracyclines (e.g. doxycycline) interfere selectively with the 30S bacterial subunit, although in the case of the latter drug, the human 40S subunit is also sensitive. The selective toxicity of tetracyclines for bacterial cells resides in the differential permeability of the membranes, and the fact that bacterial cells actively take up the antibiotic while mammalian cells exclude the drug. Different aminoglycosides bind to different sites on the ribosome, and therefore resistance to one does not necessarily imply resistance to others. Binding of the aminoglycoside to the 30S subunit appears to result in either inhibition of protein synthesis, or a misreading of the genetic code (resulting in the accumulation of toxic, non-functional and potentially fatal proteins). The effectiveness of aminoglycosides is augmented by their active uptake, a feature that is absent in anaerobic microbes, on which aminoglycosides have little action. They also penetrate poorly into mammalian cells and are thus of limited value in treating infections that involve intracellular bacteria.

Chloramphenicol, and also macrolides such as erythromycin, selectively inhibit bacterial protein synthesis by binding to the 50S subunit, and the binding sites are obviously closely orientated, as binding of one inhibits binding of the other. Other significant protein inhibitors include clindamycin (50S subunit), fusidic acid (associated factors) and mupirocin (enzymes which couple amino acids to tRNA for delivery to the ribosome).

INHIBITORS OF NUCLEIC ACID SYNTHESIS OR FUNCTION

Important antimicrobial agents in this group include the sulphonamides, quinolones, nitroimidazoles and rifampicin. Sulphonamides achieve their selective effect by interrupting the metabolic pathways that lead to the synthesis of functional nucleic acids, while quinolones exert a more direct effect and inhibit some part of DNA replication.

Most bacteria must synthesize folic acid and, unlike mammalian cells, they cannot take it up from the environment. Sulphonamides block folate synthesis at an early stage in the pathway, resulting in a failure to form the purine nucleotides and thymidine that are essential for DNA synthesis. The site of action of the quinolones (e.g. ciprofloxacin) appears to be DNA gyrase (or probably more correctly the DNA gyrase–DNA complex), which is required for the successful unwinding of the supercoiled DNA helix prior to replication and transcription. Why quinolones are so rapidly cidal is largely unknown, although activation of an autolytic process akin to that found with β-lactams has been postulated.

The basis of the selective activity of nitroimidazoles (e.g. metronidazole) against anaerobes is the fact that the low redox values found inside anaerobes create a reduced derivative of the drug which is highly antibacterial. This reduced form apparently causes strand breakage in DNA.

Rifampicin and related rifamycins inhibit mRNA formation by binding to the β-subunit of DNA-dependent RNA polymerase. This blocks the action of the enzyme at the initial stage of transcription. As this is a site of frequent mutation, the emergence of resistance to rifampicin is a relatively common occurrence.

BACTERIAL RESISTANCE TO ANTIBIOTICS

NATURAL (INTRINSIC) RESISTANCE

Some bacteria have always been naturally resistant to certain antibiotics. This so-called intrinsic or inherent resistance is chromosomally mediated, and may result from a variety of factors, including the absence of the drug's target site and failure to penetrate the outer membrane barrier found in Gram-negative bacteria. Such resistance is usually well known and predictable, and causes few therapeutic problems.

Most antimicrobial agents have to pass through the cell wall and cytoplasmic membrane (inner membrane) in order to reach their target site. Not surprisingly, the cell wall of Gram-negative bacteria, which is more complex than that of Gram-positive organisms and which contains an outer membrane (Figure 2.2), offers more of a barrier to many antibiotics, including penicillins and vancomycin, than does the Gram-positive outer layer.

The outer membrane contributes a number of important characteristics to Gram-negative bacteria. It contains endotoxins responsible for initiating the events associated with septic shock, and it serves as a permeability barrier to antimicrobial agents. In order to pass through the outer membrane, antibacterial drugs must either be water soluble and small enough to pass through the water-filled porin channels, or be able to diffuse through the lipid matrix of the membrane. Antimicrobial agents with an appropriate surface charge may promote the latter by destabilizing the lipopolysaccharide cross-linking on the surface of the outer membrane. Vancomycin is the classic example of an antibiotic that is too bulky to pass through the porin channels, and cotrimoxazole is an example of a lipid-soluble antimicrobial agent which can passively diffuse through membranes. Aminoglycosides are an example of cationic drugs which may facilitate their own uptake by disrupting the integrity of the membrane.

ACQUIRED RESISTANCE

More important to the clinician than the well-documented intrinsic resistance, is the ability of bacteria to acquire resistance after having previously been susceptible to the therapeutic agent in question. This acquired resistance may occur by spontaneous mutation or by the transfer of DNA from a resistant clone to a previously susceptible recipient. While acquired resistance may be slow to emerge (e.g. penicillin resistance in pneumococci), it more often develops rapidly, as evidenced by resistance of *Staphylococcus aureus* to benzylpenicillin, which occurred in the 1940s and 1950s. The emergence of resistance in previously susceptible microbes is clearly one of the major concerns facing the medical profession.

Spontaneous mutation. Bacteria are in a constant state of flux, and spontaneous mutations that result in resistance to most antibiotics occur with reasonable frequency *in vitro* (around 10^{-5} to 10^{-8} per cell division). A single mutation at an appropriate genetic site usually produces only a slight decrease in susceptibility, with the level of resistance rising with successive mutations at other sites. Although the presence of the antibiotic does not specifically induce resistance, it has recently been suggested that the rate of mutation in general may be influenced by the presence of antibiotic residues in humans, other animals and the environment.

In most established infectious processes (where bacterial cell numbers invariably exceed 10^8), a small number of bacteria will be present that are resistant to the antibiotics that will subsequently prove to be effective in curing that disease. The host defence mechanisms are usually capable of eliminating these resistant forms. Seldom do antibiotics alone sterilize tissues and secretions of their abnormal bacterial load. It is therefore not surprising that infectious processes are extremely difficult to treat in the absence of adequate white blood cell numbers or function (e.g. chronic granulomatous disease, where deficiency in intracellular killing by phagocytes leads to recurrent and chronic bacterial infections). A high prevalence of spontaneously occurring resistant mutants is a feature of certain bacteria (e.g. *Pseudomonas aeruginosa*), and is associated with high bacterial numbers in locations where a poor blood supply or limited antibiotic diffusion results in the bacteria being exposed to low antibiotic concentrations over long periods of time.

DNA transfer. It seems that the remarkable ability of many bacteria to acquire resistance genes by transfer of DNA from one cell to another accounts for most examples of clinically significant antibiotic resistance. Genetic transfer between cells usually involves small extrachromosomal pieces of DNA termed plasmids, although occasionally chromosomal genes may also be transferred. As well as being transferred between closely related strains, plasmids can be transferred between bacteria that belong to apparently unrelated and diverse genera. Chromosomally mediated resistance is clearly more stable and permanent than is resistance associated with plasmid genes. Plasmids may be spontaneously lost, especially in the absence of any appropriate selection pressure (e.g. the presence of an antibiotic).

The three classical methods of gene transfer are transformation, transduction and conjugation. In the former, small fragments of DNA liberated from one bacterial cell become incorporated into the genome of a second cell. This is considered to be the main mechanism responsible for penicillin resistance in pneumococci. In transduction, fragments of DNA from the donor cell are carried into the recipient cell via a bacterial virus or bacteriophage. This method appears to be significant in the transfer of plasmid DNA between staphylococci. In conjugation, plasmid DNA (or occasionally chromosomal DNA) is transferred between cells via pili which serve as conjugation tubes. This process seems to occur regularly with bacteria in the intestinal tract, and can result in the transfer of DNA between totally unrelated genera.

Of particular interest (and concern) in recent times has been the discovery of transposons, conjugative transposons and integrons. It seems that fragments of chromosomal or plasmid DNA can move from one DNA site to another within a cell (transposon) or between cells (conjugative transposon). Such 'jumping genes' or transposons are not self-replicating and must be included in a replicating DNA element (e.g. a plasmid or chromosome) for further propagation. They often involve large fragments of DNA containing multiple genes, including those coding for antibiotic resistance, and in the case of conjugative transpons, conjugal DNA transfer. The existence of transposons helps to explain how a single antibiotic resistance gene can become spread over a wide variety of unrelated replicons and between diverse strains. Transposons have been linked with the formation of multi-resistant R-plasmids, and with the acquisition of methicillin resistance in *Staphylococcus aureus* and vancomycin resistance in enterococci. Plasmids and conjugative transposons are proving to be very interactive gene-transfer elements, with an impressive capacity for interactions both with each other and with other genetic elements.

Integrons, which are usually found on transposons, specialize in creating clusters of genes, including those that confer antibiotic resistance. The integron provides an integration site for incoming gene cassettes, plus an enzyme (integrase) that mediates orientation-specific integration of these genes and a promoter that ensures expression of the resulting operon (Salyers and Amábile-Cuevas, 1997).

TOLERANCE

In 1970, a pneumococcal mutant was discovered in which the minimum bactericidal concentration (MBC) of penicillin was many times that of the minimum inhibitory concentration (MIC), an unusual event for an essentially bactericidal antibiotic. The term 'antibiotic tolerance' has been coined to describe this novel type of response in which cell-wall-inhibiting β-lactams act primarily as bacteriostatic agents. Although the mechanism of this resistance is unknown, it seems that factors which interfere with the microbe's normal autolytic enzyme system appear to be responsible. Tolerance (MBC ⩾ 32 times the MIC) is now a well-documented laboratory phenomenon in streptococci, and staphylococci, and appears to be unusually frequent among bacterial pathogens that cause valvular endocarditis.

BIOCHEMICAL MECHANISMS OF RESISTANCE

Among the clinically important strains of antibiotic-resistant bacteria, certain well-defined biochemical mechanisms of resistance are found. These are as follows:
1. enzymatic modification or destruction of the drug;
2. alteration of the target site, thereby reducing or eliminating binding of the antibiotic;
3. a reduction in intracellular accumulation of the drug, either by decreased permeability or by increased efflux of the drug;
4. acquisition of a replacement for the metabolic step inhibited by the drug (metabolic bypass).

These mechanisms are summarized in Figure 3.3.

Drug modification
Soon after the introduction of penicillin, strains of *Staphylococcus aureus* resistant to its action appeared. These were shown to elaborate a series of enzymes, termed penicillinases, now known as β-lactamases, which hydrolysed the drug to an inactive product, penicilloic acid (see Figure 3.4). This led to the development and use of the so-called penicillinase-stable (resistant) penicillins, of which flucloxacillin is an example.

Genes for the elaboration of β-lactamases can be plasmid or chromosomally encoded. The *bla* gene responsible for TEM-1 β-lactamase production is part of a transposon that can move from one plasmid to another; it has spread widely among many Gram-negative bacteria (Palzkill, 1998). While some β-lactamases, particularly in Gram-positive bacteria, are inducible (i.e. the very low levels of enzyme produced in the absence of a β-lactam are dramatically increased in the presence of the antibiotic), many Gram-negative organisms synthesize β-lactamases continuously,

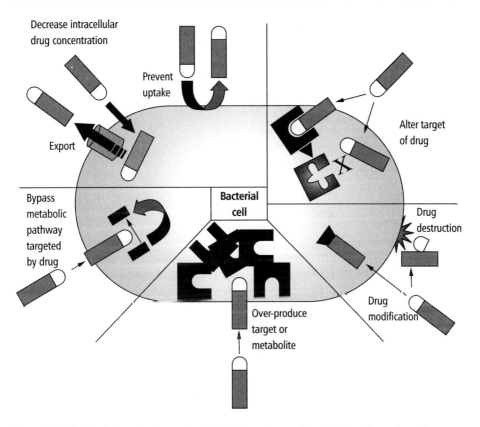

Figure 3.3 Biochemical mechanisms of antimicrobial resistance. Adapted from Murray (1997).

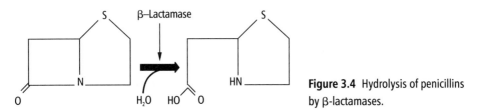

Figure 3.4 Hydrolysis of penicillins by β-lactamases.

albeit often in small amounts. Clinically important chromosomal mutations have occurred in a number of Gram-negative bacteria which originally only produced inducible enzymes. These single-step mutations have resulted in constitutive high-level β-lactamase (e.g. hyper-production of the K1 chromosomal β-lactamase by *Klebsiella oxytoca*).

The escalating world-wide use of third-generation cephalosporins during the early 1990s is thought to be related to the emergence of numerous new extended-spectrum β-lactamases (ESBL) that are capable of inactivating most penicillins, cephalosporins and monobactams, but not 7-methoxy cephalosporins (cephamycins) or the carbapenems. As a general rule, ESBL activity is plasmid mediated and thus readily transferable. Most ESBLs are mutants of the classical TEM-1 and SHV-1 enzymes. ESBL production has created potential problems concerning the treatment of infections by a variety of Enterobacteriaceae, especially those by *Klebsiella* species (see Table 3.1). In an attempt to curb the increasing numbers of new ESBLs, restrictions on the use of third-generation cephalosporins have now been implemented in many countries. ESBLs are inhibited by (i.e. susceptible to) clavulanic acid-type β-lactamase inhibitors.

Enzymatic inactivation is an important mechanism of resistance to the aminoglycosides (Miller *et al.*, 1997). Many of these aminoglycoside-modifying enzymes are coded by R-plasmids and may confer full or partial resistance. They result in a variety of changes (e.g. acetylation, adenylylation and phosphorylation) of appropriate substrates. The enzymes are often plasmid or transposon encoded, and are classified according to the precise type of modification performed and also by the site of modification on the aminoglycoside molecule (e.g. aminoglycoside-acetylating enzymes; AAC).

Alteration of the target site

This type of resistance has been described for almost all groups of antibiotics, including β-lactams, quinolones and glycopeptides, and most often arises from the selection of rare mutants from among an otherwise susceptible cell population.

In recent years, methicillin (flucloxacillin, dicloxacillin, nafcillin)-resistant strains of *Staphylococcus aureus* (MRSA) have emerged world-wide, and in some countries now represent over 30% of all hospital-derived and/or 10% of community-derived *Staphylococcus aureus* isolates. First reported in *Staphylococcus aureus* by Jevons in 1961, methicillin resistance has always been relatively common in some coagulase-negative species (e.g. *Staphylococcus epidermidis*). MRSA strains are considered to be therapeutically resistant to all β-lactams (including imipenem), and are often resistant to other potentially useful antistaphylococcal agents (e.g. rifampicin, ciprofloxacin). While multiple resistance

Table 3.1 ESBL-producing Enterobacteriaceae[a]

Microbe	Number of isolates tested	Number of ESBL-producing isolates	Prevalence (%)
Enterobacter sakazakii	6	1	17
Serratia marcescens	40	2	5.0
Klebsiella oxytoca	45	2	4.4
Citrobacter freundii	68	3	4.4
Klebsiella pneumoniae	341	12	3.5
Enterobacter aerogenes	87	1	1.1
Enterobacter cloacae	156	1	0.6
Escherichia coli	1143	2	0.2

[a]Adapted from Emery and Weymouth (1997).

seems to be a feature of hospital-associated MRSA strains, those strains that are found in the community are often susceptible to non-β-lactam anti-staphylococcal drugs. The range of non-β-lactams to which strains are susceptible tends to vary geographically.

Methicillin resistance is usually governed by an acquired chromosomal gene (*mec*A) which codes for a new penicillin-binding protein, PBP2a. This protein has a relatively low affinity for all of the common β-lactam antimicrobial agents, and in the presence of antibiotic is able to take on the critical tasks of cell wall assembly that are normally performed by the other more antibiotic-susceptible PBPs. PBP2a only functions when all other normal PBPs have been inactivated. As the level of resistance does not correlate with the amount of PBP2a produced, it is clear that other chromosomal and/or plasmid-controlled factors can be involved in methicillin resistance. The *mec*A gene is under the control of associated suppressor and inducer genes, and is located on a large (e.g. 30-kb) transposable fragment of DNA (the mec locus), which often harbours a number of other genes that are responsible for resistance to a variety of unrelated antibiotics. The mec locus is absent from methicillin susceptible *Staphylococcus aureus* (MSSA), and has possibly been acquired from coagulase-negative staphylococci.

Also increasing in incidence world-wide are strains of *Streptococcus pneumoniae* (pneumococci) that are either less susceptible (MIC 0.1–1.0 mg/L) or clearly resistant (MIC $\geq$ 2.0 mg/L) to penicillin. Such strains are often multi-resistant and appear to be mainly community based. The transition to penicillin resistance occurs following a series of steps involving incorporation (by transformation) of fragments of different genes that code for penicillin-binding proteins (i.e. the cell wall synthesis enzymes). This results in 'mosaic genes' containing parts derived from one or more other streptococcal species. The subsequent gene product – a modified normal PBP – has lowered affinity for penicillin and other β-lactams. This has led to the name 'low-affinity penicillin-binding proteins', although it should be remembered that the real function of these enzymes is to make cell wall, not to bind penicillin. PBP1a, PBP2b and PBP2x of pneumococci seem to be primarily involved in this resistance (Smith and Klugman, 1998).

A variation of making an altered or alternative target site is seen with vancomycin-resistant enterococci (VRE). In enterococci, the normal target for vancomycin is the pentapeptide peptidoglycan precursor terminating in D-alanyl-D-alanine. The glycopeptides vancomycin and teicoplanin bind to this terminal dipeptide, inhibiting further cell wall synthesis (Arthur *et al.*, 1996; Murray, 1997; French, 1998). At present, at least five phenotypes of vancomycin resistance have been described (VanA-VanE), which can be distinguished both genetically and on a number of phenotypic characteristics, including vancomycin and teicoplanin MIC levels (Murray, 1998; Casadewall and Courvalin, 1999). In the best studied of the phenotypes, VanA resistance (high-level resistance to both vancomycin and teicoplanin) is associated with acquisition of a transposon containing seven genes, all of which appear to play some role in resistance. The enzyme products of three of these genes (*vanH*, *vanA*, *vanX* or the *vanHAX* operon) are important in the formation of a new depsipeptide D-alanyl-D-lactate terminus to the pentapeptide. The genes encode a dehydrogenase (VanH) and a ligase (VanA) that synthesize D-alanyl-D-lactate, and a D,D-dipeptidase (VanX) that hydrolyses the endogenous D-alanyl-D-alanine and thereby limits the availability of precursors containing the target of glycopeptides. The D-lactate residue results in an ester rather than amide linkage of the terminal moiety (Nicas et al., 1997; Marshall et al., 1998). The net result is the loss of one of the key hydrogen bonds important in complex formation with vancomycin, and binding is at least 1000-fold weaker. The D-alanyl-D-lactate depsipeptide allows cell growth to occur in the presence of glycopeptides when all of the normal dipeptide substrate has been removed, an important function of some of the gene products. The role of the animal growth promotant, avoparcin, in the selection of enterococci expressing the VanA phenotype was a hotly debated and controversial topic in Europe during the early 1990s, and was a major influence in the banning of antimicrobial growth promotants in general in Europe from mid-1999.

Most VREs of clinical significance are strains of *Enterococcus faecium*, although inherent low-level vancomycin resistance (VanC phenotype) is a feature of rare human pathogens such as *Enterococcus gallinarum*. Low-level or intermediate vancomycin resistance (IVR) has recently been detected in MRSA strains in Japan and the USA. The genetic and biochemical mechanisms involved are unknown, although they are probably not related to *vanA* or *vanB* genes and their substrates.

Quinolone antimicrobial agents target DNA gyrase (topoisomerase II), an enzyme that is essential for DNA replication. Structural alteration to one of the two subunits comprising this enzyme can result in high-level quinolone resistance – a feature that is increasingly being seen in *Pseudomonas aeruginosa*, *Staphylococcus aureus* and *Campylobacter jejuni*.

Decreased intracellular drug accumulation

Most β-lactams reach their target site in Gram-negative bacteria by passing through the water-filled porin channels that extend across the outer membrane. These porins consist of a trimer of proteins, with passage of the antibiotic being largely governed by its physical size. Resistance to imipenem is associated with the loss of a specific porin protein (D2). The roles of other outer membrane structures (e.g. lipopolysaccharide) in limiting membrane permeability are still largely unresolved.

It is becoming increasingly apparent that at least as important as decreased outer membrane permeability is the ability of some bacteria, such as *Pseudomonas aeruginosa*, to excrete (efflux) antibiotics actively out of the cell through the inner and/or outer membranes. Mutants of *Pseudomonas aeruginosa* have been found with high-level quinolone resistance associated with the production of new outer membrane proteins in the 51–54 kDa range. These proteins apparently actively transport antibiotic out of the cell. Trends in bacterial resistance to fluoroquinolones have recently been reviewed (Acar and Goldstein, 1997). There is concern about the increasing resistance to fluoroquinolones being seen in *Pseudomonas aeruginosa*, *Klebsiella pneumoniae*, *Serratia marcescens* and *Staphylococcus aureus*.

Metabolic bypass

The bacteriostatic effect of sulphonamides is associated with their competitive inhibition of an enzyme which links p-aminobenzoic acid (PABA) and pteridine to form dihydropteroate. Resistant Gram-negative bacilli synthesize a dihydropteroate synthetase that is unaffected by the drug. Trimethoprim blocks a later step in the same metabolic pathway (see Figure 3.1) by inhibiting the dihydrofolate reductase enzymes of susceptible bacteria. Resistant strains synthesize a new trimethoprim-insensitive transposon or plasmid-encoded dihydrofolate reductase as well as the normal drug-sensitive chromosomal enzyme (Greenwood, 1995c).

CLINICAL PROBLEMS OF DRUG RESISTANCE

Most infections that are encountered in surgery are still readily amenable to treatment with traditional and proven antimicrobial agents. However, it is obvious that problems of resistance are increasing world-wide, and that appropriate steps to curtail its progression and spread must be implemented (Archibald et al., 1997; Pfaller et al., 1998a,b; Anon., 1999b). It is apparent that bacteria isolated from patients in the 1930s had virtually no resistance genes, and that strains resistant to new antimicrobial agents were not often seen in many species of bacteria until the agents had been used for some time. In general, we know little about how resistant genes have emerged, and why some appeared rapidly while others have taken decades to become apparent, and why they are so resistant to elimination (Salyers and Amábile-Cuevas, 1997). However, the emergence and propagation of resistant microbes does seem to be associated with heavy antibiotic use and a ready means of spread of the resistant microbe. Thus we can delay emergence by using less antimicrobial agents and we can retard dissemination by practising good hygiene, infection-control policies and avoidance of specific agents that select for resistance genes in contiguous populations (Johnson, 1998). In the 1995 World Health Organization (WHO) list of essential and 'reserved agents', second- and third-generation cephalosporins, the fluoroquinolones and vancomycin were considered to be the most important reserved agents (Couper, 1997). Resistance patterns clearly vary geographically, and local monitoring and management of any outbreaks must obviously be a top priority if the problem is to be minimized (Goldmann and Huskins, 1997).

Extended spectrum β-lactamases

The increasing use of third-generation cephalosporins has apparently been associated with the appearance of novel β-lactamases, including the extended spectrum TEM-1 and SHV-1 derivatives (extended-spectrum β-lactamases, ESBLs) (Medeiros, 1997). Organisms such as *Klebsiella pneumoniae* are commonly involved, although ESBLs are now appearing more frequently in *E. coli* and other enterics (see Table 3.1). Detection of ESBL-producing Enterobacteriaceae in the laboratory is not easy, and strains are often incorrectly recorded as susceptible to antibiotics such as ceftazidime.

In a survey in the USA, at least 1.5% of routine isolates of the family Enterobacteriaceae were found to express ESBLs (Emery and Weymouth, 1997). By using National Committee for Clinical Laboratory Standards (NCCLS) guidelines, 26–39% of the ESBL-producing isolates would have been reported as susceptible to ceftazidime, depending on the routine sensitivity method employed. On the other hand, using cefpodoxime, all ESBL-producing strains would have been recorded as resistant or intermediate. At least seven of nine patients infected with ESBL-producing strains were cured with therapy which included an extended-spectrum cephalosporin. Other studies have suggested that the clinical effectiveness of cephalosporins in treating ESBL-producing enterics may be variable and dose dependent (Thauvin-Eliopoulos et al., 1997). It therefore becomes unclear whether or not it is clinically necessary or cost-effective to institute testing for ESBL production on a routine basis. In cases where there is a poor or slow clinical response to third-generation cephalosporins in patients with infections by Gram-negative enteric bacilli, the possibility of participation of an ESBL-producing strain should be considered and more appropriate therapy initiated. Others clearly consider ESBL producers to be clinically significant and, as far as resistance problems are concerned, the Gram-negative equivalent of MRSA and VRE.

As a general rule, cephalosporins, penicillins and monobactams should be avoided when considering treatment options for ESBL-producing enterics. Potentially more useful agents include fluoroquinolones (e.g. ciprofloxacin), carbapenems (e.g. meropenem), which are probably the drug of choice, and β-lactam/β-lactamase inhibitor combinations (e.g. piperacillin/tazobactam), which must be given in high doses (Gold and Moellering, 1996). While β-lactamase inhibitors appear to be useful in therapy, others consider that their protective role has yet to be fully established (Jacoby, 1998). ESBL-producing strains are often also resistant to gentamicin, which may be a significant indicator of their potential presence. This information is summarized in Table 3.2.

Members of the genus *Enterobacter* have been described as 'pathogens poised to flourish at the turn of the century' (Sanders and Sanders, 1997). Most of these microbes are innately resistant to older antimicrobial agents and have the ability to develop resistance to new agents rapidly. They appear to be increasingly important in nosocomial infections, with multiple resistant strains apparently accompanying high cephalosporin use. Although enterobacters have been implicated in surgical wound infection at almost every body site, two areas have been recognized as particularly prone to involvement, namely the sternum/mediastinum and posterior spinal tissues. Important species include *Enterobacter cloacae* and *Enterobacter aerogenes*. Suggested antibiotic regimens include combinations of third-generation cephalosporins and aminoglycosides, cotrimoxazole with or without a cephalosporin or aminoglycoside, fluoroquinolones and imipenem (see Sanders and Sanders, 1997).

Like other resistant hospital microbes, enterobacters have the ability to spill over into the community, and they may be associated with clinical syndromes traditionally blamed on bacteria such as *Streptococcus pyogenes* and *Staphylococcus aureus*. One such species is *Enterobacter cancerogenus* (syn. *Enterobacter taylorae*), which has been associated with wound infections and bacteraemia in individuals environmentally exposed following multiple trauma to the head or severe crush injuries (Abbott and Janda, 1997). However, unlike many of the hospital-associated enterobacters, *Enterobacter cancerogenus* is usually susceptible to a wide range of extended-spectrum β-lactams.

Table 3.2 Treatment options for selected antimicrobial-resistant bacteria

Bacteria	Resistant to	Alternative considerations	Comment
Extended-spectrum β-lactamase (ESBL)-producing Enterobacteriaceae	Penicillins and cephalosporins, including extended-spectrum penicillins and third-generation cephalosporins; aztreonam	Meropenem, imipenem, ciprofloxacin, piperacillin/tazobactam	Often also cross-resistant to aminoglycosides such as gentamicin
Methicillin-resistant *Staphylococcus aureus* (MRSA)	*all* β-lactams; often other useful anti-staphylococcal agents (e.g. fusidic acid, rifampicin, ciprofloxacin, gentamicin)	Vancomycin	Check for susceptibility to cotrimoxazole, fusidic acid, rifampicin, clindamycin, gentamicin, ciprofloxacin and newer quinolones
Vancomycin-resistant enterococci (VRE)	Vancomycin	Imipenem plus ampicillin (strains may still be susceptible to ampicillin plus gentamicin)	Quinupristin/dalfopristin combinations for *Enterococcus faecium*; newer fluoroquinolones are under investigation
Bacteroides fragilis species group resistant to some β-lactams and clindamycin	Cefotetan, cefoxitin, clindamycin	Metronidazole, meropenem, imipenem, chloramphenicol	Resistance to metronidazole and carbapenems almost unknown; resistance to most other β-lactams and clindamycin slowly escalating
Streptococcus pneumoniae non-susceptible (intermediate) or resistant to penicillin (PRP)	Penicillins, cephalosporins	Vancomycin plus third-generation cephalosporin (e.g. ceftriaxone); high-dose amoxycillin in some situations	A third-generation cephalosporin (e.g. ceftriaxone) alone can be considered with some less susceptible strains; quinupristin/dalfopristin and new fluoroquinolones are under investigation

Methicillin-resistant *Staphylococcus aureus*

Staphylococcus aureus and especially coagulase-negative staphylococci such as *Staphylococcus epidermidis* are becoming increasingly resistant to all antistaphylococcal β-lactams (e.g. flucloxacillin, dicloxacillin, nafcillin, co-amoxyclav, cefuroxime) (Archer and Climo, 1994). Treatment options for these so-called methicillin-resistant isolates (MRSA) depend largely on the strain involved and on local sensitivity patterns (see Table 3.2). (*Note that as laboratory tests for MRSA now invariably involve oxacillin rather than methicillin, the term oxacillin-resistant* Staphylococcus aureus, *or ORSA, is possibly more correct.*) By definition and for clinical purposes, MRSA are considered to be resistant to *all* β-lactams, including imipenem (Howe et al., 1996). Many of the high-level MRSA strains (oxacillin MIC ⩾ 128 mg/L) are also multi-resistant, with glycopeptides such as vancomycin being the only reliably effective agent. Depending on the strain involved, other potential alternatives include cotrimoxazole, rifampicin, fusidic acid, gentamicin, ciprofloxacin and newer fluoroquinolones such as clinafloxacin (Lowy, 1998). These are often used in some type of combination. Mortality rates in patients with MRSA bacteraemia are significantly higher than those in patients with methicillin-susceptible *Staphylococcus aureus* (MSSA) bacteraemias (Catchpole et al., 1997), although this may simply be a reflection of age, severity of illness, comorbidity or the poor cidal activity of drugs such as vancomycin. MRSA strains with oxacillin MICs in the range 32–128 mg/L are becoming increasingly common in community settings (cMRSA). These are not usually multi-resistant, and respond to a variety of antimicrobial agents other than vancomycin (e.g. a combination of ciprofloxacin and either rifampicin or fusicid acid). They appear to be more virulent than the largely hospital-based high-level oxacillin resistant strains, and appear to be associated with typical *Staphylococcus aureus* disease (e.g. skin and soft-tissue lesions, osteomyelitis). In New Zealand, these cMRSA strains are now responsible for around 20% of all community-acquired *Staphylococcus aureus* infections in some geographical localities. A similar pattern may be occurring in Australia and the USA. An interesting observation has been that critically ill hospitalized patients may frequently be colonized and/or infected with more than one genotype of MRSA at any one time. The clinical and laboratory implications of this finding are as yet unknown. It also seems that the potential for human-to-animal spread of MRSA exists (Seguin et al., 1999).

In hospitals where MRSA are present or anticipated, empirical therapy for suspected serious staphylococcal sepsis should still utilize a penicillinase-stable penicillin (e.g. flucloxacillin) together with an appropriate agent to cover methicillin-resistant strains. The choice of the second antimicrobial agent depends on local sensitivity patterns and also to some extent on the seriousness of the infection. Alternatives include vancomycin, cotrimoxazole, gentamicin, fusidic acid, rifampicin and ciprofloxacin. More specific monotherapy or combined therapy can be initiated when sensitivity results become available. Strains of MRSA with reduced susceptibility to vancomycin are slowly appearing world-wide (Hiramatsu et al., 1997; Sieradzki et al., 1999; Smith et al., 1999). Considerations for their control have been proposed (Wenzel and Edmond, 1998).

Mupirocin-resistant *Staphylococcus aureus*

For some time, mupirocin ointment has been used to reduce nasal and skin colonization by *Staphylococcus aureus*, especially methicillin-resistant strains (MRSA). It appears to be a safe and effective procedure (Hudson, 1994; Fernandez *et al.*, 1995). Unfortunately, indiscriminate use of this topical antibiotic and its availability over the counter have resulted in the rapid emergence of staphylococcal clones resistant to its actions (Eltringham, 1997). The incidence of mupirocin-resistant *Staphylococcus aureus* has risen from 0% to over 15% in some countries in recent years. Clearly restrictions should have been imposed on the availability and use of this agent long before resistant strains emerged to their present level. Resistance appears to be plasmid mediated and readily transmissible. Controlling skin and/or nasal colonization by MRSA resistant to mupirocin is both difficult and expensive (Irish *et al.*, 1998).

The epidemiology and control of MRSA in the hospital environment is a hotly debated subject. Although spread from infected and colonized patients appears to occur directly via the hands (McDonald, 1997), environmental sources are perhaps underestimated as a reservoir of contagion (Blythe *et al.*, 1998; Stacey *et al.*, 1998). MRSA can clearly survive for several weeks on inanimate objects such as curtains and dry mops. In addition, it is now apparent that cMRSA may be regularly introduced into the hospital environment from the community in a similar manner to methicillin-sensitive *Staphylococcus aureus* strains. Prevention of cross-infection in hospitals obviously revolves around continual implementation of standard isolation procedures, including the cohorting of infected patients and strict personal hygiene (e.g. hand-washing). In view of the increasing community reservoir, screening of staff (and probably also patients) would now seem to be unjustified, although some may question this with regard to hospital areas where the impact of MRSA infection is likely to be greatest (e.g. intensive-care units) (Humphreys and Duckworth, 1997). It is debatable how valid the recently revised guidelines for the control of MRSA infection in UK hospitals will be for other countries. The 1998 Working Party Report (see Duckworth *et al.*, 1998) is certainly meticulous in its detail and recommendations, but fails to address the fact that MRSA are now becoming a community problem and may be introduced into hospitals via normal visitors as well as by colonized and/or infected patients and staff.

Glycopeptide-resistant enterococci

Vancomycin-resistant enterococci (VRE) have emerged globally as important nosocomially-derived pathogens, especially in the intensive-care setting (Arthur *et al.*, 1996; Dahlberg *et al.*, 1996; Gold and Moellering, 1996; Gorbach, 1996; Howe *et al.*, 1996; Cars, 1997; Murray, 1997). Reasons for the emergence of VRE are largely speculative (Donskey and Rice, 1999), although the increasing therapeutic use of cephalosporins to which enterococci are inherently resistant, and vancomycin in human medicine, and of similar drugs (e.g. avoparcin) in agriculture and veterinary medicine, are possible explanations. Metronidazole (and possibly other anti-anaerobe agents) also appears to favour intestinal colonization by enterococci. Clearly vancomycin must be reserved for infections involving microbes for which no alternative antibiotic is available. This does not include most cases of *Clostridium difficile*-associated diarrhoea, where adequate alternatives (e.g. metronidazole) are available. As vancomycin is not absorbed, oral therapy with this drug (as used in severe cases of antibiotic-associated diarrhoea) has the potential to select vancomycin-resistant enterococci to a much higher degree than does parenteral therapy (Murray, 1997). An interesting observation has been the association of previous metronidazole therapy as a risk factor for VRE bacteraemia (Lucas *et al.*, 1998). Enterococcal overgrowth in the gastrointestinal tract and translocation of the microbes into the bloodstream appear to be potentiated by metronidazole.

In addition to vancomycin and teicoplanin (glycopeptides), enterococci have increasingly developed high-level resistance to aminoglycosides and penicillins (enterococci have always shown low-level intrinsic resistance to both of these antimicrobial agents). This has important clinical implications, as the usual treatment for serious enterococcal infection involves the synergistic combination of penicillin plus gentamicin, or amoxycillin (or ampicillin) with or without gentamicin. High-level resistance to aminoglycosides (with the exception of streptomycin) abolishes the synergistic activity of these combinations. In fact, strains of *Enterococcus faecium* that are resistant to every useful currently available antibiotic have been described (Leclercq, 1997).

For enterococci that do not show high-level resistance to gentamicin or penicillins, ampicillin (or amoxycillin) plus gentamicin is the therapy of choice. In cases where sensitivity patterns suggest that these agents may be ineffective, vancomycin or teicoplanin are logical choices. Therapeutic options for VRE may be limited (see Table 3.2). Where strains do not reveal high-level ampicillin resistance (i.e. MIC ⩽ 32 mg/L), a combination of high-dose ampicillin plus imipenem has been suggested as a possibility, while streptogramin combinations (e.g. quinupristin/dalfopristin, Q/D) have been used experimentally for the treatment of severe infection due to multi-resistant strains of *Enterococcus faecium* (strains of *Enterococcus faecalis* are resistant to streptogramins). An interesting recent observation has been that a combination of ampicillin plus ceftriaxone seems to be superior to ampicillin alone with enterococci that show high-level gentamicin resistance (Gavalda *et al.*, 1999), despite the fact that enterococci are inherently resistant to all cephalosporins, including ceftriaxone. Some of the newer fluoroquinolones (e.g. clinafloxacin) appear to have excellent *in-vitro* activity against multi-resistant VRE, and have been suggested as possible therapeutic alternatives. Time will tell. An alarming finding has been the detection of Q/D-resistant *Enterococcus faecium* in turkeys fed sub-therapeutic levels of virginiamycin, a mixture of two streptogramins (Welton *et al.*, 1998).

Clearly the use of glycopeptides such as vancomycin and teicoplanin should be restricted to scenarios where they are the only available agent. Any suggestion that they should be used routinely for prophylaxis in patients undergoing clean orthopaedic surgical procedures must be questioned. Single-dose teicoplanin is effective and well tolerated in this situation (Mini *et al.*, 1997), but should only be considered in cases where there is a high risk of infection due to methicillin-resistant strains of *Staphylococcus aureus* and *Staphylococcus epidermidis*, or in patients with major β-lactam allergy.

Pencillin-resistant pneumococci

Pneumococci that are less susceptible or resistant to penicillin now create problems in the treatment of pneumococcal meningitis and, to a lesser extent, otitis media (Gold and Moellering, 1996; Gorbach, 1996; Howe *et al.*, 1996; Quintiliani *et al.*, 1996; Bradley and Scheld, 1997). Treatment options for such strains include third-generation cephalosporins (e.g. ceftriaxone, cefotaxime), often combined with vancomycin or rifampicin (Table 3.2). Laboratory data suggest that some of the newer extended-spectrum cephalosporins (e.g. cefpirome) may also be worthy of consideration, and are possibly superior to ceftriaxone. Imipenem is also very active against these organisms, but is contraindicated in meningitis because of the risk of seizures. In this respect, meopenem is a better choice (Bradley and Scheld, 1997; Fitoussi *et al.*, 1998). The newer fluoroquinolones are also possible considerations (Haria and Lamb, 1997). Although resistant pneumococci may appear to be susceptible to chloramphenicol in laboratory tests, this drug appears to perform poorly in clinical situations. High-dose intravenous penicillin or amoxycillin is probably effective for pneumonia unless the strain is highly resistant (Gold and Moellering, 1996; Quintiliani *et al.*, 1996). As a general rule, 'resistant' pneumococcal infection in non-meningeal sites responds equally well to third-generation cephalosporins and amoxycillin, although monotherapy with either of these for meningitis should not be used (Quintiliani *et al.*, 1996).

While decreased susceptibility to penicillin is generally accompanied by a corresponding decrease in susceptibility to third-generation cephalosporins, the latter have a greater affinity for PBPs and are potentially more potent. The recent emergence of pneumococci that display high-level resistance to cefotaxime but only intermediate susceptibility to penicillin suggests that the complete saga and implications of penicillin resistance in pneumococci have yet to unfold. In many countries, penicillin resistance (MIC $\geq$ 2 mg/L) is now approaching 5% of pneumococcal strains recovered from invasive disease, with a further 10–15% of strains showing reduced (intermediate) penicillin susceptibility (MIC 0.1–1.0 mg/L). Corresponding cefotaxime figures are around 1% in resistant strains (MIC $\geq$ 2 mg/L) and 8% in the less susceptible or intermediate range (MIC 1 mg/L). In all cases these percentages appear to be increasing rapidly.

Antibiotic-resistant anaerobes

Antibiotic resistance, including plasmid-mediated resistance, among the *Bacteroides fragilis* species group is well documented. A recent survey (Snydman *et al.*, 1996) has suggested that the rates of resistance of *Bacteroides fragilis* and non-*fragilis* species of the *Bacteroides fragilis* group (e.g. *Bacteroides thetaiotaomicron*) to some previously useful drugs are increasing. No resistance was found to metronidazole and chloramphenicol, and virtually no resistance to imipenem. Increasing levels of resistance to β-lactam combinations (e.g. piperacillin/tazobactam) were seen in the non-*fragilis* species (although 95–98% of strains were still susceptible), and to cefotetan and clindamycin in both *Bacteroides fragilis* and non-*fragilis* species (see also Patey *et al.*, 1994). Metronidazole clearly remains the drug of choice for infections involving anaerobes of the *Bacteroides fragilis* and *Clostridium* groups, although this agent shows poor activity against some Gram-positive anaerobes such as actinomycetes and propionibacteria.

There is now accumulating and compelling evidence that inadequate empirical treatment (e.g. use of drugs to which the causal microbe is resistant) is associated with a poor outcome and significantly increased mortality (e.g. Montravers *et al.*, 1996). A surgeon's knowledge of local emerging resistance patterns is now clearly an important therapeutic consideration. Table 3.2 summarizes the important aspects of antibacterial resistance and the possible therapeutic alternatives.

PREVENTION STRATEGIES

A few simple actions can dramatically reduce the emergence and spread of resistant clones (Jones and Fraise, 1997). Restrictions on the availability and use of antibiotics in human and veterinary medicine and the agricultural industry should be implemented and/or continued. There is no doubt that the use of vancomycin analogues, fluoroquinolones and streptogramins as growth promotants in animals has contributed significantly to the selection of resistant bacteria. Such animal strains have the ability to infect or colonize humans, and to serve as a reservoir of resistant genes that are readily transposable to human bacterial strains (Barton, 1998). Rotating or diversifying the type of antibiotic used is another logical measure, and it could become hospital policy. It is common sense to employ high initial doses and complete an adequate duration of therapy. Where treatment is to be long-term (e.g. in tuberculosis), or involves microbes that are known to readily become resistant (e.g. *Pseudomonas aeruginosa*), combinations of antibiotics with differing modes of action should be considered. Certainly, short-term high-dose therapy seems to carry less risk of selecting resistant cells than does low-dose prolonged treatment, although the emergence of MRSA has been correlated with the use of cephazolin for surgical prophylaxis, with subsequent amplification of strains being attributed to third-generation cephalosporin use (Schentag *et al.*, 1998).

In many cases, a reduction in the spread of MRSA and methicillin-susceptible *Staphylococcus aureus* (MSSA) in the hospital environment can be achieved by implementing strict hand-washing protocols (Cox and Conquest, 1997). Increasing the number of beds in a fixed area heightens the risk of cross-infection (Kibbler *et al.*, 1998). However, it does seem that MRSA have the ability to survive in the environment (e.g. on dry mops or on television sets) to a greater extent than has been suggested in the past for strains of *Staphylococcus aureus* (Oie and Kamiya, 1996; Blythe *et al.*, 1998; Stacey *et al.*, 1998). In addition, colonization of the skin (especially if damaged), throat or bladder by MRSA is often seen in hospitalized patients, and may prove almost impossible to eliminate. There is also clear evidence that antibiotic-resistant staphylococci (e.g. ciprofloxacin-resistant *Staphylococcus epidermidis*) are selected by antibiotic use, and are readily transferred between patients and staff. New and revised comprehensive guidelines for the control of MRSA infection in hospitals (in the UK) have recently been formulated (Duckworth et al., 1998), and the key points from these guidelines are summarized in Table 3.3.

Implementation of strict infection control measures is often seen as a non-priority area in some hospitals. This is unfortunate if emerging problems of resistance are to be recognized and measures rapidly implemented to stop their spread. Collaboration and dialogue between laboratory workers, epidemiologists, clinicians and administrators is obviously of major importance if the unrestricted spread of resistant microbes is to be curtailed. This function is usually undertaken by a hospital infection control committee, and has to be supported by responsible hospital policy and appropriate funding. While the function of this committee would include (*inter alia*) inspection of water towers and air-conditioning systems for legionellae, formulating an HIV-screening policy and monitoring nosocomial infections, other essential functions would include environmental studies to determine the degree of local contamination, monitoring of wound infections and antibiotic-prescribing habits, and management of patients with infectious disease. The infection control committee should also monitor the post-operative management of wounds, ward wound-dressing procedures, and the disposal of contaminated dressings and linen – hence the critical role of hospital administration on the committee. While it has been widely advocated that patients who

Table 3.3 Key points from the revised guidelines for the control of MRSA[a]

- Reassess previous guidelines in view of changing patterns of MRSA in many countries (e.g. increasing, mainly community-acquired problem).
- Continue to limit intra-hospital spread of MRSA and implement measures to curtail community spread as for normal *Staphylococcus aureus* infections; these are clearly cost-effective.
- Key hospital practices in limiting spread include strict adherence to hand-washing and general cleaning, and patient isolation or cohorting.
- There should be pre-admission screening of patients/staff moving from colonized hospitals to hospitals free of MRSA.
- Where MRSA is endemic, utilize limited resources in areas where the impact of spread is likely to be greatest (e.g. intensive-care units).
- In-house patient and/or staff screening is usually only indicated in high-risk areas.
- Recommended sampling sites for screening are the nose, perineum or groin, any lesions and manipulated sites. The throat should be screened in denture wearers and if nasal carriage persists despite topical therapy. The bladder may be an undetected site of long-term colonization.
- Prolonged (> 7 days) or repeated (more than twice during one admission) courses of mupirocin should be avoided in order to prevent selection of resistant strains.
- Ward closure should be recommended only as a last resort.

[a]Reproduced from Duckworth et al. (1998).

shed large quantities of MRSA should be temporarily isolated, cohorting all infected patients together may be a more acceptable and feasible alternative to single-room isolation.

However, despite such in-house procedures, MRSA often continue to survive and spread in hospitals. Reservoirs are notoriously difficult to pinpoint, and may include diagnostic and other obscure locations, and both patients and staff within an institution and those entering from another hospital. In some institutions, screening of staff and patients transferring between hospitals or following outbreaks has been suspended, with no apparent deleterious effects on the occurrence or spread of MRSA. This may not be wise in settings where there is a high prevalence of resistance, or in areas where infection may have catastrophic consequences.

In cases where appropriate surveys have been undertaken, it seems that MRSA have now become established in the community, where they may account for up to 20% of all *Staphylococcus aureus* infections that require medical attention. These strains behave like MSSA and may be brought into hospitals by overtly infected individuals (e.g. those with skin abscesses that require surgical intervention), but show little intra-hospital spread or persistence provided that cross-infection prevention policies are maintained. Such strains usually show low-level oxacillin resistance, are not multi-resistant, and clearly have different epidemiological features to the usual high-level, methicillin-resistant, multi-resistant hospital strains. The presence and importance of these community-based strains (cMRSA) clearly has implications with regard to hospital MRSA policies, and the choice of empirical therapy in patients admitted with probable *Staphylococcus aureus* sepsis. Approximately 50% of staphylococcal infections that occur in hospitals originate in the community; patients are admitted who either already have or are incubating these infections, which can then present opportunities for cross-infection to others.

The control of VRE, ESBL-producing enterics and other multi-resistant Gram-negative bacilli in hospital environments has not received the attention or press coverage of the 'killer bug', MRSA. Limiting the occurrence and spread of resistant Gram-positive cocci and Gram-negative bacilli relies on infection-control principles as outlined for MRSA, and wherever possible restriction in the use of the 'selecting' antibiotic (e.g. vancomycin for VRE, and third-generation cephalosporins for ESBL-producing Gram-negative bacilli). The emergence and importance of resistant microbes is clearly associated with heavy antibiotic use in a setting that facilitates easy microbial multiplication and spread (Archibald *et al.*, 1997; Murray 1997).

In some cases, detection of antimicrobial resistance in bacterial isolates by conventional methods (see p. 160) may not always be easy or even possible. As the genetic basis of resistance has been elucidated for many antimicrobial agent-microbe combinations, genetic methods for assessing resistance have been developed for many microbes, including bacteria such as MRSA, VRE, ESBL-elaborating Gram-negative bacilli, and resistant strains of *Mycobacterium tuberculosis* (see Cockerill, 1999).

TREATMENT STRATEGIES AND PRINCIPLES

In a 1993 European survey covering larger hospitals in France, Germany, Italy, Spain and the UK (Halls, 1993), general surgery was the second largest user of antimicrobial agents after internal medicine, and a major factor related to antibiotic use in surgery was prophylaxis. Within the surgical specialties, orthopaedic surgery followed by lower abdominal and gastric/upper gastrointestinal surgery were associated with the largest consumption of antimicrobial agents. In each case, prophylaxis accounted for at least 50% of the antibiotic therapy days (over 75% in the case of orthopaedics). General surgery was one of the largest users of injectable antibiotics. Antimicrobial costs are thus a major consideration in surgery budgets, and may escalate unnecessarily if recommended guidelines are not followed (Ryono *et al.*, 1996).

Patients who are undergoing surgery may have their defences to infection compromised by necrosis or major trauma, foreign bodies and prostheses, underlying illness or malignancy, exogenous immunosuppressive medication, and tissue and endothelial damage associated with endotoxin, cytokine and nitric oxide release. In cases where there are single or multiple abscesses, surgical intervention is the first priority. Resuscitation with the maintenance of oxygenation by adequate perfusion of injured tissue will aid repair and the resurrection of normal microbial barriers. The possible addition of immune globulin may compensate for the depletion of complement and other essential blood components. Water retention, an increase in cardiac output, and associated hypoxia and acidosis are common in critically ill patients, and need to be taken into account when formulating and prescribing antimicrobial therapy.

Antimicrobial therapy demands an initial clinical evaluation of the nature and extent of the infective process as well as knowledge of the likely causative microbe(s). With many surgery-related infections, the causal microbes are usually part of the indigenous or normal flora. The clinical assessment should be supported wherever possible by laboratory investigations aimed at establishing the microbial aetiology and the sensitivity of the microbe(s) involved to possible therapeutic regimens. In most cases, the causal microbe(s) is likely to be susceptible to a variety of antimicrobial agents, and the choice of one or more of these depends on a variety of drug and patient factors. These include the preferred route and frequency of administration, known adverse effects, the drug's pharmacological and kinetic features (including its likely penetration and activity at the site where the microbe is growing), the age and clinical state of the patient, and cost. Failure to reassess and/or alter empirical treatment on receipt of laboratory results is clearly associated with increased mortality rates (Salonen *et al.*, 1998).

PHARMACOKINETIC FACTORS

Route of administration
The aim of therapy should be to produce antimicrobial concentrations of the drug at the site of infection during the dosing interval. Many antibiotics do not

produce adequate plasma levels when given by mouth, and are available only as injectable (parenteral) preparations. Wherever possible, however, antibiotics are given by mouth, as this involves tremendous savings both in the cost of the antibiotic and in the cost and ease of its administration. Local chemical irritation of the gastric and intestinal mucosa may accompany the use of many oral antibiotics – hence the increasing use of enteric-coated or slow-release formulations. Oral therapy may present problems in burn patients with gastrointestinal ileus. Parenteral administration, especially prolonged infusion via the intravenous route, also carries the inherent risk of catheter-related sepsis. In addition, some intravenously administered antibiotics produce local irritation or phlebitis. It may be necessary to adjust the pH of an infusion by suitable buffering. Pain may also accompany intramuscular injection of drugs, which can require the concurrent use of a local anaesthetic.

As with other forms of stress, surgery is a time of fluctuating haemodynamics, physiological shifts, intense metabolic changes, and protein catabolism and anabolism. Sometimes these changes can occur within hours. There is a reasonable body of literature on the adverse effects of surgery on gastric emptying, but not on the effects of surgery on the distribution, metabolism and excretion of drugs.

Following major procedures, gastric emptying is reduced or absent, which prevents the delivery of orally administered drugs to their major site of absorption, namely the small bowel. Some antimicrobial agents (e.g. metronidazole) can be administered rectally, although this route of administration depends largely on the type of surgery employed. The use of intramuscular or (more commonly) intravenous routes of antimicrobial drug administration is common following surgery. Although costly, these guarantee adequate drug delivery provided that the circulation is intact. However, distribution is affected by changes in blood volume, alterations in circulation, increases in the extracellular fluid and changes in the circulating plasma protein levels. Little is known about alterations in drug metabolism following surgery.

In addition, the renal elimination of drugs is affected in patients post-operatively, although the effects of biliary clearance during this period are difficult to determine. Despite the lack of research into pharmacokinetics during the postoperative period, and given the immense and often sudden changes that are observed in patients following surgery, it is reasonable to recommend vigilance with regard to antimicrobial therapy during this period (Kennedy and van Riji, 1998).

The most direct method of ensuring adequate concentrations of antibiotic in the blood and tissues, is by intravenous injection. Single rapid intravenous injections result in the highest immediate concentrations, although this benefit may be offset to some degree by rapid excretion (Greenwood, 1995d). Therefore many agents are given by infusion over 15 to 20 min. Degradation of the drug in the diluting solution can occur with prolonged infusion times, which also tend to result in lower plasma levels. It seems that combinations of penicillins and aminoglycosides that are regarded as synergistic *in vivo* may inactivate each other when mixed in solution at high concentrations for any length of time. The β-lactam ring of the penicillin and amino groups of the aminoglycoside appear to interact chemically. This interaction probably has no significance at the concentrations that are achieved in the body, but penicillins and aminoglycosides should not be mixed together in intravenous solutions (see p. 89).

Rapid infusion or injection results in high plasma concentrations of the drug, but these rapidly decline as the drug diffuses into extracellular spaces and, to a lesser degree, the cells (see Figure 3.5). Most antibiotics diffuse poorly through cell membranes, and only a few reach therapeutically significant intracellular levels. After intravenous infusion, antibiotics rapidly equilibrate throughout their volume of distribution. In the critically ill and in trauma patients, volumes of distribution may vary widely, mainly decreasing tissue concentrations of the drug. Consequently, under-dosing may result.

Distribution and protein binding

The volume of distribution represents the hypothetical volume that is necessary to dilute an administered dose in order to achieve a measured plasma concentration, and it varies between antibiotics depending on the degree to which they are bound to plasma proteins (e.g. albumin). Antibiotics that are not bound to plasma proteins move freely into extravascular fluids by passive diffusion. As protein binding increases, less and less of the drug is free to move, although it seems that until binding exceeds 90% there is little change in the free drug concentration. In reality, antibiotics that are highly protein bound have a small

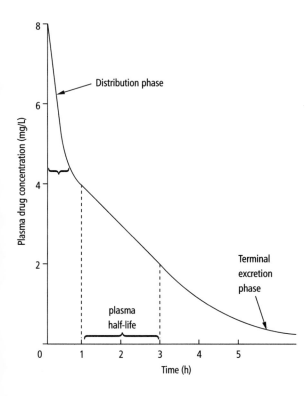

Figure 3.5 Decline in plasma concentration of a drug after its intravenous injection. Three phases may be distinguished: an initial rapid fall as the drug is distributed from the plasma; a less rapid fall during the main period of excretion and metabolism; and a terminal slow decline representing, for example, the release of the drug from binding sites. Redrawn from Greenwood (1995d).

volume of distribution equivalent to the vascular space, whereas less tightly bound agents distribute throughout the extracellular water space. With some drugs (e.g. ceftriaxone), over 90% of the administered dose is bound and is antibacterially inactive. Only the free diffusible fraction of the drug is considered to have any antibacterial effect.

In order to be therapeutically effective, adequate concentrations of the free drug must occur at the site of infection. It is clear from their proven clinical efficacy on susceptible microbes that this occurs with compounds that display high levels of protein binding. Presumably manufacturers' dosing schedules give total doses which result in appropriately high levels of unbound drug in relevant areas. In addition, it is obvious that bound drug will go wherever the protein goes, and that includes the inflammatory exudate poured into the site of infection. Once at

Table 3.4 Antibiotic doses in renal failure and for patients on dialysis

Antibiotic	Dose for normal renal function	Adjustment for renal failure[a]			Method[b]	Dose after haemodialysis
		Mild	Moderate	Severe		
Penicillins						
Penicillin G	0.5–4 mu q 4 h	NC	75%	20–50%	D	Yes
Amoxycillin	0.25–0.5 g q 8 h po	NC	q 8–12 h	q 24 h	I	Yes
Ampicillin	0.25–2 g q 6 h	NC	q 6–12 h	q 12–24 h	I	Yes
Flucloxacillin	1–2 g q 4 h	NC	NC	NC	–	SRF
Dicloxacillin	0.25–0.5 g q 6 h po	NC	NC	NC	–	SRF
Nafcillin	1–2 g q 4 h	NC	NC	NC	–	SRF
Piperacillin	3–4 g q 4–6 h	NC	4 g q 8 h	4 g q 12 h	I	Yes
Piperacillin/tazobactam	3.375 g q 6 h	NC	2.25 g q 6 h	2.25 g q 8 h	DI	Yes
Ticarcillin[c]	3 g q 4 h	NC	2 g q 4–8 h	2 g q 12 h	DI	Yes
Ticarcillin/clavulanate	3.1 g q 4 h	NC	2 g q 4–8 h	2 g q 12 h	DI	Yes
Cephalosporins						
Cephazolin	1–2 g q 8 h	NC	q 12 h	q 24–48 h	I	Yes
Cefamandole	1–2 g q 4–6 h	NC	q 8 h	q 12 h	I	Yes
Cefuroxime	0.75–1.5 g q 8 h	NC	q 8–12 h	q 24 h	I	Yes
Cefoxitin	2 g q 8 h	NC	q 12 h	q 24–48 h	I	Yes
Cefotetan	2–3 g q 12 h	NC	q 24 h	q 48 h	I	Yes
Cefotaxime	2 g q 8 h	NC	q 12–24 h	q 24 h	I	Yes
Ceftriaxone	1–2 g q 12 h	NC	NC	NC	–	Usual
Ceftazidime	2 g q 8 h	NC	q 24–48 h	q 48 h	I	Yes
Ceftizoxime	2 g q 8 h	NC	q 12–24 h	q 24 h	I	Yes
Cefpodoxime	0.4 g q 12 h	NC	q 24 h	q 48 h	I	Yes
Cefepime	2 g q 12 h	NC	1–2 g q 24 h	0.5 g q 24 h	D/I	Yes
Aminoglycosides						
Amikacin[d]	15 mg/kg/day	c.75% q 24 h	c.50% q 24 h	c.25% q 24–48 h	DI	Yes
Gentamicin, tobramycin[d]	5–7 mg/kg/day	c.75% q 24 h	c.50% q 24 h	c.25% q 24–48 h	DI	Yes

Table 3.4 Continued

Antibiotic	Dose for normal renal function	Adjustment for renal failure[a] Mild	Moderate	Severe	Method[b]	Dose after haemodialysis
Others						
Ciprofloxacin	500–750 mg q 12 h po 400 mg q 12 h IV	NC	50–75%	50%	D	Yes
Sparfloxacin	400 mg on day 1, then 200 mg qd	NC	400 mg on day 1, then 200 mg q 48 h	ND	I	ND
Clindamycin	300–900 mg q 6 h IV	NC	NC	NC	–	Usual
Chloramphenicol	0.25–1 g q 6 h	NC	NC	NC	–	Usual
Fusidic acid	500 mg q 8 h	NC	NC	NC	–	Usual
Imipenem/cilastatin	0.5–1 g q 6 h	NC	50% q 8–12 h	25–50% q 12 h	DI	Yes
Meropenem	0.5–1 g q 8 h	NC	50% q 8–12 h	50% q 24 h	DI	ND
Metronidazole	250–750 mg q 8 h	NC	NC	NC	–	Usual
Rifampicin	600 mg/day	NC	NC	NC	–	Usual
Teicoplanin	6–12 mg/kg/day	NC	q 48 h	q 72 h	I	SRF
Vancomycin	15 mg/kg q 12 h	q 24 h	q 1–4 days	q 4–7 days	I	SRF

NC, no change; ND, no data; SRF, dose as for severe renal failure.
[a]Mild, creatinine clearance 50–80 mL/min;
  moderate, creatinine clearance 10–50 mL/min;
  severe, creatinine clearance < 10 mL/min.
[b]I, interval extension; D, dose reduction.
[c]Ticarcillin needs adjustment of dosage after peritoneal dialysis.
[d]Monitoring of levels is recommended.

this site, it appears that the concentration of free (active) drug is still defined by the equilibrium between free and bound drug (i.e. as the unbound fraction diffuses away, more of the plasma-bound drug dissociates, and the equilibrium between the bound and free compound is maintained) (Greenwood, 1995d).

Not surprisingly, the amount of free drug (i.e. the agent's volume of distribution) is important in prophylactic settings, where delivery of the antibiotic to the surgical site is governed by its extravascular fluid concentration. In addition, the level of free drug at the site of infection is higher with antibiotics that are not highly protein bound. Therefore high levels of protein binding are generally considered to be an undesirable characteristic, although they may result in more prolonged persistence of antibiotics that are cleared by the kidney (only free drug is filtered out).

Half-life

The time required for the concentration of drug in the plasma to fall by 50% is called the plasma half-life (see Figure 3.5). Half-lives vary considerably among antibiotics (e.g. 30 mins for benzylpenicillin and 8 h for ceftriaxone). In addition, drug that is distributed into the tissues may be bound there, and this drug, as distinct from that in the plasma, may be relatively slowly remobilized and excreted (Greenwood, 1995d).

The relationship between half-life and protein binding is expressed as the area under the concentration–time curve (AUC) for free drug (see Solomkin, 1988). Not surprisingly, antibiotics with short half-lives have small AUCs and must be given frequently in order to maintain effective levels at the site of infection. High levels of protein binding clearly decrease the AUC for free drug.

Breakdown and excretion

The route of elimination is the other important pharmacokinetic variable that influences the choice of antibiotic, and it can vary considerably even within a given group or class of antibiotics. For example, most members of the β-lactam group are excreted in the urine, where they reach very high levels. Imipenem is hydrolysed by a renal peptidase to form an inactive and toxic metabolite in the urine. The action of the enzyme is inhibited by cilastatin, which is usually administered in equal amounts with imipenem. This combination results in high urine levels of active drug and prolongs its circulating half-life. Co-administration of probenecid also reduces renal excretion of β-lactams and results in a consequent prolongation of the plasma half-life.

In contrast, some β-lactams (e.g. ceftriaxone), are excreted in significant amounts in the bile (see Table 4.4). This results in high antibiotic concentrations in the bile and potentially the small intestine. Obviously such drugs are good candidates for prophylaxis and treatment of biliary tract infections, as drug concentrations usually far exceed inhibitory levels of the bacteria involved. Unfortunately, high drug levels in the bile may affect its viscosity (e.g. biliary sludging occurs with some cephalosporins) or disturb the balance of the intestinal flora and result in the development of problems such as antibiotic-associated colitis (see p. 206).

Cerebrospinal fluid penetration

Most antibiotics do not cross the blood-brain barrier unless the cerebral capillary junctions open, i.e. penetration of many antibiotics into the cerebrospinal fluid (CSF) is poor in the absence of inflammation. In addition, capillary permeability may revert to normal before all viable bacteria are eradicated, resulting in lowered drug levels and incomplete sterilization of the CSF. Antimicrobial agents which show reasonable CSF penetration in the presence of inflammation include cefuroxime, ceftriaxone, ceftazidime, gentamicin, chloramphenicol, metronidazole and some fluoroquinolones.

Renal failure

In patients with impaired renal function, the dose of those antibiotics that are excreted predominantly by the kidneys should be reduced. In general, nephrotoxic drugs should be avoided in patients with impaired renal function. For patients on dialysis, the manufacturer's literature should be consulted. More common drugs that require no or only a minor reduction in dosage in cases of renal insufficiency include cefotaxime, ceftriaxone, the macrolides, flucloxacillin, rifampicin and metronidazole. Those that require a significant dosage reduction include aminoglycosides, cephalosporins and penicillins not listed above, aztreonam, imipenem, vancomycin, cotrimoxazole and ciprofloxacin. Tetracyclines (other than doxycycline), nalidixic acid and norfloxacin should be avoided in patients with renal

Table 3.5 Categories of antibacterial agent safety in pregnancy

	Australian Drug Evaluation Committee[a]	Food and Drug Administration (USA)[b]
Category A	Drugs that have been taken by a large number of women of childbearing age, without any proven increase in the frequency of malformations or other direct or indirect harmful effects on the fetus having been observed	Controlled studies in women demonstrate no fetal risk
Category B	Drugs that have been taken by only a limited number of pregnant women and women of childbearing age, without an increase in the frequency of malformations or other direct or indirect harmful effects on the fetus having been observed	Animal studies demonstrate no fetal risk, but there are no human trials
Group B1	Studies in animals have not shown evidence of an increased occurrence of fetal damage	or
Group B2	Studies in animals are inadequate or may be lacking; however, available data show no evidence of an increased occurrence of fetal damage	Animal studies demonstrate a risk that is not corroborated in human trials
Group B3	Studies in animals have shown evidence of an increased occurrence of fetal damage, the significance of which is considered to be uncertain in humans	
Category C	Drugs that, because of their pharmacological effects, have caused or may be suspected of causing harmful effects on the human fetus or neonate without causing malformations. These effects may be reversible. Product information should be consulted for further details	Animal studies demonstrate a fetal risk, but there are no human trials or Neither human nor animal studies are available
Category D	Drugs that have caused, are suspected to have caused, or may be expected to cause an increased incidence of human fetal malformations or irreversible damage. These drugs may also have adverse pharmacological effects. Product information should be consulted for further details	Evidence exists for fetal risk in humans, but benefits may outweigh risk
Category X	Drugs that have such a high risk of causing permanent damage to the fetus that they should not be used in pregnancy or when there is a possibility of pregnancy	Evidence exists for fetal risk in humans, and benefit is clearly outweighed by risk

[a]Australian Drug Evaluation Committee (1996) *Medicines in pregnancy*, 3rd edn. Canberra: Australian Government Publishing Service.
[b]Food and Drug Administration categories as cited by Bartlett JG. (1996) *Pocket book of infectious disease therapy*. Baltimore MD: Williams and Wilkins; or by Sanford JP, Gilbert DN, Sande MA. (1996) *Guide to antimicrobial therapy*, 26th edn. Dallas, TX: Antimicrobial Therapy Inc.

failure. Mechanisms of renal excretion and hepatic metabolism are poorly developed in newborns, who are susceptible to chloramphenicol toxicity. Elderly patients often have impaired renal function and are also particularly susceptible to aminoglycoside toxicity. The doses of common antibacterial agents to be used in renal failure and for patients on dialysis are shown in Table 3.4.

Pregnancy and lactation

A reduction in plasma concentration of penicillins may occur during pregnancy. The fetus may be susceptible to the toxic effects of aminoglycosides, while excretion of sulphonamides in breast milk may be harmful to neonates. The antibiotics that are probably safe during pregnancy are penicillins, including amoxycillin/clavulanic acid (co-amoxyclav), cephalosporins, aztreonam, imipenem and most macrolides (but not erythromycin estolate), while those that are contraindicated include all tetracyclines, quinolones and possibly metronidazole (at least in the first trimester). Most other antibacterial agents present a potential risk and should only be used when absolutely essential. No antimicrobial agent is completely contraindicated in all circumstances in pregnancy, but the categories listed in Tables 3.5 and 3.6 may allow the choice of a safer alternative or raise the question of deferring treatment.

With breast-feeding mothers, tetracyclines (which affect the teeth), chloramphenicol (which causes bone-marrow suppression) and sulphonamides (which

Table 3.6 Relative safety of antibacterial agents in pregnancy[a]

Antibacterial agents	Category Australian	FDA (USA)	Comment
Aminoglycosides			
Amikacin	D	D	Congenital deafness has followed streptomycin use, therefore any aminoglycoside should only be used when essential. Single-dose spectinomycin appears to be safe for gonorrhoea in pregnancy
Gentamicin	D	C	
Tobramycin	D	D	
Cephalosporins			
Cefotaxime	B_1	B	All cephalosporins listed here are regarded as being safe
Cefotetan	B_1	B	
Cefoxitin	B_1	B	
Cefpirome	B_2	B	
Ceftazidime	B_1	B	
Ceftriazone	B_1	B	
Cefuroxime	–	B	
Cephazolin	B_1	B	
Penicillins			
Amoxycillin	A	B	All penicillins are regarded as being safe
Amoxycillin/clavulanate	B_1	B	
Ampicillin	A	B	
Benzylpenicillin	A	B	
Flucloxacillin	B_1	B	
Piperacillin	B_1	B	
Piperacillin/tazobactam	B_1	B	
Ticarcillin	B_2	B	
Ticarcillin/clavulanate	B_2	B	
Other β-lactams			
Aztreonam	B_1	B	Probably safe but inadequately studied
Imipenem/cilastatin	B_3	C	Maternal intolerance in some pregnant animals – caution advised

Table 3.6 Continued

Antibacterial agents	Category		Comment
	Australian	FDA (USA)	
Macrolides			
Erythromycin (except estolate)	A	B	Erythromycin estolate is associated with an increased risk of cholestatic hepatitis in pregnant women. Other erythromycins are routinely used for chlamydial infection in pregnancy
Quinolones			
Ciprofloxacin	B_3	C	Quinolones, including nalidixic acid, cause arthropathy and cartilage damage in juvenile experimental animals. Experience with nalidixic acid suggests that it is safe in practice. The newer fluoroquinolones are relatively contraindicated, probably only because less experience has accrued
Norfloxacin	B_3	C	
Tetracyclines			
Doxycycline	D	D	All tetracyclines are contraindicated during the second and third trimesters because of the possibility of retardation of fetal skeletal development and enamel hypoplasia with discoloration of teeth. Intravenous use has been associated with hepatotoxicity and nephrotoxicity in pregnant women, especially in those with renal insufficiency or if overdosed
Other antibacterial agents			
Choloramphenicol	A	–	Possibility of grey baby syndrome when given near term, also idiosyncratic aplastic anaemia
Clindamycin	A	–	Appears safe
Fusidic acid	C	–	Risk of kernicterus when given perinatally, otherwise safe
Metronidazole	B_2	B	Mutagenic in bacteria and carcinogenic in mice after long-term use, therefore usually avoided in first trimester. A recent meta-analysis suggests that it is safe (*American Journal of Obstetrics and Gynecology* (1995) **172**, 525–9)
Rifampicin	C	C	Cause of skeletal malformation in animals and of postnatal haemorrhage in humans. If given in late pregnancy, the mother and neonate should be given vitamin K
Teicoplanin	B_3	–	
Vancomycin	B_2	B	No studies: use with caution (oral vancomycin for *Clostridium difficile* disease is safe as it is not absorbed)

[a]Adapted from Lang, S. (ed.) (1997) *Guide to pathogens and antibiotic treatment 1998*. Auckland: Adis International with permission.

compete for bilirubin binding to albumin) should be avoided. Metronidazole may give the milk a bitter taste. Whenever possible the following antibacterial agents should be avoided: azithromycin, clarithromycin, fluoroquinolones, nitrofurantoin and sulphonamides.

Cidal vs. static action

When host defences are impaired, complete killing or lysis of bacteria may be required in order to achieve a successful outcome. In this situation it may be detrimental to combine bactericidal and bacteriostatic agents. For instance, with a combination of tetracycline (static) and penicillin (cidal on actively growing cells), the final outcome is in theory mainly dependent on the static agent, and antibiotic killing should not occur. In addition, infectious agents that are capable of survival within phagocytic cells are protected from the action of many antibiotics. Some antibiotics (e.g. rifampicin, macrolides, quinolones) may achieve effective intracellular antibacterial concentrations without detrimentally influencing the metabolism and activity of phagocytic cells.

DRUG INTERACTIONS

The major pharmacokinetic mechanisms involved in the interactions of antibiotics are as follows:
1 changes in gastrointestinal absorption of the antibiotic; and
2 interference with the process of hepatic metabolism (Robinson, 1998).

Gastrointestinal absorption

Changes in drug absorption may be attributed to several mechanisms, including the presence of food, complexing of the interacting drugs, and changes in the gut bacterial flora. As a general rule, antibiotics are usually absorbed within 30 min of ingestion of a dose. Drugs which are to be taken on an empty stomach (e.g. flucloxacillin) should therefore be taken 30 min before starting a meal, 30 min after a light meal or 2 h after a heavy meal. Dairy foods (e.g. milk, cheese, yoghurt) can significantly reduce the levels of fluoroquinolones (e.g. ciprofloxacin).

Antibiotics may combine with other drugs or with elements in food in the gut to form insoluble, poorly absorbed complexes. Examples include fluoroquinolones complexing with several positively charged (cationic) elements (e.g. magnesium, aluminium, bismuth, calcium, iron, zinc).

Some antibiotics are partly metabolized by bacteria that are normally present in the intestinal tract. Changes in the nature or composition of this normal flora that are brought about by antibiotic use may therefore affect the blood levels of these drugs. The oestrogen in oral contraceptives is inactivated in the liver and excreted back into the intestinal tract via the bile. Microbes in the intestines reactivate these conjugated oestrogens, which are subsequently reabsorbed. Although reliable data is lacking, it is thought that most antibiotics may result in reduced blood levels of oestrogen and oral contraceptive failure. Additional contraceptive precautions should therefore be taken until the course of antibiotics is finished and oral contraceptive efficacy has been re-established (usually about 7 days after finishing antibiotics).

Changes in hepatic metabolism

Antibiotics may inhibit enzymes that are important in the metabolism of other drugs, resulting in raised levels of the drug in question or a prolongation of its clinical effect. On the other hand, a few antibiotics may cause induction of liver enzymes and reduced levels of co-administered drugs. Some examples of more significant surgery-related drug interactions (see also Table 3.7) include the following:
1 the ability of rifampicin to decrease (by induction of liver enzymes) the half-life of many drugs, including anticoagulants, prednisone and azole antifungal agents;

2 the ability of chloramphenicol and metronidazole (by a reduction in liver enzymes) to increase the anticoagulant activity of compounds such as warfarin, and to promote bleeding;
3 the ability of ciprofloxacin (by inhibition of cytochrome P_{450} metabolism) to lead to the accumulation of toxic concentrations of theophyllines;
4 the ability of some broad-spectrum cephalosporins to increase the effects of anticoagulants which rely on the inhibition of prothrombin synthesis by vitamin K manufactured by intestinal microbes;
5 the potentiation of the neuromuscular blocking effects of aminoglycosides by anaesthetic agents;
6 the additive renal toxicity of combined aminoglycoside and vancomycin therapy;
7 competitive inhibition and enhanced nephrotoxicity when cyclosporin is given with aminoglycosides.

It is well documented that certain antibiotics, such as aminoglycosides and β lactams, are chemically and physically incompatible and should not be mixed. For maximum effect, they should be administered 20–30 min apart. It does not appear to matter in what order the drugs are given. Antagonism may occur with certain combinations of drugs, especially when bacteriostatic drugs inhibit cell division or protein synthesis required for a second drug's bactericidal activity. The actual relevance of this potential interaction *in vivo* is not known (see also Table 3.8 for more detailed lists of adverse reactions and complications).

Table 3.7 *In vivo* incompatibilities of selected antimicrobial agents[a]

Antibiotic(s)	Interacting agent(s)	Adverse reaction
Aminoglycosides	Curare-like agents	Neuromuscular blockade
Chloramphenicol Sulphonamides Isoniazid	Phenytoin	Phenytoin toxicity
Ciprofloxacin Erythromycin	Theophylline	Agitation, convulsions
Metronidazole	Alcohol	Nausea and vomiting (disulfiram effect)
Rifampicin	Oral contraceptives	Decreased efficacy
Sulphonamides Nalidixic acid	Anticoagulants	Increased anticoagulation
Tetracyclines	Oral antacid preparations and oral iron	Decreased tetracycline absorption

[a]Adapted from Greenwood, D. (ed.), (1995) *Antimicrobial chemotherapy*, 3rd edn. Oxford: Oxford University Press.

Table 3.8 Relative frequency of selected adverse reactions to antimicrobial agents[a]

Antimicrobial agent	Frequent	Infrequent
Aminoglycosides	Nephrotoicity (reversible) Ototoxicity (non-reversible)	Rashes Neuromuscular blockade

Cephalosporins	Hypersensitivity rashes
	Candida overgrowth
	Clostridium difficile overgrowth
		Nephrotoxicity (cephaloridine)
		Anaphylaxis
		Haematological toxicity
		  (e.g. positive antiglobulin test,
		    hypoprothrombinaemia
		    (cefamandole),
		    neutropenia)
		Biliary sludging (ceftriaxone)
		Hepatotoxicity
		Neurotoxicity
		Serum sickness reaction (cefaclor)

Chloramphenicol	Dose-related bone-marrow toxicity	Aplastic anaemia Grey baby syndrome Optic neuritis
Clindamycin	Rash Diarrhoea	Hepatitis Pseudomembranous colitis
Cotrimoxazole and sulphonamides	Rashes (sulphonamide)	Megaloblastic anaemia
Erythromycin	Gastrointestinal intolerance	Cholestatic jaundice Deafness
Fluoroquinolones	Gastrointestinal intolerance Rashes	Confusion/convulsions/dizziness Photosensitivity Destructive arthropathy/arthralgia Tendonitis and rupture
Fusidic acid	Gastrointestinal intolerance (oral)	Hepatotoxicity (intravenous)

Imipenem	Hypersensitivity reactions	Neuropathy
		  (e.g. convulsions or
		    epileptic seizures)

Metronidazole	Gastrointestinal intolerance Metallic taste Disulfiram reaction with alcohol	Peripheral neuropathy Urinary discoloration Rash
Penicillins	Hypersensitivity reactions – mainly rashes *Candida* overgrowth *Clostridium difficile* overgrowth	Haematological toxicity Encephalopathy Interstitial nephritis Hepatoxicity (e.g. cholestasis (with flucloxacillin)) Platelet dysfunction (ticarcillin)

Table 3.8 Continued

Antimicrobial agent	Frequent	Infrequent
Rifampicin	Hepatotoxicity Liver enzyme induction Red discoloration of body fluids	Hypersensitivity 'Influenza syndrome' (intermittent treatment) Haematological toxicity
Tetracyclines	Gastrointestinal intolerance *Candida* overgrowth Dental staining and hypoplasia in childhood	Photosensitivity Nephrotoxicity Staphylococcal enterocolitis
Vancomycin[b]	'Red man' syndrome Fever Phlebitis	Nephrotoxicity Ototoxicity

[a]Adapted from Greenwood, D. (ed.), (1995) *Antimicrobial chemotherapy*, 3rd edn. Oxford: Oxford University Press
[b]Adverse effects are reportedly less common with the more purified preparations available since the 1980s.

ANTIMICROBIAL SIDE-EFFECTS

Hypersensitivity

Among the antibiotics, the β-lactams display the greatest potential to produce hypersensitivity or allergic reactions. The major determinants of this are the penicilloyl derivatives, which are formed by the breakdown of the β-lactam ring associated with penicillins. Penicilloyl derivatives bound to serum and tissue proteins account for over 90% of the immunologically active penicillin metabolites (Anon., 1996a). Skin tests are an inaccurate way of diagnosing penicillin hypersensitivity, and are not without risk. 'Desensitization' is rarely justified.

Hypersensitivity reactions range from mild rashes to life-threatening anaphylaxis (type 1 reactions). Because of the similar structure of the penicillins, hypersensitivity to one agent is usually accompanied by allergy to the whole group. In addition, structural similarity between penicillins and cephalosporins is associated with some degree of cross-reaction between the two groups. However, the incidence of allergy to cephalosporins in patients with a history of penicillin allergy is probably less than 10%. Cephalosporins should therefore only be used with extreme caution in patients with known anaphylactic reactions to penicillins, and the same applies to imipenem and probably also to meropenem (Anon., 1996a). Clinical cross-reactivity between penicillins and monobactams does not appear to be significant, and patients with penicillin allergy can probably be given aztreonam with minimal risk of anaphylaxis. The maculopapular erythematous rash that occurs in patients who are receiving amoxycillin (or ampicillin)-type β-lactams and who have concurrent viral infections (e.g. glandular fever, cytomegalovirus infection) does not appear to have an allergic basis, and is unlikely to reoccur after the viral illness has resolved. Alternatives to β-lactams in patients with serious β-lactam allergy include vancomycin, teicoplanin, clindamycin, erythromycin and aminoglycosides.

Rashes

Skin rashes are probably the most frequent adverse reaction associated with the use of antimicrobial agents. Most reactions are caused by hypersensitivity (see above), but other reactions may occur (e.g. maculopapular, vesicular and bullous eruptions, exfoliation, and erythema multiforme and nodosum). Photosensitivity has emerged recently as a major concern with quinolones, especially in areas of the world that receive bright sunlight. A lupus syndrome is a rare event with penicillins.

Over-rapid infusion of vancomycin can result in release of histamine, which in turn may result in acute flushing, tachycardia and hypotension ('red man' syndrome), or in sudden vomiting. Tetracyclines bind avidly to developing bones and teeth, which can become discoloured and hypoplastic. In addition to joint and cartilage problems in children, fluoroquinolones may cause tendonitis and even tendon rupture, particularly of the Achilles tendon.

Changes in normal flora

Even so-called 'narrow-spectrum' antibiotics can have pronounced effects on the normal flora (e.g. the effects of penicillin on the coccal and anaerobic flora of the mouth). Single-dose prophylaxis has also been shown to influence the flora, which may take several days to return to normal (Sanderson, 1993). A rather alarming finding in this respect has been the selection of distinct strains of *Staphylococcus aureus* by cephazolin prophylaxis (see p. 114 and Kernodle *et al.*, 1998a,b). However, in general the antibiotics which most influence the normal flora are broad-spectrum agents such as tetracyclines, co-amoxyclav and many of the newer extended-spectrum β-lactams. Use of such agents may result in overgrowth of yeasts such as *Candida albicans* and the development of overt oral or vaginal thrush. Broad-spectrum drugs may also result in the overgrowth of resistant strains of *Staphylococcus aureus* in the gut, with a resultant enterocolitis (this was a much feared complication in surgical patients given oral pre-operative antibiotics, but is seldom seen nowadays).

Gastrointestinal effects

Also of importance is the syndrome of antibiotic-associated diarrhoea or colitis (pseudomembranous colitis) caused by toxin-producing strains of *Clostridium difficile*. Strains of *Clostridium perfringens* are rare causes of this condition. Almost all known antibiotics have been associated with *Clostridium difficile* overgrowth, although those most commonly involved appear to be clindamycin, co-amoxyclav and perhaps the cephalosporins. The microbe appears to elaborate two distinct toxins, namely toxin A, which is an enterotoxin and primarily responsible for the diarrhoea, and toxin B, which is a cytotoxin and has a lethal effect on cultured tissue cells. *Clostridium difficile* seems to be part of the normal flora in up to 10% of adults, producing disease when much of the remaining flora is reduced by antibiotic use. The causative anaerobe appears to be capable of hospital spread (possibly via spores), and may present a significant cross-infection problem (see p. 207). In such circumstances, *Clostridium difficile* may be recovered from the stool of over 50% of patients in a ward.

Hepatitis

The liver is a major organ of drug metabolism and excretion. In recent years, considerable concern has arisen in Australia, The Netherlands and Sweden with regard to the propensity of flucloxacillin to evoke serious hepatic disease (e.g. cholestasis). This usually occurs within 6 weeks of the commencement of treatment. Hepatic injury by other penicillinase-stable penicillins (e.g. dicloxacillin, nafcillin, co-amoxyclav) appears to be less common (Vial et al., 1997). Because of these problems, measures to restrict the use of flucloxacillin for minor staphylococcal infections have been instituted. Acute liver injury is seen with a variety of other antimicrobial agents, including ceftriaxone, clindamycin, fusidic acid, the macrolides and probably fluoroquinolones (see Vial et al., 1997). Tetracycline should be avoided in patients with pre-existing liver disease or during pregnancy, as under these conditions liver necrosis may prove fatal.

Neural effects

It has long been known that ototoxicity is an important side-effect of aminoglycoside therapy. It appears to result from excessive drug accumulation, and not necessarily with high peak levels as is often stated. Ototoxicity is irreversible, and increased susceptibility to this adverse effect has been linked to a mutation in mitocondrial DNA (see p. 91). Massive parenteral doses of penicillin may induce encephalitic reactions, while other β-lactams such as imipenem are contraindicated in meningitis because of their ability to induce convulsive or epileptic reactions. β-lactams given to patients with renal impairment and neurological disease may produce focal or grand mal epilepsy. High doses of quinolones may also induce convulsions. Metronidazole may produce a reversible peripheral neuropathy, while neuromuscular blockade is a rare but serious adverse reaction with aminoglycosides, which produce this reaction by a curare-like anticholinesterase effect and by competing with calcium. This is most likely to be seen following the use of muscle relaxants during anaesthesia (Finch, 1995).

Renal effects

As the kidneys are the major route of drug excretion, it is not surprising that nephrotoxicity in some form or other is seen with a number of antibiotics. Benzylpenicillin and occasionally other penicillins may produce a hypersensitivity interstitial nephritis, while of the cephalosporins, cephaloridine and cephalothin are the most nephrotoxic (this has limited their use). However, the aminoglycosides are the antibiotics most frequently associated with nephrotoxicity, which is potentiated by underlying renal disease or the simultaneous use of other nephrotoxic agents (e.g. vancomycin, cyclosporin or intravenous radio-contrast medium as used in computerized tomography). Although often debated, little variation in nephrotoxicity appears to occur between the different aminoglycosides, although it is said to be least with netilmicin and greatest with gentamicin. Proximal tubular epithelial cells appear to become engorged with the drug and burst, leaving the basement layer intact. The toxicity is therefore reversible (unlike aminoglycoside ototoxicity).

Bone-marrow effects

Bone-marrow toxicity may be either selective, affecting only one cell line, or non-selective, producing pancytopenia and marrow aplasia. Immune-mediated haemolysis may occur, as may bleeding from platelet dysfunction or thrombocytopenia. Eosinophilia may represent a hypersensitive reaction (Finch, 1995). While chloramphenicol has achieved notoriety for inducing bone-marrow depression, and the sulphonamide-containing drugs for inducing haematological side-effects, several β-lactams have also been associated with haematological toxicity. Penicillins and cephalosporins have been associated with Coombs' antibody-positive disease, the likes of ampicillin and flucloxacillin with a selective white-cell depression, and a vitamin K-dependent bleeding disorder with cephalosporins possessing a methyl-thiotetrazole side-chain (e.g. cefamandole, cefotetan). Although the latter is uncommon, bleeding has been seen in elderly or malnourished patients undergoing major surgery. It can be treated and/or prevented by vitamin K supplementation.

Fever

While persistent fever may be a result of inappropriate diagnosis or therapy, it seems that antibiotics themselves are capable of causing fever. Suggested reasons for this include direct toxicity, allergic phenomena and liberation of inflammatory-inducing endotoxin and other substances from killed bacteria. Prior or concurrent administration of anti-inflammatory agents may help to alleviate such problems. Recent studies have shown that down-regulation of lymphocyte and macrophage responses by treatment with corticosteroids significantly alleviates the outcome of staphylococcal infections when combined with appropriate antibiotics (Tarkowski and Wagner, 1998).

COMBINATION ANTIMICROBIAL THERAPY FOR SURGICAL INFECTIONS

Combination therapy has been used for as long as antimicrobial agents have been available. Well-established indications for prescribing more than one antibacterial agent at a time include the following:
1 as empirical therapy for serious infections before culture and/or susceptibility results are available;
2 in polymicrobial infections;
3 in an attempt to promote synergy;
4 to minimize the emergence of resistant strains during therapy.

However, combined therapy may result in a variety of problems, including excessive costs, adverse drug reactions, possible antagonism, the selection of multi-resistant strains and superinfection (Anon., 1997b). In addition, it often provides a false sense of security. Some trials have shown that with the currently available antibiotics, combined short-term antibiotic therapy of identified microbes in fact has no real advantage over monotherapy.

The combination of two different groups of antibiotics may result in bactericidal activity that is significantly greater than the sum of either agent alone – this is known as synergy. If the combination results in a decreased effect, it is said to be antagonistic, while any result which is less than synergistic but clearly not antagonistic is described as additive. However, it must be remembered that much of our knowledge of these phenomena is derived from laboratory rather than clinical studies.

Regardless of the susceptibility of enterococci to penicillin alone, the combination of benzylpenicillin or amoxycillin plus an aminoglycoside (e.g. gentamicin) significantly improves the outcome of patients with enterococcal endocarditis and other significant enterococcal infections, where the causative strain does not show high-level gentamicin resistance. The postulated theory is that penicillin damage to the cell wall allows increased uptake of the aminoglycoside and enhanced killing. The emergence of high-level gentamicin resistance in enterococci negates the therapeutic use of such a combination.

Other more commonly used combinations include the following:
1 aminoglycosides (e.g. tobramycin) and either an antipseudomonal penicillin (e.g. piperacillin) or a cephalosporin (e.g. ceftazidime) for serious *Pseudomonas* sepsis;
2 a β-lactam plus a β-lactamase inhibitor (e.g. amoxycillin plus clavulanic acid, or piperacillin plus tazobactam) which in reality is a combination of two β-lactams with one possessing high affinity and resistance to β-lactamase degradation;
3 a penicillinase-stable penicillin (e.g. flucloxacillin) plus either rifampicin or fusidic acid for loculated infections by *Staphylococcus aureus* (e.g. osteomyelitis);
4 quinupristin plus dalfopristin for glycopeptide-resistant *Enterococcus faecium* infections;
5 a combination of metronidazole plus ciprofloxacin for the treatment of intra-abdominal infections. This is possibly microbiologically superior to the usual triple regimen of metronidazole, gentamicin and amoxycillin, but is no different to imipenem alone.

Some of the more significant surgery-related antibiotic combinations are summarized in Table 3.9 (see also Rybak and McGrath, 1996).

THE POST-ANTIBIOTIC EFFECT (PAE)

The phenomenon of persistent suppression of bacterial growth after a short exposure to antimicrobial agents, has been referred to as the post-antibiotic effect (PAE). The PAE is the time in hours required by an organism to demonstrate viable growth after exposure to and removal of an antibiotic. This phenomenon is exhibited by virtually all classes of antimicrobial agents, but especially aminoglycosides and quinolones. It was one of the reasons behind the initial studies of once-daily aminoglycoside dosing. While aminoglycosides and quinolones demonstrate PAEs of around 2–4 h for both Gram-negative and Gram-positive bacteria, most β-lactams and also vancomycin demonstrate PAEs only against Gram-postive bacteria. The exception appears to be the β-lactam imipenem, which exhibits a pronounced PAE against *Pseudomonas aeruginosa*,

Table 3.9 Important surgery-related antibiotic combinations

Clinical situation	Bacteria involved	Antibiotic suggestions[a]
Aspiration pneumonia (within a hospital)	*Staphylococcus aureus*, oral anaerobes, Gram-negative bacilli	Flucloxacillin + gentamicin ± metronidazole or co-amoxyclav + gentamicin
Intra-abdominal sepsis	*Bacteroides fragilis*, coliforms, enterococci, pseudomonads	Ciprofloxacin + metronidazole ± amoxycillin or gentamicin + metronidazole + amoxycillin
Methicillin-resistant *Staphylococcus aureus*	MRSA	Combination of the following depending on local sensitivity patterns – cotrimoxazole, fusidic acid, ciprofloxacin, rifampicin, gentamicin
Necrotizing fasciitis	*Staphylococcus aureus*, pyogenic streptococci, Gram-negative bacilli (including pseudomonads), obligate anerobes	Cefuroxime + ciprofloxacin + metronidazole or cefotaxime + ciprofloxacin + clindamycin
Osteomyelitis	*Staphylococcus aureus*	Flucloxacillin[b] + fusidic acid or rifampicin
Pelvic inflammatory disease	Gonococci, chlamydiae, anaerobes, coliforms, mycoplasmas	Co-amoxyclav + doxycycline or cefoxitin + doxycycline or erythromycin
Pseudomonas septicaemia	*Pseudomonas aeruginosa*	Tobramycin + piperacillin or tobramycin + ceftazidime
Severe hospital-acquired septicaemia	Gram-negative bacilli, pneumococci, *Staphylococcus aureus*, others	Gentamicin + flucloxacillin[b]

[a] Monotherapy with agents such as imipenem or piperacillin/tazobactam may be equally effective.
[b] Substitute vancomycin ± rifampicin if MRSA is common.

particularly *in vivo*. This may be in some way related to leucocyte activity. Synergistic PAEs have been observed between combinations of β-lactams and aminoglycosides, and by the addition of rifampicin to a number of other antibiotics, although the clinical relevance of all this is as yet undetermined (Rybak and McGrath, 1996).

chapter 4

SELECTED ANTIBACTERIAL AGENTS

The systematic search for new antibiotics began in the 1940s with large-scale screening programmes for compounds that inhibited bacterial growth. These studies were fuelled by the revelations in the 1930s that the dye prontosil red (composed of two components, one of which was sulphanilamide) and culture filtrates of a fungus belonging to the genus *Pencillium* possessed antibacterial properties. Over the next 30 years or so, many microbes – both fungi and bacteria – were empirically screened for metabolites with antibacterial properties. During this period virtually all of the current antibiotic classes with clinical utility were discovered (see Figure 4.1 and Knowles, 1997).

Since the mid- to late 1960s, the rate of discovery of new antibacterial classes has tailed off dramatically. Countering this, however, has been the development of new antibiotics through synthetic modification of existing antibiotics and/or increasing knowledge of their specific molecular target sites. This new approach,

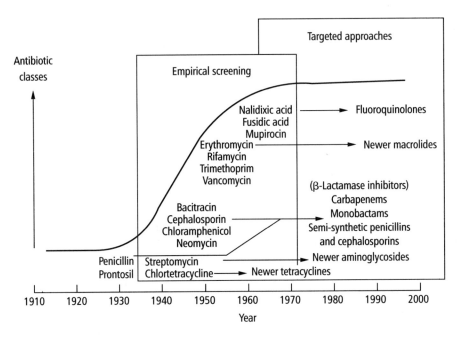

Figure 4.1 The discovery of some surgically important antibiotic classes, either by empirical screening or by targeted approaches. Adapted from Knowles (1997).

known as targeted screening, is based on developing detailed molecular structure–antibacterial activity relationships (Knowles, 1997).

The number of bacteria that are more commonly responsible for surgery-related infections is well defined, although the list is slowly increasing as a consequence of improved bacteriological techniques and the sepsis related to the cohort of severely ill and debilitated patients now seen in surgical critical care. Important pathogens include staphylococci (e.g. *Staphylococcus aureus, Staphylococcus epidermidis*), streptococci (e.g. *Streptococcus pyogenes, Streptococcus anginosus*), enterococci (e.g. *Enterococcus faecalis*), various Gram-negative enteric bacilli (e.g. *E. coli, Klebsiella pneumoniae*, enterobacters), non-enteric Gram-negative bacilli (e.g. *Pseudomonas aeruginosa*), and a wide variety of obligate anaerobes (e.g. the *Bacteroides fragilis* group, clostridia, fusobacteria, prevotellae and peptostreptococci). Possible antibiotic considerations for these bacteria are summarized in Table 4.1.

In therapy, the β-lactam antimicrobial agents are by far the most commonly used. Aminoglycosides and quinolones offer alternatives for infections involving Gram-negative bacilli, while a wide variety of antimicrobial agents possess anti-anaerobe activity. The most useful of these are metronidazole and, to a lesser extent imipenem, clindamycin and penicillin. Antimicrobial agents that are becoming increasingly important in the therapy of infections due to more resistant Gram-positive cocci include vancomycin, the newer fluoroquinolones (e.g. clinafloxacin), fusidic acid and rifampicin.

β-LACTAMS

GENERAL

The penicillins, cephalosporins, monobactams, carbapenems and the β-lactamase inhibitors comprise the cell-wall-active β-lactam group of antimicrobial agents. All of them possess a 4-membered β-lactam ring (hence the name) whose integrity is essential for antibacterial activity. In all but the monobactams, this structure is fused to a 5- or 6-membered associated ring that is variously capped by a carbon, sulphur or oxygen atom (see Figure 4.2). The character of the ancillary ring and of the side-chains designated R1 and R2 determine the differences in antibacterial spectrum, resistance to β-lactamase enzymes, pharmacokinetics and some adverse effects (Nathwani and Wood, 1993).

Of the β-lactams, the most commonly used subgroups are the closely related penicillins and cephalosporins. In the former, the β-lactam ring is fused to a 5-membered thiazolidine ring, while in the latter it is attached to a dihydrothiazine structure (see Figure 4.2). The various semi-synthetic penicillins now in use have been formed by adding different side-chains at the 6-position of the penicillin nucleus, 6-aminopenicillanic acid (6-APA). The nucleus was obtained by removing the acyl side-chain from the original biological penicillin G (benzylpenicillin). In a similar manner, semi-synthetic cephalosporins have been prepared from cephalosporin C; in this case, the nucleus remaining after removal of the

Table 4.1 Antibiotic considerations for common pathogens and/or bacterial groups

Microbe	Antimicrobial agent	
	First choice(s)	Alternatives
Obligate anaerobes		
Bacteroides fragilis	Metronidazole	Imipenem, meropenem, clindamycin, piperacillin/tazobactam, co-amoxyclav, cefoxitin, cefotetan, chloramphenicol
Clostridia	Metronidazole, penicillin	(as above), vancomycin
Fusobacteria	Metronidazole	(as for *Bacteroides fragilis*)
Peptostreptococci	Penicillin, metronidazole	(as for clostridia)
Porphyromonas species	Metronidazole	(as for *Bacteroides fragilis*)
Prevotella species	Metronidazole	(as for *Bacteroides fragilis*)
Gram-positive cocci		
Enterococci	Amoxycillin ± gentamicin	Vancomycin, clinafloxacin, imipenem, co-amoxyclav, quinupristin/dalfopristin
Staphylococcus aureus	Flucloxacillin or similar penicillinase-stable penicillin[b]	Vancomycin, cotrimoxazole, fusidic acid, co-amoxyclav, cephalexin, cefuroxime, clinafloxacin, ciprofloxacin, rifampicin, clindamycin, imipenem
Staphylococcus epidermidis	Vancomycin, netilmicin	Teicoplanin, flucloxacillin (or similar agent), ciprofloxacin, clindamycin, imipenem
Streptococcus anginosus *Streptococcus pyogenes*	Penicillin (possibly second- or third-generation cephalosporin)[a]	Ceftriaxone, cefuroxime, amoxycillin, clindamycin, flucloxacillin, erythromycin, vancomycin, imipenem
Gram-negative bacilli		
E. coli and related enteric coliforms	Aminoglycosides (e.g. gentamicin)	Fluoroquinolones (e.g. ciprofloxacin), third-generation cephalosporins (e.g. ceftriaxone), meropenem, piperacillin/tazobactam
Pseudomonas aeruginosa	Tobramycin + piperacillin	Other aminoglycosides, ceftazidime, meropenem/imipenem, ciprofloxacin, aztreonam, ticarcillin (possibly combinations)

[a]Cephalosporins such as cefuroxime and ceftriaxone.
[b]Vancomycin for MRSA (or non-β-lactam combination depending on local sensitivity patterns).

side-chain was 7-aminocephalosporanic acid (7-ACA). In addition, the extra carbon atom in the dihydrothiazine ring offers the possibility of further modification at other positions.

The best marker of β-lactam antibacterial activity is the time for which the plasma concentration exceeds the microbe's MIC. Optimal activity is therefore achieved by continuous infusion or by employing frequent dosing intervals, or by using agents with a long half-life. Unlike the situation with aminoglycosides, increased bacterial killing is not seen with increasing concentrations of β-lactams above the MIC (Turnidge, 1998).

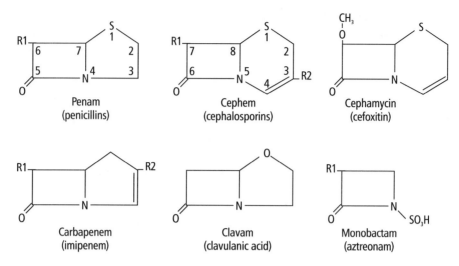

Figure 4.2 Basic molecular structures of β-lactam antibiotics (examples are shown in parentheses). Adapted from Greenwood (1995a).

As already discussed (see p. 28 and p. 30), the target sites of the β-lactams are the enzymes (penicillin-binding proteins or PBPs) associated with the cross-linking of the newly formed linear glycan strand to the existing cell wall peptidoglycan. Resistance to β-lactams in Gram-positive bacteria is either due to the presence of novel or altered normal PBPs, or more commonly to the elaboration of enzymes, β-lactamases, which open (hydrolyse) the β-lactam ring as shown in Figure 3.4 (p. 35). On hydrolysis, the molecule becomes biologically inactive. Genes for β-lactamase production are usually plasmid located and inducible. In Gram-negative microbes, resistance to β-lactams can be associated with decreased intracellular accumulation of drug (outer membrane permeability or efflux), by changes to the PBP target sites, or by β-lactamase production. Of growing concern in recent years has been the emergence world-wide of novel extended-spectrum β-lactamases (ESBLs). These enzymes are probably linked to the increasing use of extended-spectrum cephalosporins.

Most β-lactam antibiotics are poorly absorbed from the gastrointestinal tract, and their absorption is delayed by the presence of food. Certain specially selected drugs (e.g. penicillin V, cephalexin, amoxycillin) are reasonably well absorbed orally. Most β-lactams are widely distributed in the body and cross the placenta. However, only a few reach significant concentrations in the bile, cerebrospinal fluid or breast milk. Most have short plasma half-lives, and a large proportion show significant protein binding (e.g. 96% with flucloxacillin). Excretion is mainly by renal tubular secretion, and blood levels may be increased in renal failure or by administration of probenecid.

PENICILLINS (PENAMS)

Natural penicillins

The original preparations of penicillin were apparently mixtures of four closely related compounds, namely penicillins F, G, K and X. Penicillin G (benzylpenicillin), often simply referred to as 'penicillin', was chosen for further study as it demonstrated the most desirable properties and could be produced from a strain of *Pencillium chrysogenum* in large – almost pure – amounts. Three preparations of benzylpenicillin are currently available, namely aqueous, procaine and benzathine (see Table 4.2). Because benzylpenicillin is unstable in gastric acid, it does not reach reliable levels when given orally, and must be administered intravenously (the usual route for the aqueous form) or intramuscularly as with the long-acting procaine or benzathine formulations. Oral probenecid given at the same time as penicillin will slow kidney excretion and prolong serum levels of the drug.

Until recently, benzylpenicillin was considered to be the treatment of choice for many streptococcal and several anaerobic infections (e.g. clostridial, oral anaerobes including *Prevotella* and *Porphyromonas* species) as well as some infections associated with human or animal bites (e.g. *Eikenella corrodens, Pasteurella multocida*). However, an increasing number of important pathogens now elaborate β-lactamases that are destructive to benzylpenicillin (e.g. *Staphylococcus aureus, Bacteroides fragilis*) (see Nathwani and Wood, 1993). Newer antibiotics with improved activity and pharmacokinetic properties are available for treatment of infections by many of these microbes.

The first major success in improving the pharmacological properties of penicillin was achieved with phenoxymethylpenicillin (penicillin V), which is acid stable and orally achieves better and more reliable serum levels than benzylpenicillin. The antibacterial spectrum of penicillin V is similar to that of penicillin G.

Antistaphylococcal penicillins

By the end of the 1950s, most hospital strains of *Staphylococcus aureus* were resistant to benzylpenicillin because of their ability to elaborate β-lactamases more specifically referred to as staphylococcal penicillinases. This led to the search for semi-synthetic compounds that are resistant to penicillinase hydrolysis. Several penicillinase-stable (resistant) penicillins were developed (e.g. nafcillin, methicillin, oxacillin, cloxacillin, dicloxacillin, flucloxacillin) and have been in use as antistaphylococcal penicillins since the early 1960s. All of them except methicillin can be given orally. Some of them, including flucloxacillin, are highly bound to serum proteins. Of the available agents, flucloxacillin achieves higher serum levels than the others after oral administration, and is probably the preferred agent. In recent years, however, concern has arisen in Australia about the association of flucloxacillin with severe and debilitating cholestatic hepatitis (Turnidge, 1995), with patients over 55 years old and those taking the drug for more than 14 days being most at risk. Dicloxacillin may be less likely to cause this problem. Perhaps alternatives (e.g. cephalexin, cefuroxime axetil) should now be considered for minor skin and soft-tissue infections.

Broad-spectrum penicillins

Extension of the activity of penicillin to encompass Gram-negative bacilli was achieved by adding an amino group to the side-chain to form aminopenicillins such as ampicillin and its derivative amoxycillin. Although slightly less active than

Table 4.2 Comparison of some common penicillins

Generic name	Administration	Half-life (h)	Comment	Renal clearance	Protein binding (%)	Usual adult dose (g)
Benzylpenicillin (penicillin G)			Susceptible to β-lactamase			
Aqueous	IV, IM	0.5	Serum level c. 20 mg/L	+	55	0.6–2.4 q 4 h[b]
Procaine	IM	–	Serum level c. 2 mg/L Given once daily	+		1–1.5 daily
Benzathine	IM	–	Serum level < 0.2 mg/L Given weekly or monthly	+		0.9 (1.2 mu) monthly
Phenoxymethylpenicillin (penicillin V)	PO	1.0	Serum level 2–5 mg/L Antibacterial activity similar to penicillin G	+	80	0.25–0.5 q 6 h
Flucloxacillin	IV, IM, PO	0.5	Stable vs. staphylococcal β-lactamase but intrinsic activity c. 10% that of penicillin G vs. susceptible staphylococci (and streptococci)	+	92	0.25–0.5 q 8 h PO; 1–2 q 4 h IV[c]
Nafcillin	IV, IM, PO	0.5		–[d]	87	1–2 q 4 h IV, IM
Dicloxacillin	PO	0.5		+	97	0.25–0.5 q 6 h PO
Amoxycillin	IV, IM, PO	1.0	The aminopenicillins amoxycillin and ampicillin have better activity than penicillin G against Gram-negative bacteria, but are susceptible to β-lactamase	+	17	0.5 q 8 h PO; 1–2 q 6 h IV
Ampicillin	IV, PO	1.0		+	17	0.25–0.5 q 6 h PO; 1–3 q 4–6 h IV
Pivampicillin	PO	1.0	Ampicillin is better absorbed when given as pivampicillin	+		0.5 q 12 h
Piperacillin	IV, IM	1.0	More extensive Gram-negative activity, including *Pseudomonas aeruginosa*, but susceptible to β-lactamase	+[a]	50	3–4 q 4–6 h IV
Ticarcillin	IV	1.2		+	50	3 q 3–6 h IV

Table 4.2 Continued

Generic name	Administration	Half-life (h)	Comment	Renal clearance	Protein binding (%)	Usual adult dose (g)
Amoxycillin/clavulanic acid (co-amoxyclav)	IV, PO	1.3	Addition of β-lactamase inhibitor extends spectrum to β-lactamase-producing staphylococci, haemophili, moraxellae, gonococci and enteric Gram-negative bacilli otherwise resistant to amoxycillin, ampicillin, ticarcillin and piperacillin	+		0.5 q 8–12 h PO; 1.2 q 6–8 h IV
Ampicillin/sulbactam	IV	1.0		+		1.5–3 q 6 h
Ticarcillin/clavulanic acid	IV	1.0		+		3.1 q 4–6 h
Piperacillin/tazobactam	IV, IM	1.0		+		3.375 q 6 h IV (Zosyn); 4.5 q 8 h IV (Tazocin)

[a] Some biliary excretion.
[b] 1 million units equivalent to around 0.6 g; can give up to 12 g/daily.
[c] Cholestatic hepatitis is reported to be common in Australia – use only in severe infection.
[d] Hepatic metabolism.
IM, intramuscular; IV, intravenous; PO, oral.
Adapted from Lang, S. (ed.) (1997) *Guide to pathogens and antibiotic treatment* 1998. Auckland: Adis International, with permission.

benzylpenicillin against Gram-positive cocci, they do have much improved activity against enteric Gram-negative bacilli and *Haemophilus influenzae*. Amoxycillin, but not ampicillin, is well absorbed when given orally, but both are susceptible to staphylococcal β-lactamases. Esterified pro-drugs of ampicillin (e.g. pivampicillin) show improved absorption when given orally. Ampicillin is released during absorption by the activity of non-specific tissue esterases.

Extended-spectrum (antipseudomonal) penicillins

None of the penicillins mentioned so far exhibit useful antipseudomonal activity. *Pseudomonas aeruginosa* is an increasingly important pathogen in immunocompromised and burns patients, and is notoriously resistant to many antimicrobial agents. However, a number of extended-spectrum or antipseudomonal penicillins have been produced, such as the carboxypenicillin, ticarcillin, and the ampicillin derivatives, namely azlocillin, mezlocillin and piperacillin (ureidopenicillins). These must be administered by injection. Drugs such as ticarcillin and piperacillin are often used in combination with aminoglycosides against serious Gram-negative sepsis, including that due to *Pseudomonas aeruginosa*. Neither piperacillin nor ticarcillin is stable to staphylococcal β-lactamases. Some indication of the development of the penicillins is given in Figure 4.3, while Table 4.3 outlines their antimicrobial spectrum.

Adverse reactions

The main disadvantage of using penicillins is the risk of allergy – the most frequent complication of therapy is penicillin hypersensitivity (0.5–2.0% of patients). The penicilloyl and penicillanic derivatives are the major determinants of penicillin allergy. The penicilloyl derivative, which is produced through opening of the β-lactam ring, thereby allowing amide linkage to body proteins (as haptens), is the most important antigenic component (see Chambers and Neu, 1995b). Mild reactions may consist only of a morbilliform or scarlatiniform rash on the trunk and extremities, which can disappear spontaneously during treatment. Such skin reactions are more common after amoxycillin, and may be associated with viral diseases such as infectious mononucleosis. Urticarial or oedematous skin rashes, with or without fever, and sometimes accompanied by joint swelling, laryngeal oedema, conjunctivitis or other symptoms, should be taken more seriously. Anaphylactic shock is fortunately rare, but is extremely serious, and lethal in around 10% of cases. Other adverse effects include diarrhoea (including pseudomembranous colitis), haematological abnormalities, renal toxicity, cholestatic hepatitis and (rarely) seizures.

Alternative antibiotics for patients with penicillin allergy vary according to the type of hypersensitivity involved. The incidence of allergy to cephalosporins in patients with a history of penicillin allergy is probably less than 10%. Cephalosporins should therefore only be used with extreme caution in patients with known anaphylactic reactions to penicillins; the same applies to imipenem and probably also to meropenem (Anon., 1996a). Clinical cross-reactivity between penicillins and monobactams does not appear to be significant, and patients with penicillin allergy can probably be given aztreonam with minimum

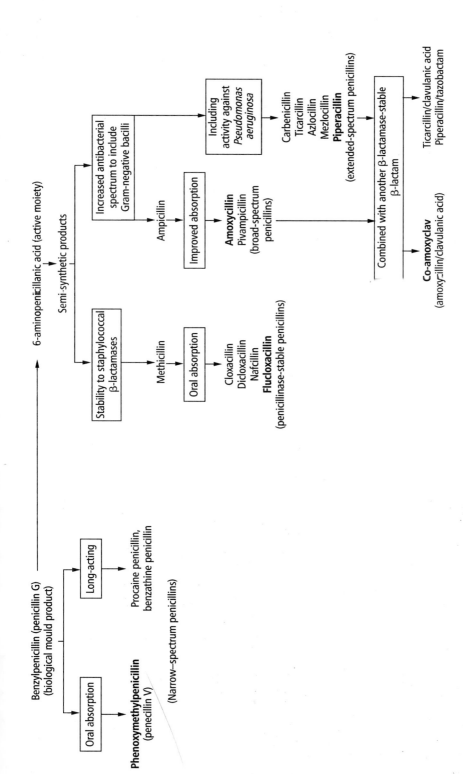

Figure 4.3 Development of penicillins. Adapted from Greenwood (1992).

Table 4.3 Useful antibacterial spectra of commonly used penicillins

Penicillin/group	Therapeutic activity						
	Staphylococcus aureus	*Streptococci*[b]	Enterococci	Gram-negative enteric bacilli	*Pseudomonas aeruginosa*	*Bacteroides fragilis*	Clostridia
Benzylpenicillin	−	++	−[a]	−	−	−	++
Phenoxymethylpenicillin	−	++	−	−	−	−	+
Flucloxacillin and related anti-staphylococcal agents[e]	++	+	−	−	−	−	+
Amoxycillin; ampicillin	−	++	+	V	−	−	+
Piperacillin, ticarcillin group[d]	−	+	V	+	++[c]	−	+
Co-amoxyclav	+	++	+	V	−	+	+
Piperacillin/tazobactam; ticarcillin/clavulanic acid	+	+	V	++	++[c]	+	+

[a]Synergy with gentamicin may make this an acceptable agent.
[b]Decreasing susceptibility of streptococci to penicillins in general.
[c]Usually in combination with an aminoglycoside (e.g. tobramycin).
[d]Includes carbenicillin, azlocillin and mezlocillin.
[e]Includes cloxacillin, dicloxacillin and nafcillin (oxacillin, methicillin).
+, usually susceptible; ++, very effective drug; −, not susceptible; V, variable sensitivity.

risk of anaphylaxis. In cases where the reported penicillin allergy is mild (e.g. rash), cephalosporins can probably be administered with reasonable safety (see p. 61). If in doubt, a non-β-lactam antibiotic should be used. More classical examples include erythromycin (or a similar macrolide), clindamycin and vancomycin for streptococcal and staphylococcal disease, aminoglycosides for infections involving Gram-negative enteric bacilli, and metronidazole for anaerobic infections.

CEPHALOSPORINS (CEPHEMS)

Cephalosporin C, on which the cephalosporins are based, is a β-lactam in which the β-lactam ring is fused to a 6-membered dihydrothiazine ring to give the cephem nucleus (see Figure 4.2). Acid hydrolysis of this compound results in the formation of 7-aminocephalosporanic acid (7-ACA), the basic structure for the future development of semi-synthetic cephalosporins. Unlike penicillins, in which substitutions are only practicable at the 6 position, 7-ACA can be modified at several positions, especially 7 and 3 (see Figure 4.2). Modifications at the 7-acyl group tend to alter the antimicrobial spectrum, including β-lactamase stability, while changes or substitutions at position 3 predominantly cause changes in metabolic and pharmacokinetic (e.g. bioavailability) parameters (Karchmer, 1995).

Important characteristics of cephalosporins that are reflected in their use include the overall resistance of the group to inactivation by β-lactamases (including staphylococcal penicillinases), their ability to reach areas of the cell where the penicillin-binding-protein target sites are located, and their affinity for these enzymes. As a group, the cephalosporins exhibit no therapeutic activity against enterococci, and most of them are also devoid of useful anti-anaerobe activity.

The cephalosporins are usually classified according to their generations (e.g. first, second and third generation), which relate both to the sequence of their discovery and to their increasing activity against Gram-negative bacilli. New agents (fourth generation) are continually being added whose antibacterial spectra combine some of the advantages of the first-, second- and third-generation compounds. Examples of more commonly available or used cephalosporins together with their important characteristics are listed in Table 4.4, and Figure 4.4 depicts the events which have occurred in the development of the cephalosporins since the initial work with cephalosporin C in 1955.

First-generation cephalosporins

In general, movement from first- to third-generation drugs is accompanied by decreasing Gram-positive (staphylococcal, streptococcal) activity and increasing activity against Gram-negative bacilli. First-generation cephalosporins (see Table 4.4) are very active against Gram-positive cocci (except enterococci), and show variable activity against Gram-negative cocci and bacilli. They lack therapeutic activity against enterococci and anaerobes (see Table 4.5). Of these agents, cephalexin is extremely well absorbed, and can be considered as an alternative to

Table 4.4 Comparison of more common cephalosporins

Generic name	Route of administration	Half-life (h)	Protein binding (%)	Renal clearance	Significant biliary excretion	Usual adult dose (gm)	Daily adult dose for severe infection (gm)
First generation							
Cephalothin	IV	0.6	71	+		1–2 q 4–6 h	6–12
Cephazolin	IV, IM	1.8	80	+	+	1 q 8 h	3–6
Cephradine	PO	0.7	10	+		0.5 q 6 h	
Cephalexin	PO	0.9	10	+		0.5–1 q 6 h	
Second generation							
Cefaclor	PO	0.8	25	+		0.25–0.5 q 8 h	
Cefuroxime	IV, IM	1.3	35	+		1.5 q 8 h	4.5–6
Cefuroxime axetil	PO	1.3	35	+		0.25–0.5 q 12 h	
Cefamandole	IV, IM	0.8	75	+	+	1–2 q 4–6 h	6–12
Cefoxitin	IV, IM	0.8	70	+		2 q 4–6 h	6–12
Cefotetan	IV, IM	3.5	90	+	+	2–3 q 12 h	4–6
Cefmetazole	IV	1.1	65	+		2 q 6 h	8
Third generation							
Cefotaxime	IV, IM	1.0	35	+		2 q 6–8 h	6–12
Ceftriaxone	IV, IM	8.0	83–96	+	+	1–2 q 12 h	2–4
Ceftazidime	IV, IM	1.8	17	+		2 q 8 h	6
Cefoperazone	IV, IM	2.0	87–93	+	+	2 q 8–12 h	6–12
Ceftizoxime	IV, IM	1.7	30	+		2 q 8–12 h	6–12

Table 4.4 Continued

Generic name	Route of administration	Half-life (h)	Protein binding (%)	Renal clearance	Significant biliary excretion	Usual adult dose (gm)	Daily adult dose for severe infection (gm)
Cefpodoxime proxetil	PO	2.2	40	+		0.4 q 12 h	
Cefixime	PO	3.7	67	+ (plus other)		0.4 q 24 h (or 0.2 q 12 h)	
Ceftibuten	PO	2.5	63	+		0.2 q 12 h (or 0.4 q 24 h)	
Fourth generation							
Cefpirome	IV, IM	2.0	10	+		1–2 q 12 h	
Cefepime	IV	2.1	20	+		1–2 q 12 h	

IM, intramuscular; IV, intravenous; PO, oral.
Adapted from Lang, S. (ed.) (1997) *Guide to pathogens and antibiotic treatment 1998*. Auckland: Adis International, with permission.

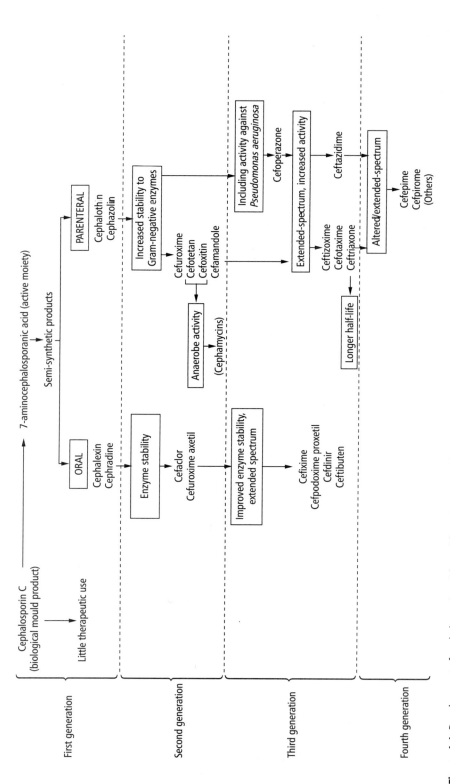

Figure 4.4 Development of cephalosporins. Adapted from Greenwood (1992).

flucloxacillin for skin and soft-tissue staphylococcal infections. Potential nephrotoxicity has limited the use of some first-generation compounds (e.g. cephaloridine, cephalothin).

Second-generation cephalosporins
Second-generation cephalosporins (see Table 4.4) fall into two groups, namely the true cephalosporins (e.g. cefamadole, cefuroxime) and the 7-methoxy-cephalosporins or cephamycins (e.g. cefoxitin, cefotetan, cefmetazole). Esterified oral preparations are available for some drugs (e.g. cefuroxime axetil). Compared to earlier agents, the true cephalosporins exhibit enhanced activity against staphylococci and streptococci, as well as improved activity against *Haemophilus influenzae*, *Moraxella catarrhalis* and some Enterobacteriaceae (Table 4.5). In comparison to first- and other second-generation cephalosporins, the cephamycins show inferior activity against staphylococci and streptococci, but an enhanced antibacterial effect against selected Enterobacteriaceae; this latter point is especially true for cefotetan, which also has a more prolonged half-life. Cephamycins are also active against *Bacteroides* species, including the *Bacteroides fragilis* species group, although low serum and tissue levels and the increasing emergence of resistant strains has limited their use with serious sepsis. Cefoxitin seems to be the most active against anaerobes. The increased activity of second-generation compounds in general against Gram-negative bacteria is probably related to enhanced penetration through the cell wall outer membrane, improved β-lactamase stability, and increased affinity for the PBP target sites.

Third-generation cephalosporins
The third-generation cephalosporins have excellent antibacterial activity against most Gram-negative enteric bacilli, with specific drugs (e.g. ceftazidime, cefperazone) active against *Pseudomonas aeruginosa* (see Table 4.5). With the possible exception of ceftazidime, most of them are also extremely effective against streptococci and to a lesser degree against *Staphylococcus aureus*. They also show excellent activity against a variety of other pathogens (e.g. the gonococcus). Like all compounds with a cephem nucleus, they lack activity against enterococci. Included in this group are a number of orally available compounds with improved β-lactamase stability (e.g. cefixime, ceftibuten, cefpodoxime proxetil and the related cefdinir). The first two of these have poor activity against staphylococci, and ceftibuten has poor antipneumococcal activity. A recent Canadian study (Blondeau *et al.*, 1997) suggested that over 80% of Enterobacteriaceae studied were still susceptible to ceftriaxone. Cephalosporins with an acetyl group at position 3 (e.g. cefotaxime) undergo *in vivo* metabolism to a disacetyl form which may be synergistic with the parent compound to increase its antibacterial effect against microbes such as *Staphylococcus aureus*.

In recent years, the increasing use of third-generation cephalosporins (e.g. ceftazidime) has been associated with the emergence of novel plasmid-encoded extended-spectrum β-lactamases (ESBLs) that are active against many extended-spectrum β-lactam antimicrobial agents. In addition, a number of chromosomally encoded β-lactamases that mediate resistance to all cephalosporins have been

Table 4.5 Useful antibacterial spectra of commonly used cephalosporins

Cephalosporin/group	Therapeutic activity						
	Staphylococcus aureus	Streptococci[b]	Enterococci	Gram-negative enteric bacilli	Pseudomonas aeruginosa	Bacteroides fragilis	Clostridia
Cephalexin	+	+	–	V	–	–	V/–
Cephazolin	+	+	–	V	–	–	V/–
Cefuroxime	+	++	–	V/+	–	–	V/–
Cefamandole	+	+	–	V/+	–	–	V/–
Cefoxitin; cefotetan	V/+	V	–	+	–	V/+	+
Cefotaxime; ceftriaxone	V/+	++	–	++	–	–	V/–
Ceftazidime	V/+	V	–	++	+[a]	–	–
Cefpodoxime proxetil	V/+	+	–	+	–	–	–
Cefixime	V/–	++	–	++	–	–	–
Cefepime	V/+	++	–	++	+[a]	–	V/–
Cefpirome	V/+	++	–	++	V/+[a]	–	V/–

+, usually susceptible; ++, very effective drug; –, not susceptible; V, variable susceptibility.
[a]Often used in combination with an aminoglycoside (e.g. tobramycin).

linked to the increasing number of infections now being encountered with enterobacters and *Citrobacter freundii* (Karchmer, 1995). Treatment options for cephalosporin-resistant Gram-negative bacteria include imipenem, ciprofloxacin and β-lactam/β-lactamase inhibitor combinations such as piperacillin/tazobactam (see p. 87).

Fourth-generation cephalosporins
The fourth-generation cephalosporins (represented by cefepime and cefpirome) have broad antibacterial activity that encompasses Gram-negative bacteria including *Pseudomonas aeruginosa*, and Gram-positive bacteria such as staphylococci and streptococci (Fung-Tomc, 1997). Compared to third-generation compounds, enhanced activity against Gram-negative bacilli (e.g. indole-positive *Proteus* species, *Citrobacter freundii*, *Enterobacter* species, *Serratia marcescens*, *Pseudomonas aeruginosa*) is conferred by more rapid penetration across the outer membrane (probably related to the presence of a positively charged quaternary nitrogen at position 3, which is thought to orient the molecule towards the entrance of the porin channel), and poor affinity for and increased stability against common chromosomal β-lactamases. However, it is recommended that Gram-negative bacilli elaborating plasmid-mediated extended-spectrum β-lactamases (ESBLs) should be regarded as clinically resistant to all cephalosporins, including fourth-generation compounds. This is possibly an over-simplification, as it seems that the effectiveness of cephalosporins in the treatment of infections caused by ESBL-producing enterics may be highly dependent on dosing regimens (Thauvin-Eliopoulos *et al.*, 1997). At present, cefepime is the only fourth-generation cephalosporin approved for clinical use in the USA (Wynd and Paladino, 1996; Fung-Tomc, 1997), although trials with cefpirome are well advanced. The latter reveals potential for use with pneumococci that are less susceptible to penicillins (Wiseman and Lamb, 1997; Anon., 1998b). A comparison of the antibacterial activity of common cephalosporins is shown in Table 4.5.

Other 'cephalosporins'
Loracarbef is an orally administered carbacephem antibiotic, with a carbon instead of a sulphur atom at position 1 in the cephem nucleus. Otherwise its molecular structure is similar to that of the second-generation cephalosporin, cefaclor. Compared to cefaclor, loracarbef shows modestly increased activity against some β-lactamase-producing respiratory pathogens (e.g. *Haemophilus influenzae*). It is not stable to ESBLs.

Moxalactam is essentially considered to be a third-generation cephalosporin, although it has an oxygen rather than a sulphur atom at position 1, i.e. it is an oxycephem. A methoxy group at position 7 confers increased β-lactamase stability and anaerobic activity on this compound.

Pharmacokinetics
As already stated, absorption from the gastrointestinal tract occurs with only a few cephalosporins. Absorption of esterified forms may be enhanced if they are taken with food, while the bioavailability of cefpodoxime is decreased by H_2-antagonists. Of the newer agents, only ceftibuten shows absorption comparable to that of the earlier cephalexin.

In general, tissue and CSF penetration by cephalosporins is excellent, while those with significant biliary excretion (e.g. ceftriaxone) reach high levels in the bile. However, most cephalosporins are excreted via the kidneys and achieve very high concentrations in the urine. Accumulation of commonly used third-generation agents is likely to occur in the setting of combined hepatic and renal failure, and dose adjustment will therefore probably be required (Karchmer, 1995).

Cephalosporins do not exhibit concentration-dependent killing of bacteria, nor do they demonstrate a marked post-antibiotic effect. Thus the time for which the cephalosporin concentration exceeds the MIC of the pathogen at the site of infection is the important pharmacokinetic parameter for enhanced efficacy. Continuous infusion therapy, shorter dosing intervals or the use of agents with extended half-lives (e.g. ceftriaxone) would appear to be logical considerations with regard to cephalosporin therapy.

Adverse reactions

Compared to most other antimicrobial agents, cephalosporins have a favourable toxicity profile. Thrombophlebitis is a common complication (1–5% of patients) of most intravenously administered cephalosporins. However, as with the penicillins, the most common systemic adverse event seen with cephalosporin use is hypersensitivity, although this is rare compared to the frequency seen with penicillins. Reactions range from skin rash (1–3% of patients) to immediate or accelerated IgE-mediated anaphylactic or urticarial reactions. Serum sickness reactions are most often seen with oral cefaclor use in children.

The specific haptens associated with cephalosporin hypersensitivity are not known. The rate of cephalosporin reactions among patients with a history of penicillin allergy has been estimated to be in the range 3–7% (Karchmer, 1995). In general, cephalosporins can be used with reasonable safety in patients who exhibit mild penicillin allergy (see p. 74). Other adverse events (Karchmer, 1995) include haematological reactions such as hypoprothrombinaemia (inhibition of vitamin K synthesis), antibiotic-associated diarrhoea (probably highest in those with biliary excretion), sludge in the gall-bladder and common bile duct with ceftriaxone, nephrotoxicity (which has limited the use of cephaloridine), and disulfiram-like alcohol reactions with cephalosporins such as cefotetan.

OTHER β-LACTAMS

Carbapenems

These differ from the penicillins in that methylene replaces sulphur in the 1 position (see Figure 4.2). Carbapenems are derivatives of thienamycin, a natural product of the actinomycete *Streptomyces cattleya*. Two carbapenems (imipenem and meropenem) are currently in use, while a third (biapenem) is undergoing investigation.

Imipenem Imipenem has a very wide spectrum of activity (see Table 4.6), being lytic for most aerobic and anaerobic bacteria, and it has an important role in the empirical treatment of life-threatening infections. More significant resistant bacteria are MRSA, *Enterococcus faecium*, *Clostridium difficile*, *Stenotrophomonas* (*Xanthomonas*) *maltophilia* and *Burkholderia* (*Pseudomonas*) *cepacia*. It is a consideration for strains of *Streptococcus pneumoniae* that are less susceptible or resistant to penicillin. Like penicillin, it is only bacteriostatic to *Enterococcus faecalis*. The excellent antibacterial activity of imipenem against most Gram-negative and Gram-positive bacteria appears to be related to its ability to bind strongly to high-molecular-weight PBPs.

Imipenem exhibits a post-antibiotic effect against Gram-negative bacilli similar to that seen with aminoglycosides and quinolones. It resists hydrolysis by most β-lactamases, including those that hydrolyse third-generation cephalosporins. However, a few microbes (e.g. *Stenotrophomonas maltophilia*) elaborate a β-lactamase that is capable of hydrolysing imipenem and other carbapenems. In general, resistance to imipenem is attributable to permeability factors (e.g. the loss of outer membrane porin protein D2). This appears to be an increasing problem with *Pseudomonas aeruginosa*.

Because of its instability in gastric acid, imipenem must be given by injection. It is removed from the circulation by glomerular filtration and hydrolysed by a renal peptidase to form an inactive and toxic compound in the urine. The action of this peptidase can be inhibited by cilastatin, which is usually administered in equal amounts with imipenem. Without cilastatin, metabolites of imipenem are nephrotoxic. In patients with acute renal failure managed by continuous veno-venous haemofiltration, higher imipenem doses (e.g. at least 2000 mg 24-hourly) are required, which inevitably result in a marked accumulation of cilastatin. It is clear that fixed equal concentrations of imipenem and cilastatin are not satisfactory for patients with acute renal failure, for whom higher than normal doses of imipenem are required. It has been suggested that in such patients an appropriate dose of each component should be given separately (e.g. a 4:1 ratio of imipenem to cilastatin), or that a carbapenem (e.g. meropenem) that does not require the addition of cilastatin should be used (Tegeder *et al.*, 1997).

Apart from the adverse reactions that are common to most β-lactams, the most serious problem associated with imipenem use is seizures – an infrequent ($< 2\%$ of patients) side-effect that is most often seen in patients with underlying central nervous system pathology (e.g. meningitis) and in those with impaired renal function (Chambers and Neu, 1995a).

Meropenem Meropenem has a similar spectrum of activity to imipenem, although it is perhaps marginally more active against Gram-negative bacilli (possibly due to improved penetration through the outer membrane) and slightly less active against Gram-positive cocci. It may be effective against some imipenem-resistant strains of *Pseudomonas aeruginosa* (Anon., 1996b). In contrast to imipenem, it is stable to the renal peptidase, dihydropeptidase-1.

The toxicity profile of meropenem is similar to that of imipenem, except that there is growing evidence that it has a lower incidence of adverse gastrointestinal events, and is less epileptogenic and less nephrotoxic (Chambers and Neu, 1995a; Wiseman *et al.*, 1995). Meropenem can be given by intravenous bolus injection or by infusion. Imipenem/cilastatin can only be given by intravenous infusion or deep intramuscular injection. Both have similar half-lives of about 1 h.

Table 4.6 Comparison of the therapeutic activity of selected aminoglycosides, β-lactams and fluoroquinolones

Antimicrobial agent	Therapeutic activity						
	Staphylococcus aureus[b]	Streptococci	Enterococci	Gram-negative enteric bacilli	*Pseudomonas aeruginosa*[c]	*Bacteroides fragilis*	Clostridia
Gentamicin	+	–	–[a]	++	+	–	–
Tobramycin	+	–	–	++	++	–	–
Netilmicin	++	–	–	++	+	–	–
Ceftriaxone	V/+	++	–	++	–	–	V/–
Ceftazidime	V/–	V/–	–	+	+	–	–
Cefpirome	V/+	++	–	++	+	–	–
Flucloxacillin	++	+	–	–	–	–	V/+
Amoxycillin	–	+	+	V/–	–	–	++
Co-amoxyclav	++	+	+	V	–	+	++
Piperacillin; ticarcillin	–	+	+	+	+	–	+
Piperacillin/tazobactam	++	+	+	++	+	+	++
Imipenem; meropenem	++	+	+	++	++	++	++
Aztreonam	–	–	–	+	+	–	–
Ciprofloxacin	+	V	V/–	++	V/+	–	V/–
Clinafloxacin	++	+	+	++	V/+	+	+

[a]Synergistic with penicillin or amoxycillin.
[b]β-Lactams ineffective against MRSA.
[c]Synergy with combined aminoglycoside/β-lactam regimens.
+, usually susceptible; ++, very effective drug; –, not susceptible; V, variable sensitivity.

Monobactams

Aztreonam is a modified monocyclic β-lactam compound (see Figure 4.2) obtained from *Gluconobacter violaceum* (Chambers and Neu, 1995a). It does not show significant activity against Gram-positive or anaerobic bacteria, being incapable of binding to the PBPs of such species. In Gram-negative bacilli, it binds primarily to PBP3, resulting in long filaments of non-viable bacteria.

Aztreonam has been successfully used in the treatment of a variety of infections involving Gram-negative bacteria (e.g. cystitis, peritonitis). Possibly because of emerging resistance, it seems to be one of the least favoured of the anti-Gram-negative bacillus drugs, and is usually regarded as a second-line agent. Because of its lack of cross-reactivity with other β-lactams, aztreonam can be safely used in patients with serious allergy to penicillins or cephalosporins. It is a potent inhibitor of chromosomal cephalosporinases (Bush group 1) from bacteria such as *Pseudomonas aeruginosa*, and appears to enhance the activity of cefepime against strains of this bacterium that express these enzymes. Aztreonam combined with cefepime may be a unique double β-lactam combination that is highly effective against *Pseudomonas aeruginosa* (Lister et al., 1998).

β-lactamase inhibitors

A number of β-lactams exist which have weak antibacterial properties and are potent inhibitors of many plasmid-encoded (e.g. types II, III, IV and V β-lactamases, staphylococcal penicillinases, ESBLs) and some chromosomal (type 1) β-lactamases. These so-called β-lactamase inhibitors include clavulanic acid and several penicillanic acid sulphone derivatives (e.g. sulbactam and tazobactam) (see Figure 4.2). These inhibitors can be regarded as basically similar, although minor differences in properties do occur. Tazobactam has only weak enzyme-inducing activity and is the only inhibitor that is active against some of the potent class I chromosomal β-lactamases (Bryson and Brogden, 1994). β-Lactamase inhibitors apparently form stable suicide complexes with the bacterial β-lactamase.

β-Lactamase inhibitors have been formulated with an active β-lactam, usually a penicillin, as a companion agent (Chambers and Neu, 1995a). The antibacterial activity of the resulting β-lactam combination is largely determined by the companion antimicrobial agent, which may have its β-lactamase susceptibility reversed by the concurrent presence of the inhibitor. Some Gram-negative bacilli (e.g. *Pseudomonas aeruginosa*) produce chromosomal β-lactamases that are unaffected by these β-lactamase inhibitors.

Amoxycillin-clavulanate The most commonly used β-lactam combination is co-amoxyclav (amoxycillin-clavulanate, 'Augmentin'), which has proved to be a useful antimicrobial agent in a variety of settings. Nausea and diarrhoea are significant adverse effects of this combination when given orally. In the surgical area, co-amoxyclav has been used predominantly as an anti-Gram-positive coccal and anti-anaerobe agent for minor skin and soft-tissue infections. It has also been advocated as a useful single agent for surgical prophylaxis, and in the treatment of post-operative respiratory infections. Co-amoxyclav should not be used alone in life-threatening sepsis that is likely to contain Gram-negative enteric bacilli as a significant component.

Ampicillin-sulbactam The combination of ampicillin and sulbactam ('Unasyn') has the same spectrum of antibacterial activity as co-amoxyclav. It is invariably used only as a parenteral formulation, and in many countries it has been largely replaced by co-amoxyclav, which has the added benefit of oral administration.

Ticarcillin-clavulanate Compared to co-amoxyclav, ticarcillin-clavulanate ('Timentin') has improved Gram-negative bacillus activity, although it is prudent to combine it with another agent (e.g. gentamicin) in seriously ill patients. Intra-abdominal and gynaecological infections, as well as soft-tissue infections and osteomyelitis, have been successfully treated with ticarcillin-clavulanate combinations (Chambers and Neu, 1995a).

Piperacillin-tazobactam Of the commonly used and/or available β-lactam combinations, piperacillin-tazobactam ('Tazocin', 'Zosyn') would seem to possess the widest antibacterial spectrum, the latter being similar to that of imipenem (Perry and Markham, 1999). Increasing numbers of studies have demonstrated its usefulness as the sole antimicrobial agent in a number of clinical situations involving mixtures of bacteria (e.g. intra-abdominal infections). The most commonly reported adverse effect has been diarrhoea, which may occur in up to 10% of patients (Bryson and Brogden, 1994; Schoonover et al., 1995). As bacteria that are capable of elaborating chromosomally mediated class I enzymes unaffected by β-lactamase inhibitors seem to be on the increase, the utility of β-lactam–β-lactamase inhibitor combinations as empirical monotherapy for life-threatening sepsis of unknown aetiology is perhaps questionable.

Investigational
New β-lactams with an increased affinity for PBP2a (i.e the enzyme that is largely responsible for methicillin resistance in *Staphylococcus aureus*) and potential activity against MRSA are currently under investigation (Nicas et al., 1997).

AMINOGLYCOSIDES

GENERAL PROPERTIES

Streptomycin, a metabolite of the actinomycete *Streptomyces griseus*, was the first aminoglycoside to be reported (in 1944). At least 11 other aminoglycosides have subsequently been discovered, those ending with the suffix -*mycin* being derived from species of the genus *Streptomyces,* and those with the suffix -*micin* being derived directly or indirectly from members of the genus *Micromonospora*. Commonly used and/or currently available aminoglycosides include gentamicin, tobramycin, amikacin and netilmicin. The most common agent, gentamicin, is a mixture of three closely related constituents, namely C_1, C_{1a} and C_2, elaborated by *Micromonospora* species (Greenwood, 1995b).

Aminoglycosides are inexpensive, cationic compounds that exhibit concentration-dependent killing of susceptible species. Another important feature of the group is their additive or synergistic activity with penicillins and cephalosporins, especially against Gram-positive cocci. A combination of penicillin and gentamicin displays synergistic activity against enterococci, and is a widely used therapeutic option, despite the fact that enterococci are intrinsically resistant to low concentrations of gentamicin. Synergy is also seen with combinations of ampicillin and gentamicin. Clearance of the bacteraemic phase of Gram-positive infections is also reported to be augmented by such combinations. The prevalence of aminoglycoside resistance (usually by enzymatic inactivation) has remained low, as has the emergence of resistance during therapy (Gilbert, 1995). The type of aminoglycoside-modifying enzyme that is present in any locality tends to mirror the use of particular aminoglycosides (e.g. AAC(6') enzymes following amikacin rather than gentamicin use) (Davies and Wright, 1997).

It is of interest, particularly in view of their apparent synergistic activity *in vivo*, that the cationic aminoglycosides interact chemically with β-lactam antibiotics. The reaction results in a nucleophilic opening of the β-lactam ring with acylation of an amino group of the aminoglycoside and associated loss of antibacterial activity (Gilbert, 1995). As this reaction requires several hours for completion, the clinical impact would appear to be minimal. However, penicillins and aminoglycosides should not be mixed in the same solution prior to infusion. For greatest synergy, the two drugs should be administered 20 to 30 min apart – it seems to make no difference which drug is infused first.

ANTIBACTERIAL SPECTRUM

The *in vitro* spectrum of activity of aminoglycosides compared to selected β-lactams and fluoroquinolones is shown in Table 4.6. In general, aminoglycosides are effective against most Enterobacteriaceae (e.g. *E. coli*, klebsiellae, enterobacters, citrobacters, *Proteus* species), *Pseudomonas aeruginosa*, and *Staphylococcus aureus*, including some strains of MRSA. Aminoglycosides have no predictable activity against *Stenotrophomonas* (*Xanthomonas*) *multophilia* and *Burkholderia* (*Pseudomonas*) *cepacia* or any anaerobes. While gentamicin is the agent usually employed, there is some evidence that tobramycin may be more effective against *Pseudomonas aeruginosa*. Similarly, netilmicin appears to be more effective than gentamicin or tobramycin against β-lactam-resistant strains of *Staphylococcus epidermidis*. Resistance to one aminoglycoside does not necessarily imply resistance to others; in this respect amikacin is the most likely to remain active.

PHARMACOKINETICS

Aminoglycosides are highly soluble in water but are unable to cross lipid cellular membranes readily. They enter Gram-negative bacteria by binding electrostatically and rapidly to the outer membrane. This passive binding results in disruption of the cell wall's normal permeability function, and allows entry of the drug, which is then actively transported across the cytoplasmic membrane and bound to the ribosomal target site(s). Intracellular aminoglycoside concentrations soon greatly exceed the external concentrations as the drug is irreversibly trapped. Uptake into the cell requires aerobic oxidative metabolism – hence the inactivity of aminoglycosides against anaerobes and enterococci. Aminoglycosides are degraded in acidic environments, including pus. As already mentioned, aminoglycosides are rapidly bactericidal, with a rise in the rate of killing as the antibiotic concentration is increased, regardless of the inoculum density (i.e. unlike the β-lactams, aminoglycosides exhibit concentration-dependent bactericidal activity) (Dew and Susla, 1996; Lacy et al., 1998). In the past, high drug concentrations have been avoided because of potential ototoxicity. However, transient high concentrations (e.g. 15–20 mg/L) are now considered to be advantageous.

Two other important properties of aminoglycosides which have influenced the move to 24-h dosing are their well-documented post-antibiotic effect (PAE) and their adaptive resistance (Dew and Susla, 1996). In the former, persistent suppression of bacterial growth occurs for some hours after the antibiotic has been removed. Among other parameters, the PAE appears to be related to the initial level of drug and the length of time for which the bacteria are exposed to it (Lacy et al., 1998).

Adaptive resistance is the term used to describe the ability of bacterial cells to survive a second supposedly cidal pulse of drug in the presence of pre-existing non-inhibitory levels of the drug. The mechanism responsible for this phenomenon is unknown, but clearly the next dose of aminoglycoside should be given when maximum wash-out of previous doses has occurred. Administering aminoglycosides once daily may also minimize the risk of under-dosing, which can occur in critically ill or traumatized patients and those who have fluid retention and/or high-output cardiac failure. Another important feature of once daily aminoglycoside dosing is its decreased tendency to induce renal and vestibular problems.

CLINICAL USE/TOXICITY

Considerable debate surrounds the administration of aminoglycosides. Logically, the shorter the infusion period, the higher the peak concentration and the higher the rate of killing. The aim is to achieve a peak drug concentration to MIC ratio of at least 10 as soon as possible in the therapeutic regimen (Kashuba et al., 1998). In cases where larger (e.g. 5–7 mg/kg) single daily doses are prescribed, it would appear reasonable to infuse over 20 to 30 min in order to minimize the risk of a rapid rise in serum concentration that might precipitate neuromuscular blockade (Gilbert, 1995). Blood–tissue equilibrium occurs about 60 min after the commencement of the infusion. After intramuscular administration, equilibrium again occurs after about 60 min, although this may be delayed in patients with hypotension and impaired tissue perfusion. Instillation as a bladder irrigation or aerosol nasal spray is accompanied by no detectable blood levels.

As would be expected of drugs with low protein binding and high water solubility, aminoglycosides are freely distributed in the vascular space, and to a lesser extent in the interstitial spaces of most tissues. Aminoglycosides do not cross biological membranes readily (due to their size, polycationic charge and lipid insolubility), with the exception of renal tubular cells and presumably inner ear cells, which have an inherent transport mechanism. Renal proximal convoluted tubular cells concentrate aminoglycosides, and when cells become overloaded they burst but leave the basement membrane intact. This is the basis for the renal toxicity of aminoglycosides. Almost all of a parenteral dose of aminoglycoside is excreted unchanged by the kidney.

Aminoglycosides bind to specific receptors on tubular cells. This is followed by endocytosis. The decreased renal toxicity of once vs. multiple daily dosing appears to be related to saturation of the receptor sites once as opposed to two to three times with multiple dosing regimens. For a given daily dose of drug, the magnitude of toxicity is greatest when the dose is divided into multiple small increments, and least when it is given as a single dose. It also seems that several days of drug administration are necessary before functional or anatomical evidence of toxicity occurs. In addition, aminoglycoside renal tubular necrosis is reversible, and cells can regenerate despite continued administration of the drug (Gilbert, 1995). Although it is often stated that tobramycin is less nephrotoxic than gentamicin, there is little clinical evidence to support such a view.

Not unexpectedly, a number of other potentially nephrotoxic agents (e.g. vancomycin) have been shown to amplify aminoglycoside renal toxicity. On the other hand, extended-spectrum penicillins tend to lower the risk of kidney injury, and this may be related to the high sodium content of the penicillin salts. Other factors that are supposedly associated with an increased risk of toxicity include old age, pre-existing renal disease, hypotension, hepatic dysfunction, and treatment for 3 days or longer (Gilbert, 1995). Certainly they represent situations in which trough levels must be obtained to ensure the adequacy of renal clearance. High trough concentrations reflect poor renal clearance and a need to adjust the timing of the dosing regimen.

Unlike renal toxicity, ototoxicity is usually irreversible, and may first manifest itself within a day or so of starting treatment (see below), or more often some time after the end of treatment. Risk factors appear to be similar to those for nephrotoxicity, and are a reflection of excessive accumulation over time, rather than of peak levels (see Gilbert, 1995). Transient high concentrations do not appear to be a risk factor.

Acute aminoglycoside-induced deafness has been associated with a specific genetic mutation called AI555G, which is transmitted from one generation to the next in some families. Patients with this mutation have a greatly increased sensitivity to aminoglycosides. The mutation appears to occur in mitochondrial DNA. The way in which the information is expressed in the DNA is similar in both mitochondria and bacteria, which may help to explain why mitochondria are susceptible to aminoglycosides.

Dosing schedules (see p. 93) will only be summarized here. With multiple daily dosing (two or three times daily), an initial loading dose independent of renal function is employed. Some authorities advocate maintaining loading dose levels for the first two or three doses (i.e. the first 24 h), before settling on the maintenance dose level. As the peak serum level obtained is dependent on the volume of distribution, adjustment for excess obesity is important. The distribution is less in adipose/fatty tissues. In practice, where the actual body weight is more than 30% above the ideal weight, the ideal body weight is added to 40% of the excess weight, and the total thus obtained is used as a basis for calculating the loading dose. Thus the loading dose is calculated from the ideal body weight plus 0.4 times the difference between the total body weight and the ideal body weight. In addition, the volume of distribution is increased in patients with severe burns, ascites and other oedematous states, and in cases where blood loss occurs. To ensure that peak concentrations corresponding to therapeutic levels are obtained (i.e. at least 5 mg/L), serum levels should be measured on about the second day. Suggested loading doses for gentamicin and tobramycin are in the range 2–3 mg/kg (see Table 4.8 and Gilbert, 1995; Thomson et al., 1996).

Maintenance doses of gentamicin and tobramycin in adults with normal creatinine clearance levels are about 1.7 mg/kg 8-hourly. In cases where trough levels are greater than 2 mg/L, excessive tissue accumulation is occurring and renal toxicity is likely. Rather than lowering the maintenance dose (which may result in sub-therapeutic levels), it is better to lengthen the time interval between doses. It is desirable to measure peak and trough serum levels after the first or second maintenance dose and to adjust the timing of the maintenance dose, accordingly. If serum creatinine levels are stable, subsequent serum aminoglycoside measurements become less important, but should be repeated every couple of days. If renal function changes, the dosing schedule should be changed and peak and trough serum levels repeated after initiation of the new regimen. Details of how dosing schedules can be adjusted for various clinical situations have been given by Gilbert (1995), and can be found in most hospital antibiotic guidelines/formularies. Recent studies have endorsed early recommendations that routine monitoring of blood aminoglycoside levels in paediatric patients (3 months to 15 years of age) with *normal renal function* is probably unnecessary (see Robinson, 1997).

With once daily administration, each dose is in reality a loading dose (Marra et al., 1996). Dosing regimens for gentamicin and tobramycin are of the order of 5 mg/kg, and this results in peak serum levels far in excess of the 5 mg/L therapeutic minimum. With once daily treatment, peak levels will be high and need not be measured. Twenty-four-hour trough levels should be maintained below 0.5 mg/L. As this level is difficult to measure accurately, it is suggested that serum levels should be measured 6 to 14 h after the infusion. Details of 'safe' levels at various times post-infusion are shown in Tables 4.7 and 4.8. Some institutions have computer programs which, when provided with the relevant details (e.g. the actual dose used, time of blood sample, creatinine clearance values), will list the likely 24-h trough level. When these are found to exceed the suggested safe level (i.e. 0.5 mg/L), increasing the time between doses is mandatory. As with multiple daily dosing schedules, trough levels should be measured around the second day of treatment and monitoring should be continued in the face of changing renal function. Kidney accumulation and toxicity can occur with once daily aminoglycoside regimens. As already mentioned, continuing to measure peak levels is only of academic interest, unless there are clinical reasons to suspect an unusually large volume of distribution. Critically ill patients clearly benefit from aggressive initial aminoglycoside therapy (Cornwell et al., 1997).

A new regimen for administering aminoglycosides to haemodialysis patients has been proposed (Matsuo et al., 1997). In this modification of the once daily regimen, the drug is administered over 60 min by drip infusion just before the

Table 4.7 Suggested 'safe' aminoglycoside serum levels with once daily dosing schedules

Time post-inoculation (h)	Level (mg/L)[a]
6	4–7
8	3–5
10	2–3
12	1–2
14	0.5–1

[a]Figures shown apply to gentamicin, tobramycin and netilmicin; double these levels for amikacin.
For additional details see Bailey et al. (1997).

Table 4.8 Aminoglycoside dosage and administration in adults with normal renal function

Aminoglycoside	Dose	Desired levels (mg/L)
Gentamicin and tobramycin	Once daily 5 mg/kg Divided doses 2 mg/kg load then 1.7 mg/kg 8-hourly	< 1 at 18 h Peak 5–10 Trough < 2
Netilmicin	Once daily 6–7 mg/kg Divided doses 2 mg/kg 8-hourly	< 1 at 18 h Peak 5–10 Trough < 2
Amikacin	Once daily 15 mg/kg Divided doses 7.5 mg/kg 12-hourly	< 2 at 18 h Peak 25–30 Trough < 10

Adapted from Lang, S. (ed.) (1997) *Guide to pathogens and antibiotic treatment 1998*. Auckland: Adis International, with permission.

dialysis session. Peak concentrations are reached at the beginning of the session, with a rapidly decreasing concentration occurring by the start of haemodialysis. The accumulation of drug in the body is minimized by this regimen, which achieves the aims of short-duration, concentration-dependent bacterial killing. Further studies are required to assess its clinical efficacy.

It is unfortunate that aminoglycosides, which are cheap, safe and effective, share the potential for nephrotoxicity, ototoxicity and (rarely) neuromuscular blockade. Together with their low therapeutic index, this has tended to restrict their use in some countries, although the introduction of once daily dosing regimens has gone some way towards addressing these problems. In many countries, aminoglycosides are still regarded as first-line drugs for treating serious infection by Gram-negative bacilli.

QUINOLONES

GENERAL PROPERTIES

The first member of the quinolone class of antimicrobial agents was nalidixic acid, identified by Lesher and his associates in 1962 as a by-product of chloroquine synthesis. However, it was the identification in the 1980s of the fluorine- and piperazinyl-substituted quinolone derivatives that initiated an escalating resurgence of interest in this class of compounds (Hooper, 1995). Compared to nalidixic acid, these substituted compounds or fluoroquinolones display substantially greater potency, an expanded antibacterial spectrum, and improved pharmacokinetic properties and body distribution.

All of the quinolone derivatives currently in clinical use have a similar dual ring structure. Their potency is greatly improved by the presence of fluorine and piperazine moieties, and the latter are also linked to oral bioavailability. In recent years, attempts to enhance activity against Gram-positive bacteria have seen further modification of the basic structure, and the formulation of a series of drugs with varying antibacterial spectra and properties as outlined in Tables 4.9 and 4.10 (see also Hooper, 1995; Anon., 1998a).

Quinolones inhibit bacterial DNA synthesis, an event that is followed by rapid cell death, although at high drug levels (which may also inhibit protein synthesis) cell killing appears to be reduced (Hooper, 1995). Resistance to quinolones occurs by changes in the target site (DNA gyrase complexes), and by decreased intracellular drug accumulation. In the clinical setting, resistance has occurred in bacteria such as *Pseudomonas aeruginosa, Campylobacter jejuni* and *Staphylococcus aureus*, and more recently in *Neisseria gonorrhoeae*. In general, resistance to one fluoroquinolone implies resistance (or at least decreased susceptibility) to most others. Studies in patients, most often with ciprofloxacin, have suggested that the area under the concentration curve/MIC ratio is the most important predictor of both clinical and microbiological cure with fluoroquinolones (Lode *et al.*, 1998).

ANTIBACTERIAL SPECTRUM

Fluoroquinolones can basically be regarded as the oral equivalents of aminoglycosides. Unlike nalidixic acid and earlier quinolones such as norfloxacin, the newer fluoroquinolones are widely distributed in tissues and body fluids, and have additional activity against an increased range of bacterial species (e.g. most Gram-negative bacilli, including *Pseudomonas aeruginosa* and *Haemophilus* species, Gram-negative cocci, including *Neisseria* species and *Moraxella catarrhalis*, and staphylococci) (see Tables 4.6 and 4.9). Some of the newer quinolones (e.g. trovafloxacin) may be synergistic with other agents (e.g. ceftazidime) against the resistant *Stenotrophomonas maltophilia* (Visalli *et al.*, 1998). They are useful in the treatment of osteomyelitis, peritonitis and necrotizing fasciitis. In general, the concentrations of most quinolones in the urine are

Table 4.9 Comparison of selected fluoroquinolones

Generic name	Oral route frequency (h)	Half-life (h)	Staphylococcus aureus	Streptococci	Enterococci	Gram-negative enteric bacilli	Pseudomonas aeruginosa[b]	Bacteroides fragilis	Clostridia
Norfloxacin[c]	12	3.3	V/+	–	–	+	V/–	–	–
Fleroxacin	24	10	V/+	–	–	+	V/–	–	–
Ciprofloxacin	12	4–6	+	V	V/–	+	V/+	–	V/–
Levofloxacin[a]	8–12	6–8	+	V/+	V	+	V/+	V/+	V
Grepafloxacin	24	12	+	V/+	V/+	V/+	V	–	V/–
Sparfloxacin	24	20	+	+	V	+	V	V	V
Moxifloxacin	24	11–15	+	+[e]	V	+	V	+/V	+
Clinafloxacin	24	5[f]	+	+[e]	+[d]	+	+	+	+

[a]More active isomer of earlier ofloxacin, some suggest 24-h dosing.
[b]Increasing levels of resistance with most fluoroquinolones – check local pattern.
[c]Poor systemic distribution.
[d]Some activity against vancomycin-resistant enterococci (see Haria and Lamb, 1997).
[e]Activity against penicillin-resistant strains.
[f]Of parent compound.
+, usually susceptible; –, not susceptible; V, variable sensitivity.

Table 4.10 Features of some newer quinolones compared with those of ciprofloxacin

Feature	Ciprofloxacin	Grepafloxacin	Levofloxacin	Sparfloxacin	Trovafloxacin*
Formulation	IV and PO	PO	IV and PO	PO	IV[a] and PO
Usual dosage (mg/day)	800–1500	300–600	500	200[b]	200
Frequency of administration	Twice daily	Once daily	Once daily	Once daily	Once daily
Spectrum of antibacterial activity					
Gram-positive	+	++	+++	+++	+++
Gram-negative	+++	+	++	++	+++
Atypical organisms[c]	+	++++	+++	+++	+++
Anaerobes	−	−	−	−	+++
Adverse effects					
Most frequent/serious	Convulsions, tendon damage, nausea	Nausea and taste disturbance	Nausea	Photosensitivity[d]	Prolongation of QT_c interval
Other	Headache, diarrhoea, dizziness, rash	Headache, diarrhoea, dizziness, rash	Diarrhoea, vaginitis, abdominal pain	Gastrointestinal, CNS effects	Nausea, headache, vomiting, vaginitis, diarrhoea
Drug interactions					
Chelating agents[e]	√	√	√	√	√
Theophylline	√	√	X	X	X
Warfarin	√	?	X	X	X

[a]The IV formulation of trovafloxacin is alatrofloxacin, a prodrug of trovafloxacin.
[b]A single dose of 400 mg is recommended as a loading dose on day 1.
[c]Includes organisms such as *Mycoplasma pneumoniae*, and *Legionella pneumophila*.
[d]The risk of photosensitivity requires strict avoidance of exposure to sun or artificial ultraviolet light during treatment, and avoidance for 5 days after therapy ends. Treatment should be promptly discontinued if signs or symptoms appear.
[e]Sucralfate, iron, magnesium, calcium and aluminium-containing antacids.
IV, intravenous; PO, oral; +, minimal efficacy; ++, moderate efficacy; +++, marked efficacy; ++++, maximum efficacy; −, no efficacy; √, occurs; X, effect does not appear to occur; ?, unknown.
*Trovafloxacin was withdrawn from sale in June 1999.
Adapted from Anon. (1998a).

sufficient to provide substantial therapeutic ratios of urinary drug concentration to the MIC of most pathogens.

All fluoroquinolones are available as oral formulations (with bioavailability ranging from 50% to almost 100%) as well as parenteral formulations, with once daily dosing schedules recommended for some of the newer agents with more prolonged half-lives. A word of caution should be given here concerning bioavailability following oral administration. When these drugs are co-administered with antacids, or with nutritional supplements given by nasogastric tube, their absorption may be severely reduced and unpredictable. Quinolones allow out-patient therapy for some patients who would otherwise require prolonged hospitalization and parenteral antibiotics.

EXAMPLES

Ciprofloxacin remains the most potent generally available fluoroquinolone, with excellent activity against many important Gram-negative bacteria. It is also effective against many strains of *Staphylococcus aureus*, although it is not as potent as some of the newer fluoroquinolone drugs (e.g. trovafloxacin). Earlier, fluoroquinolones lack therapeutically useful activity against streptococci and anaerobes, although attempts to enhance such activity have seen the development of newer agents such as levofloxacin, gatifloxacin, grepafloxacin, sparfloxacin, maxifloxacin, clinafloxacin and trovafloxacin (see Table 4.9). These are still unavailable in most countries. Trovafloxacin appears especially promising, and has a wide antibacterial spectrum including MRSA (Citron and Appleman, 1997; Haria and Lamb, 1997; Bayer et al., 1998). It also penetrates well into the CSF, and is possibly synergistic with ceftriaxone against penicillin-resistant pneumococci (Nicolau et al., 1998). Sparfloxacin has shown promising effects against pneumococcal disease in clinical trials (Aubier et al., 1998), while laboratory data suggest an extended antibacterial spectrum for grepafloxacin (Wagstaff and Balfour, 1997), levofloxacin (Wexler et al., 1998) and clinafloxacin (Ednie et al., 1998). The latter appears to have similar activity to trovafloxacin. The growing world-wide threat of penicillin-resistant pneumococci and vancomycin-resistant enterococci has provided the impetus for the development of such quinolones. Some fluoroquinolones show therapeutic activity against mycobacteria and atypical respiratory pathogens such as *Mycoplasma pneumoniae* and *Chlamydia pneumoniae* (see Table 4.10). Most of the newer agents appear to penetrate well through the biofilms encountered in some infectious processes (e.g. alginate-producing pseudomonads) (Ishida et al., 1998).

Although they are active against staphylococci, fluoroquinolones should not be regarded as first-line anti-staphylococcal agents. They should be reserved for situations where sensitivity tests suggest that the more common anti-staphylococcal agents will be ineffective (e.g. MRSA), or where antibiotic penetration may be of concern (e.g. in patients with osteomyelitis). In all cases, fluoroquinolones should be combined with another anti-staphylococcal drug (e.g. fusidic acid) in order to prevent the emergence of quinolone-resistant strains.

The rapidly escalating levels of resistance in bacteria such as *Pseudomonas aeruginosa* and *Staphylococcus aureus* to fluoroquinolones means that the latter should not be used empirically as the only agents in life-threatening infections in which these bacteria are likely to be involved.

ADVERSE EFFECTS

Fluoroquinolones are well-tolerated drugs with varying adverse effects (Anon., 1998a), the most common of which are gastrointestinal (e.g. nausea, vomiting, abdominal discomfort). Other less common adverse effects involve the central nervous system (mild headache, dizziness), skin (including photosensitization) and joints. Arthropathy with cartilage erosion and non-inflammatory effusions have been seen in the weight-bearing joints of young animals that have been given quinolones. Although this observation limited the earlier use of quinolones in children, it seems that such joint problems (e.g. arthralgia, joint swellings) are uncommon and reversible. Safety in pregnancy has not been established for any of the quinolones, and because they are excreted in breast milk they should not be given to nursing mothers. Depending on the fluoroquinolone in question, significant interactions may occur with a variety of drugs (e.g. theophylline, caffeine, non-steroidal anti-inflammatory drugs (Anon., 1998a). A major advantage of trovafloxacin over ciprofloxacin has been the absence of photosensitivity and interactions with warfarin and theophylline, although there is concern about the frequency of dizziness, nausea and headache with trovafloxacin (Anon., 1998a). (*Since the preparation of this text, trovafloxacin has been withdrawn from the market (June 1999) because of unacceptable toxic effects. This follows the earlier removal of temafloxacin (1992) and even more recently (October 1999) of grepafloxacin.*)

A novel approach in recent times has been the chemical linking of quinolones to cephalosporins. These compounds may display characteristics of both classes of antibiotics or basically function as cephalosporins. In both cases they can release the quinolone by spontaneous or β-lactamase cleavage of the combining ester linkage (Hooper, 1995).

ANTI-ANAEROBE AGENTS

GENERAL PROPERTIES

Apart from metronidazole, numerous other antimicrobial agents include anaerobe activity in their spectrum of activity (Finegold, 1995a,b; Finegold and Wexler, 1996). Examples are clindamycin, chloramphenicol, most penicillins, β-lactam combinations such as co-amoxyclav, imipenem, meropenem, and 7-methoxy cephalosporins (cephamycins) such as cefoxitin and cefotetan (see Table 4.11). However, unlike the situation with metronidazole, resistance to these other agents is now becoming more common (Rasmussen *et al.*, 1997). Anaerobes may account for 20% or more of all hospital-acquired bacteraemias (Goldstein, 1996).

METRONIDAZOLE

The nitroimidazole compound, metronidazole, is by far the most effective antianaerobe drug. Although it was first used in 1959 for the treatment of *Trichomonas* vaginitis, it is now known to be effective against infections involving anaerobic bacteria as well as a variety of protozoan parasites. Metronidazole diffuses readily into all tissues, including the CNS, and is generally well tolerated. Excellent blood and tissue levels occur following oral or intravenous administration, and it may also be given rectally. Initial therapy in seriously ill patients must be given intravenously. Non-parenteral routes can result in unacceptable time delays in reaching therapeutic levels, although a switch to an oral or rectal route is feasible when conditions warrant this. The latter are therapeutically acceptable and have considerable cost savings. Resistance to metronidazole appears to develop rarely, and is very uncommon in the important obligate

Table 4.11 Characteristics of some important anti-anaerobe agents[c]

Antimicrobial agent	Therapeutic activity					
	Aerobic bacteria	Bacteroides fragilis	Clostridium perfringens	Other clostridia	Other anaerobes[a]	Actinomycetes
Metronidazole	−	+	+	V/+	V/+	−
Imipenem	+	+	+	+	+	+
Chloramphenicol	V/+	+	+	+	+	+
Co-amoxyclav[b]	V/+	+	+	+	+	+
Piperacillin/tazobactam	+	+	+	+	+	+
Clindamycin	V	V	+	V	V	V/+
Penicillin	V	−	+	V/+	V	+
Cefoxitin/cefotetan[d]	V	V	+	V	V/+	V/+

[a]Bacteria of other genera such as *Prevotella*, *Porphyromonas*, *Bacteroides* other than *Bacteroides fragilis* group, *Peptostreptococcus*, and *Fusobacterium*.
[b]Activity similar with piperacillin/tazobactam and ticarcillin/clavulanic acid combinations.
[c]See Finegold and Wexler (1996) for additional details.
[d]Cefoxitin seems to be superior to cefotetan.
+, usually susceptible; −, not susceptible; V, variable sensitivity.

anaerobes (see p. 45). Several other similar nitroimidazoles are available in tablet form only (e.g. ornidazole, tinidazole). There is some evidence that they are better tolerated than metronidazole.

Metronidazole is a potent bactericidal agent. Its selective activity against anaerobes is due to the fact that the antibacterial form of the drug is a reduced derivative produced intracellularly at the low redox values that are only attainable by obligate anaerobes. It is thought to act by inducing strand breakage in DNA.

Metronidazole is considered to be the drug of choice for treating most anaerobic infections, including those by clostridia. With the latter, metronidazole would seem to be at least as effective as the traditional penicillin (see Finegold and Mathisen, 1995). Anaerobic infections where metronidazole is not recommended include actinomycosis and infection by the skin anaerobe *Propionibacterium acnes*. Metronidazole has excellent and predictable activity against *Bacteroides fragilis* and other important *Bacteroides* species, and the other common anaerobes that are found in the alimentary tract (e.g. *Prevotella* species, *Porphyromonas* species, peptostreptococci, fusobacteria, *Bilophila wadsworthia* and *Clostridium perfringens*). Bacteria that are also sensitive to metronidazole include *Gardnerella vaginalis* (a potential cause of bacterial vaginosis) and *Helicobacter pylori*. Table 4.11 lists the relevant features of metronidazole.

CLINDAMYCIN

Clindamycin is a chemical modification of lincomycin with increased antibacterial potency and absorption after oral delivery. It interferes with the peptide-elongation phase of protein synthesis, and the ribosomal binding site is probably similar to that of macrolides, as resistance to erythromycin affects clindamycin as well. It is an excellent anti-staphylococcal, anti-streptococcal and anti-anaerobe agent. Clindamycin is very effective against *Clostridium perfringens* and is possibly superior to penicillin and metronidazole. Although it is one of the most active antibiotics against *Bacteroides fragilis*, resistance in this species and in other *Bacteroides* species is now in the range 15–25% and increasing; similar figures apply to a number of other significant anaerobes (e.g. clostridia other than *Clostridium perfringens*, and peptostreptococci) (see Lee *et al.*, 1996). It seems that cross-resistance between clindamycin and the more commonly used but chemically unrelated macrolides may be in part responsible for this escalating resistance problem. Laboratory data suggest that clindamycin is also very active against *Bilophila wadsworthia* (Mochida *et al.*, 1998), most strains of which are β-lactamase-positive.

Apart from its use in the treatment of anaerobic sepsis, clindamycin is a serious consideration for empirical therapy of infections that are likely to involve staphylococci and/or streptococci (e.g. necrotizing fasciitis (see p. 214). In the case of the latter, clindamycin is possibly superior to penicillin (Stevens *et al.*, 1998). Where sensitivity patterns allow, clindamycin may also be used to treat MRSA infections. Its anti-staphylococcal (and anti-streptococcal) activity has perhaps been underestimated and unappreciated in recent years. It remains a consideration for non-central-nervous-system infections that are likely to be polymicrobial in origin, as an alternative to penicillin or metronidazole in treating *Clostridium perfringens* infections, and as an alternative to β-lactams in the treatment of staphylococcal infections. Clindamycin had a notable early association with pseudomembranous enterocolitis (*Clostridium difficile* overgrowth), and this reputation has been responsible for variations in its world-wide use. The characteristics and properties of clindamycin are listed in Table 4.11.

CHLORAMPHENICOL

Chloramphenicol is one of the 'forgotten' antimicrobial agents. It was one of the first therapeutically useful antibiotics to appear from the systematic screening of soil actinomycetes in the wake of the discovery of streptomycin in the 1940s (Greenwood, 1995c). Although it is a naturally occurring compound, the relatively simple molecular structure can readily be synthesized.

Chloramphenicol acts by inhibiting the peptidyl-transferase reaction (in which the peptide bond is formed) on 70S ribosomes. Its wide spectrum of activity embraces most aerobic and anaerobic bacteria, including chlamydiae. While the action of chloramphenicol against enterobacteria is bacteriostatic, it may also display potent cidal activity against other bacteria, including Gram-positive cocci. It is not a purely bacteriostatic agent as is widely quoted. Chloramphenicol also possesses the important properties of diffusing readily into cerebrospinal fluid, and of penetrating into cells and through necrotic tissues (Greenwood, 1995c).

Given these desirable properties, it is unfortunate that chloramphenicol has one grave drawback, namely potentially fatal depression of the bone marrow. This extremely rare side-effect, and the propensity of the drug to induce grey baby syndrome in neonates when the dose is not properly adjusted, have been sufficient reasons to relegate chloramphenicol to the role of a reserve drug for special purposes. It is sometimes still used to treat typhoid fever and meningitis, and remains an option for brain and other abscesses. It may be thrown a lifeline if the incidence of penicillin-resistant pneumococci continues to escalate.

GLYCOPEPTIDES

GENERAL PROPERTIES

Vancomycin and teicoplanin are glycopeptide antibiotics with activity against many Gram-positive bacteria. They act by binding to the D-alanyl-D-alanine terminus of the pentapeptide cell wall precursor, thereby blocking further cell wall synthesis. This is at an earlier stage of cell wall formation than that which is affected by the β-lactams. Glycopeptides have found a niche in the treatment of infections by Gram-positive cocci that are resistant to penicillins and cephalosporins. Glycopeptides are too bulky to pass through the outer membrane of the cell wall of Gram-negative bacteria, thus restricting the spectrum of activity to Gram-positive bacteria. As glycopeptides do not show concentration-dependent killing, it seems likely that the time above the MIC is the most important parameter for the efficacy of these agents.

Although a number of Gram-positive bacteria are known to be intrinsically resistant to glycopeptides, acquired resistance is now being seen in enterococci (see p. 37) and more recently in staphylococci. Many strains of coagulase-negative staphylococci that are resistant to β-lactams apparently contain subpopulations with decreased susceptibility to teicoplanin and vancomycin (Sieradzki et al., 1998). The supposedly 'vancomycin-resistant MRSA' strains widely reported from Japan and the USA in mid-1997 showed only intermediate resistance (MIC of 8 mg/L) of undetermined mechanism.

VANCOMYCIN

Vancomycin has been widely used in the treatment of infections by MRSA (see p. 42), and increasingly so in the treatment of other Gram-positive infections where the causal bacteria are resistant to commonly used β-lactams or the patient is allergic to β-lactams. Compared to the anti-staphylococcal β-lactams (e.g. flucloxacillin), vancomycin does not appear to be as effective against *Staphylococcus aureus* in clinical situations. This appears to be related to reduced cidal activity (Wood and Wisniewski, 1994; Hoellman et al., 1998), which can be improved by the addition of gentamicin (Houlihan et al., 1997). Vancomycin has also been used for the treatment of *Clostridium difficile* colitis, although such oral therapy (vancomycin is given intravenously for all other infections, as it is not absorbed from the gastrointestinal tract) is now restricted to all but the most severe of cases. It is also expensive. Oral vancomycin clearly results in colonization of the bowel by vancomycin-resistant enterococci in a significant proportion of recipients (Murray, 1997). Vancomycin has also been used for prophylaxis in cases where infection by MRSA is a potential problem.

Vancomycin has been available since the 1960s, and early preparations contained impurities that gave the drug a reputation for toxicity. Newer, more highly purified preparations are said to be much safer, but renal toxicity and ototoxicity still occur (Greenwood, 1995a). The drug is given by slow intravenous infusion in order to avoid the 'red man' syndrome (which occurs in more than 50% of healthy volunteers who are given vancomycin, but at a lower incidence when the drug is being used therapeutically). The clinically relevant features of vancomycin are listed in Table 4.12.

Since no definitive relationship exists between vancomycin serum concentrations and either clinical outcome or adverse effects, considerable debate surrounds the utility of monitoring serum vancomycin concentrations. It seems that routine serum monitoring may only be warranted in specific populations (e.g. those receiving concurrent aminoglycosides or other potentially nephrotoxic drugs, those receiving higher than normal doses of vancomycin, patients undergoing haemodialysis, and patients with rapidly changing renal function) (Leader et al., 1995). It has been claimed that the concentration of vancomycin which is required for clinical efficacy may be lower than was previously believed, and can be achieved with lower doses than those currently recommended. Such lower doses might reduce the likelihood of drug accumulation and subsequent toxicity, and in turn reduce the need for frequent serum assays (Saunders, 1995). Against this must be considered the potential of treatment failure!

However, until more information is available it seems premature to abandon serum monitoring in patients receiving prolonged (> 48 h) therapy in whom there is opportunity for accumulation. The currently used trough (pre-dose) range of 5–10 mg/L has proved effective and safe, although further studies may show lower levels to be efficacious. Trough levels of 12–15 mg/L should be regarded as an indication that accumulation is occurring, although some clinicians regard trough levels approaching 15 mg/L as acceptable in serious staphylococcal infections where the microbes may not be fully sensitive to vancomycin. Although questions remain concerning the optimal serum concentrations of vancomycin for prediction of efficacy and minimal toxicity, the case for routine measurement of peak (post-dose) concentrations is weak, as it provides no additional information over determining pre-dose concentrations alone (Catchpole and Hastings, 1995; Saunders, 1995).

The usual leading dose is 15 mg/kg based on lean body weight regardless of renal function. The usual adult maintenance dose is 500 mg given 6-hourly, or 1 g given 12-hourly. The calculated creatinine clearance should be divided by the patient's weight in order to convert the units to mL/min/kg. The dosage should be recalculated at least every second day.

TEICOPLANIN

Teicoplanin is a naturally occurring mixture of several closely related compounds, with a spectrum of activity similar to that of vancomycin, although some coagulase-negative staphylococci are less susceptible to teicoplanin. Unlike vancomycin, teicoplanin can be administered by intramuscular as well as intravenous injection. It also has a much longer half-life, with the potential for

Table 4.12 Some features of vancomycin, teicoplanin, fusidic acid and rifampicin

Antimicrobial agent	Usual adult dose (g)/ frequency	Half-life (h)	Therapeutic activity[c]				Comment
			Staphylococcus aureus (MSSA)	MRSA	Coagulase-negative staphylococci	Penicillin-resistant pneumococci	
Vancomycin[b]	0.5 q 6 h[a]	6–8	+	+	+	+	Must be given by slow IV infusion; cannot be given IM
Teicoplanin[b]	0.4 q 8 h[a]	40–70	+	+	V, +	+	Can be given over 3–5 min IV or IM; can be given once daily; probably less toxic than vancomycin
Fusidic acid	0.5–1 q 6–8 h IV 0.5 q 8 h PO	14	+	V/+	+	V/–	Usually combined with a second anti-staphylococcal agent
Rifampicin	0.45–0.6 q 12–24 h IV	2–5	+	V/+	V/+	+	In combination with a second antimicrobial agent

[a] After loading doses.
[b] Bacteriostatic against enterococci, but cidal if given with gentamicin (possible toxicity).
[c] +, usually effective; –, not usually effective; V, variable efficacy.
MSSA, methicillin-susceptible *Staphylococcus aureus*; MRSA, methicillin-resistant *Staphylococcus aureus*; IV, intravenous; IM, intramuscular; PO, oral.

once daily dosing and out-patient therapy, and it appears to be less likely to give rise to adverse reactions (Brogden and Peters, 1994; Greenwood, 1995a). Teicoplanin-resistant bacteria are not necessarily resistant to vancomycin, and vice versa.

FUSIDIC ACID

Fusidic acid is the only therapeutically useful member of a group of naturally occurring antibiotics that have a steroid-like molecular structure. It acts by preventing the translocation step in bacterial protein synthesis (Greenwood, 1995b), and it is bactericidal.

Although it is active *in vitro* against a variety of bacteria, fusidic acid is primarily regarded as an anti-staphylococcal agent. *Staphylococcus aureus* is particularly susceptible, and in some European countries fusidic acid has been used for years as the main-line anti-staphylococcal agent. It is interesting to note that the incidence of MRSA appears to be minimal in such countries.

Fusidic acid administered orally or parenterally penetrates well into loculated and infected tissues, including bone, and is favoured by many authorities for the treatment of staphylococcal osteomyelitis (see p. 229). It has often been stated that one of the potential drawbacks of using fusidic acid is that in any large staphylococcal population a few cells exist which are resistant to the drug and which may proliferate during therapy. This phenomenon and its consequences appear to have been overstated. However, it is usual to administer fusidic acid in combination with a second anti-staphylococcal agent (e.g. flucloxacillin) in an attempt to minimize the overgrowth of resistant clones. The important properties of fusidic acid are listed in Table 4.12. The only common side-effects are gastric irritation when taken by mouth, and skin rashes. Hepatitis commonly occurs with the intravenous formulation.

RIFAMYCINS

The clinically useful rifamycins, of which rifampicin (known as rifampin in the USA) is the most significant, are semi-synthetic derivatives of rifamycin B, which is one of a group of structurally complex antibiotics produced by *Streptomyces mediterranei* (Greenwood, 1995c). Rifamycins were apparently named after the 1957 French film, *Le Riffi* (see Farr, 1995).

While rifampicin has become established as one of the major anti-tuberculous drugs, it also exhibits potent bactericidal activity against a range of other bacteria, including staphylococci and legionellae (Farr, 1995). It is an exceptionally good anti-staphylococcal agent (see Table 4.12). It has also found a niche in the prophylaxis of close contacts of meningococcal disease, and in the elimination of throat carriage of meningococci in patients with overt disease. One of the drawbacks of rifampicin is the frequency with which mutants emerge. For this reason it is normally used in combination with other drugs.

Rifampicin inhibits DNA-dependent RNA polymerase at the β-subunit, which prevents chain initiation but not elongation. Mammalian mitochondrial RNA synthesis is not impaired at clinically achievable concentrations (Farr, 1995). Much of the usefulness of rifampicin as an anti-staphylococcal and anti-tuberculous agent is due to its solubility in acidic aqueous solutions and organic solvents, and its remarkable diffusion through lipids. It penetrates readily through cellular membranes and into necrotic tissues (Bamberger et al., 1997).

Rifampicin is well absorbed by the oral route (although it can also be given intravenously), and serious side-effects are uncommon. A minor but potentially alarming side-effect, resulting from the fact that rifampicin is strongly pigmented, is the production of red urine and other body secretions. Even contact lenses may be discoloured. Induction of intestinal and liver enzymes by rifampicin can lead to reduced bioavailability, increased excretion and antagonism of some other drugs that are handled by the liver, including oral contraceptives. Women using oral contraceptives who are given rifampicin should therefore be advised to adopt an alternative means of contraception for 2 to 3 months. Rifampicin has also caused interactions (reduced bioavailability and/or decreased serum half-life) with prednisone, digitoxin, warfarin, some azole antifungal agents and the sulphonylureas (Farr, 1995). It is contraindicated in pregnancy (except for the treatment of tuberculosis).

Of the other rifamycins, rifabutin has proved to be effective in preventing *Myobacterium avium* complex disease in AIDS patients with lowered CD_4 counts, while rifapentine offers greater convenience than rifampicin in the treatment of pulmonary tuberculosis in non-HIV-positive patients (Anon., 1999c). The reputed antiviral and antifungal potential of rifamycins would appear to have minor therapeutic implications.

TRIMETHOPRIM AND COTRIMOXAZOLE

Trimethoprim is a dihydrofolate reductase inhibitor that interferes with folate synthesis in susceptible bacteria. These include many Gram-positive and Gram-negative aerobes, with the notable exception of *Pseudomonas aeruginosa*. Trimethoprim is usually reserved for the prophylaxis and treatment of urinary tract infections.

Cotrimoxazole is a combination of trimethoprim and sulphamethoxazole. The latter also blocks folate synthesis (at a different stage to trimethoprim), and in combination with trimethoprim is synergistic for many bacteria. Cotrimoxazole is used for the treatment and prophylaxis of urinary tract infections and the treatment of a variety of respiratory tract infections (e.g otitis media, sinusitis). It is the drug of choice for the treatment of nocardiosis, the treatment and prophylaxis of *Pneumocystis carinii* pneumonia, and possibly the treatment of infections by *Stenotrophomonas maltophilia* and some strains of MRSA.

Allergic reactions, including Stevens–Johnson syndrome, are invariably attributable to the sulphur-component of cotrimoxazole. Such reactions are common in patients with HIV infections. In addition, elderly patients are at risk from bone-marrow depression. This can be treated with folinic acid, which does not impair the antibacterial activity of cotrimoxazole. Trimethoprim is relatively contraindicated in the first trimester of pregnancy, and sulphamethoxazole is contraindicated in the last trimester.

MACROLIDES

Macrolides are primarily bacteriostatic antibiotics with a varied spectrum of activity. This group includes erythromycin, roxithromycin, clarithromycin, azithromycin (which is in reality an azalide), spiramycin and the antiparasitic drug ivermectin. As they achieve high intracellular levels, particularly in phagocytic cells, there is not always a strong correlation between serum levels and clinical efficacy.

Erythromycin is a weak base that is readily inactivated by gastric acid. It is poorly soluble in water and poorly absorbed orally. These disadvantages have been largely overcome by various enteric formulations and the use of erythromycin salts. Erythromycin is the drug of choice for infections by *Mycoplasma pneumoniae* and *Legionella* species, and provides an alternative to doxycycline for chlamydiae. It can also be considered as a second-line agent for streptococcal and *Staphylococcus aureus* infections, and provides an alternative to penicillin in penicillin-allergic patients, and to tetracyclines during pregnancy.

Roxithromycin has similar antibacterial activity to erythromycin, but a longer half-life, allowing once or twice daily dosing. It also appears to be somewhat better tolerated than erythromycin.

Clarithromycin has a fairly similar antibacterial spectrum to erythromycin, but with significantly increased activity against *Haemophilus influenzae* and the *Mycobacterium avium–intracellulare* complex. It is also now commonly used as part of a regimen for the eradication of *Helicobacter pylori*. Clarithromycin is given twice daily for most infections.

Azithromycin (an azalide with a 15-membered basic ring structure, compared to the usual 14 in other macrolides) appears to be more potent than erythromycin, with increased activity against chlamydiae and ureaplasmas. Its long half-life permits once daily dosing, and it has been shown to be effective as a single dose for non-gonococcal urethritis. Azithromycin has also been used to treat cerebral toxoplasmosis in patients with AIDS.

The macrolides are among the safest of the antibiotics, with the most common side-effects being gastrointestinal in nature (around 5% of patients). Intravenous erythromycin may cause phlebitis, while the potential to induce cholestatic jaundice has restricted the use of erythromycin estolate, especially during pregnancy. The macrolides may interfere with the hepatic metabolism of other drugs (e.g. theophylline, warfarin).

A range of ketolide antimicrobial agents is currently under development. These are members of the macrolide group, and they display activity against a variety of aerobic and anaerobic bacteria (Ednie *et al.*, 1997a,b,c; Schülin *et al.*, 1997). Some compounds may be useful against erthromycin-resistant pneumococci (Barry *et al.*, 1998).

TETRACYCLINES

Tetracyclines are currently marketed only in tablet or capsular form, or as topical preparations. Of the various tetracyclines that are available, doxycycline is probably the preferred agent. It has a long half-life, allowing once or twice daily dosing.

Tetracyclines are broad-spectrum drugs with useful activity against a variety of aerobic and anaerobic bacteria. They are often used empirically for minor respiratory, skin and soft-tissue infections, including inflammatory acne. Tetracyclines are regarded as the therapy of choice for infections by a number of unusual bacteria, including chlamydiae, ureaplasmas, rickettsiae, *Borrelia* species and, in combination with other antibiotics, *Brucella* species, *Helicobacter pylori* and *Burkholderia (Pseudomonas) pseudomallei*. Doxycycline has also been used in the treatment and prophylaxis of malaria.

All tetracyclines are contraindicated in children and during pregnancy because they may discolour developing teeth, and perhaps also depress skeletal growth in infants. Photosensitivity is relatively common, as are biological side-effects such as *Candida* overgrowth.

INVESTIGATIONAL ANTIMICROBIAL AGENTS

In addition to some of the antimicrobial agents mentioned above (e.g. ketolides, newer fluoroquinolones), a number of other drugs are currently under investigation. These include streptogramins such as quinupristin/dalfopristin (Synercid or Q/D). Information from phase III trials and laboratory investigations has revealed that Q/D has low (i.e. in the therapeutically susceptible range) MIC values against most strains of staphylococci (both methicillin-susceptible and resistant strains of *Staphylococcus aureus* and *Staphylococcus epidermidis*), pneumococci (including penicillin-resistant strains) and *Enterococcus faecium* (both vancomycin-susceptible and resistant strains). Strains of *Enterococcus faecalis* have high MICs and are regarded as non-susceptible (Pechère, 1996; Nicas et al., 1997). In clinical trials, Q/D seems to be well tolerated, with the major adverse effects being arthralgia and myalgia (both occurring in around 10% of patients). Infusion via a peripheral vein results in local thrombosis/phlebitis in a few patients, and there is no evidence of phlebitis when Q/D is infused via a central vein. Mild elevation of conjugated bilirubin has been observed in about 20% of patients. As Q/D inhibits cytochrome P_{450}, it is not surprising that it has been found to inhibit the metabolism of cyclosporine and calcium-channel blockers. When administered at 7.5 mg/kg 8-hourly to severely ill patients with vancomycin-resistant enterococcal infections, the overall success rate in bacteriologically evaluable patients has been over 65%. The results are awaited of further trials concerning the use of Q/D and related compounds in patients with serious infections by resistant Gram-positive cocci.

Oxazolidinones (e.g. eperezolid, linezolid) are said to be the first new class of antibacterial agents for over a decade (Ford et al., 1997; Nicas et al., 1997). They appear to inhibit protein synthesis on the 30S subunit, and they show activity against most Gram-positive bacteria and mycobacteria.

Other investigational antibiotics which have been predominantly studied as a result of increasing vancomycin resistance (Moellering, 1998) include the oligosaccharide everninomycins, the tetracycline-like glycylcyclines (Nicas *et al.*, 1997) and the peptide antibiotics (Hancock and Chapple, 1999).

TOPICAL ANTIBACTERIAL AGENTS

Historically, topical antimicrobial agents have an established place in the treatment of several non-surgical infectious diseases, including conjunctivitis, otitis externa, extreme acne and discrete areas of impetigo (see Table 4.13). Their role in the surgical setting is less clear-cut, and in some cases it is controversial. Mupirocin has been used to eradicate nasal and skin carriage of *Staphylococcus aureus,* including MRSAs (Eltringham, 1997; MacGowan, 1998). Unfortunately, this has been accompanied by increasing levels of resistance to this agent. Antiseptics which include povidone iodine, chlorhexidine, hypochlorite, acetic acid, 70% alcohols and benzoyl peroxide are useful for decreasing the numbers of bacteria on the skin. Some authorities recommend the use of topical antibacterial agents for prophylaxis against bacterial colonization of central and peripheral intravascular catheter sites, although the efficacy of such practices is unproven. Prevention of infection in burn wounds is discussed below.

It is clearly logical to avoid topical use of valuable systemic antimicrobial agents for two reasons – first, the patient may become allergic, and secondly, the microbe may become resistant. In many cases, selection of strains that are resistant to the antibiotic being used is accompanied by resistance to a number of related or unrelated antibiotics. Selection of strains that are resistant to topical clindamycin (Dalacin T) carries with it cross-resistance to macrolides (e.g. erythromycin) and streptogramins, and may be partly responsible for the increasing clindamycin resistance levels that are now being seen in some important anaerobes. Examples of more common topical antibiotics, their antibacterial spectrum and potential applications are listed in Table 4.13 (see also Tunkel, 1995).

Prevention of infection in burn wounds

Since the 1960s, topical therapy has been reintroduced in burn wound management to prevent burn wound sepsis. Burn wounds are initially sterile, but within a few hours the surface is colonized by Gram-positive cocci, followed within 4–5 days by Gram-negative organisms. Burn wound sepsis is defined as being present when organisms exceed 100 000 per gram of burned tissue. At this level there is active invasion of the subadjacent unburned tissue. Systemic antibiotics are unable to reach full-thickness burned areas for at least 3 weeks, i.e. until a neovasculature is established. Thus topical therapy has reduced the overall mortality from burns from 38–45% to 14–24%.

Table 4.13 Characteristics of some common topical antibiotics[a]

Antibiotic[b]	Antibacterial spectrum	Useful for infections involving:
Bacitracin (often formulated with other agents, e.g. polymyxin)	Gram-positives	Skin
Chloramphenicol	Most Gram-positives and Gram-negatives except pseudomonads	Ear, eye
Ciprofloxacin	Gram-negatives including pseudomonads	Eye
Clindamycin	Staphylococci, anaerobes, streptococci	Skin
Erythromycin	As for clindamycin	Skin
Framycetin	Staphylococci, *Neisseria*, Gram-negative enteric bacilli	Ear, eye, skin[c]
Fusidic acid	Staphylococci	Eye, skin
Gentamicin	Gram-negative bacilli, staphylococci	Eye, skin
Metronidazole	Anaerobes	Skin (vagina)
Mupirocin	Staphylococci, streptococci	Skin (nasal cavity)
Neomycin	Staphylococci, Gram-negative bacilli	Eye
Neomycin + bacitracin + polymixin	As above plus streptococci, pseudomonads	Skin
Tetracycline	Similar to chloramphenicol	Ear, eye, skin

[a] See also list for burn wounds.
[b] Agents often used in combinations.
[c] As Sofra–Tulle impregnated dressing.

Topical antibacterial agents for the treatment of burn wounds include the following:
1 mafenide or sulphamylon cream (11.1%), which is effective against a wide spectrum of organisms and penetrates well to deeper tissues, but the initial application is painful and there is a 5% incidence of skin rashes. It is a carbonic anhydrase inhibitor and leads to metabolic acidosis;
2 silver nitrate (0.5%), which is active against a wide spectrum of organisms and is useful for debriding the eschar. It has the disadvantages of causing black

discolouration, poor penetration and leaching out important minerals (especially sodium) by virtue of hypotonicity;
3 silver sulphadiazine, which has the advantages of being painless and rarely causing hypersensitivity reactions. It is effective against a wide spectrum of organisms and penetrates reasonably well. However, it is not particularly effective in long-term bacterial control;
4 furazolium hydrochloride, which is sometimes used, but has little effect on pseudomonads;
5 iodine PVP complex (Betadine), which is also sometimes used. It has little effect on staphylococci, but is reasonably effective against pseudomonads. Betadine requires prolonged skin contact for optimal activity;
6 mupirocin, which is effective against staphylococci and streptococci but should probably be restricted to difficult problems.

chapter 5

ANTIBACTERIAL PROPHYLAXIS IN SURGERY

GENERAL PRINCIPLES

The absence of wound or surgical site infection is one goal of operative therapy. In practice, this is clearly unattainable, despite improved techniques and the use of antimicrobial prophylaxis. The principles of prophylaxis are established, and the benefits are enormous in terms of both cost and patient comfort. For most procedures, the rationale behind prophylaxis is known (see Wittmann and Condon, 1991; Slama, 1992; Nichols, 1994; Gyssens et al., 1996; Gyssens, 1999), and provided that a few simple rules are followed, minimal post-operative infection rates are attainable. The results of a myriad of trials show that many alternatives to suggested regimens exist, all with essentially similar outcomes. Antibiotic cost therefore becomes a major consideration, in the context that therapy and bed-stay costs of established peri-operative infections are far greater, even when indications for prophylaxis are extended to include common elective operations. It is accepted that the majority of post-surgery infections arise from intra-operative contamination of the surgical site with microbes resident in the operative field.

In all surgical procedures some microbial contamination of the operation site is inevitable. Elective procedures on microbial colonized viscera obviously carry a greater risk of post-operative infection than do elective 'clean' procedures. Experimental studies have shown that large numbers of facultative anaerobes or aerobes when present alone are required to establish intra-abdominal infection. It is even more difficult to initiate infection with obligate anaerobes alone. In comparison, lower numbers of facultative and obligate anaerobes together (e.g. *E. coli* and *Bacteroides fragilis*) will consistently result in infection, and it seems that once the dynamics of the mixture has been established, the microbes appear in almost equal numbers in wound or cavity exudates. Therefore in elective visceral surgery, an empty viscus with minimal normal flora has the least potential to produce post-operative infection.

From a historical viewpoint, the earliest effective reduction in the occurrence of surgical site infections followed the use of carbolic acid sprays in operations involving compound fractures and amputations by Joseph Lister in the 1860s. However, the studies of John Burke almost 100 years later (Burke, 1961) on experimental staphylococcal dermal lesions in guinea pigs, and their maximum suppression by antibiotics given prior to the introduction of bacteria into the wound, are generally regarded as the first clear evidence of the value of antibiotic prophylaxis, something which in fact should possibly be credited to Gabriel Seley. In the late 1930s to early 1940s, Seley was able to link peri-operative sulphonamides with a reduction in sepsis following colonic and gastric surgery (Wenzel, 1993). A widely quoted early clinical trial showing the benefits of prophylactic antibiotics is that of Bernard and Cole (1964). In a prospective appropriately controlled study involving 145 patients undergoing gastrointestinal operations, these two investigators were able to demonstrate a significant reduction in post-operative infections (from 27% down to 8%) by the use of peri-operative prophylactic antibiotics. A review of trials (involving colon surgery) published from 1965 to 1980 also supported these observations (Baum et al., 1981). The importance of including antibiotics with activity against obligate anaerobes and more oxygen-tolerant facultative anaerobes in prophylaxis for intra-abdominal operations became clear in the late 1970s (see Morris et al., 1980). While it is now apparent that antibiotic prophylaxis can reduce surgical site infection rates by as much as tenfold, it must be remembered that the risk of an adverse event with antibiotic use may be around 2%, and the perceived benefit to the patient of any attempted prophylaxis should therefore be greater.

In 1964, the National Research Council ad hoc Committee on Trauma introduced their classical incisional wound classification system (clean, clean contaminated, contaminated, dirty), which has repeatedly been shown to correlate with post-operative surgical site infection rates. Using results derived from the large (almost 63 000 wounds) Foothills Hospital project (Cruse and Foord, 1980), it was possible to show in the 1970s that, without prophylaxis, wound infection rates ranged from 1.5% for clean procedures to 40% for dirty operations.

Apart from contamination at the time of surgery, the risk of wound infection has also been shown to be associated with the duration of surgery. In the Foothills Hospital studies, it was found that the wound infection rate roughly doubled for every hour of surgery. In another study (Garibaldi et al., 1991), stepwise logistic regression analysis revealed that an operative period of over 2 h was associated with an increased wound infection rate. Surgery lasting for more than 2 h was associated with a wound infection rate of 14%, compared to 3.3% for similar operations of shorter duration. In addition, data from the original SENIC (Study on the Efficacy of Nosocomial Infection Control) report also suggested that operations lasting over 2 h had increased wound infection rates. Prolongation of an operation is a significant risk factor for subsequent surgical site infection.

The SENIC project initiated in 1980 was also able to identify a number of other potential modulators of surgery-associated infections (Haley et al., 1985). Using data generated from 58 498 patients who were admitted to hospital in 1970 and underwent one or more operations, Haley and his associates were able to develop a simplified index for predicting a patient's risk of getting a surgical site infection (both incisional wound and deep infections). The results were then applied to an additional 59 352 surgical patients who were seen at the same hospitals during 1975–1976, in order to verify the predictive power of the index. The four identified risk factors were an abdominal operation, a contaminated or dirty procedure according to the traditional wound classification system, surgery lasting over 2 h, and comorbidity as measured by three or more diagnoses while in hospital.

This SENIC risk index has subsequently been modified further. The number of discharge diagnoses has been replaced by the American Society of Anesthesiologists (ASA) pre-operative assessment score as a measure of comorbidity (Culver et al., 1991), and the first two of the SENIC risk factors combined into one category – degree of surgical site contamination. Culver and his associates were able to validate the modified index on 84 691 procedures in 44 hospitals between 1987 and 1990. The three independent risk factors that influence surgical site infection rates are thus wound category (i.e. degree of contamination), duration of operation exceeding 2 h, and comorbidity as measured by an ASA score of 3–5. The infection rate almost doubles for each of these factors, (e.g. from around 2% if no factors are present to around 16% if all three apply). Thus only one risk factor need be present to counterbalance the problems of adverse reactions associated with antibiotic use.

However, one further variable to be considered is the fact that most ($\geq$80%) surgical site infections occur weeks to months after the patient has left hospital (Sands et al., 1996). Obviously, better methods need to be developed in order to secure a valid, reliable and simple registration of relevant infectious surgical site complications. It seems probable that the figures quoted above, especially for clean procedures, are significantly low and underestimate the problem.

The other major area of debate concerns what constitutes a (significant) nosocomial surgical wound infection (i.e. whether cellulitis or the discharge of pus are minimal criteria), or in fact whether this should be enlarged to incorporate deeper infection (i.e. surgical site infections). A consensus definition of what constitutes the surgical 'site' seems to include the cutaneous incisional site plus any deeper surgical area (e.g. large bowel anastomosis), but whether the pelvis, paracolic gutters or subdiaphragmatic spaces are included, with respect to abdominal surgery, has yet to be clarified. In addition, any inter-hospital surveys and comparisons of infection rates must be adjusted to include at least the proven risk factors (McLaws et al., 1997).

When given as a single dose, antibiotic prophylaxis supposedly has a minimal effect on the selection of resistant microbes. To some extent, this depends on the antibiotic used. For instance, following cardiac surgery cefuroxime tends to permit the emergence of resistant coagulase-negative staphylococci at a greater frequency than does cefamandole. However, there is no doubt that even a single dose of an antibiotic can have a profound effect on the normal microbial flora, and that this may take several days to return to normal. Whether such changes actually matter clinically is largely unknown (Sanderson, 1993). The potential of prophylaxis, especially if prolonged, to select resistant microbes is also a significant consideration, especially in patients who have been hospitalized for any length of time and/or have received pre-operative courses of antibiotics. Ideally it would be logical for antibiotics used in prophylaxis to be different to those normally used for the treatment of established sepsis. In many countries, cephalosporins fulfil this requirement.

Although the reasons for breakthrough infections due to apparently susceptible strains are not fully understood, *Staphylococcus aureus* isolates that elaborate type A staphylococcal β-lactamase have been associated with wound infections complicating the use of cephazolin in surgical patients. This particular β-lactamase inactivates cephazolin relatively efficiently, conferring partial resistance to this cephalosporin. Recent studies involving almost 300 isolates of *Staphylococcus aureus* associated with deep surgical wound infections developing after clean surgical procedures have demonstrated a distinct subpopulation of borderline methicillin-susceptible strains (BSSA-5) that elaborated large amounts of type A staphylococcal β-lactamase (Kernodle *et al.*, 1998a). Cephazolin use in prophylaxis was considered a significant risk factor for BSSA-5 infection. The ability of strains of *Staphylococcus aureus* to elaborate differing β-lactamases is potentially a significant factor with surgical site infections following prophylaxis (Kernodle *et al.*, 1998b). These findings clearly have significant implications in relation to antibiotic prophylaxis and its potential to favour the emergence of 'resistant' microbes.

For most elective procedures it now seems that prophylaxis is used without obvious reference to the known risk factors. This is reasonable when microbial contamination is a readily predictable factor. Thus prophylactic antibiotics are undoubtedly beneficial in all clean contaminated procedures (e.g. colon resection). The benefits of prophylaxis in clean procedures (e.g. heart surgery) where subsequent infection may have horrendous repercussions are also accepted. With 'dirty' procedures or traumatic wounds more than 4 h old, prophylaxis by definition cannot be applied and empirical treatment is warranted. Areas where debate occurs include appendectomy, biliary surgery, a number of clean procedures, and during invasive investigational procedures.

APPENDECTOMY

Appendectomy is an emergency procedure involving an infected organ, so antibiotic use is not strictly prophylactic. While the level of infection is low in non-perforated cases, the situation changes markedly if perforation has occurred. In the absence of perforation, the use of prophylactic regimens and/or principles appears to be satisfactory, bearing in mind the likely dominance of obligate anaerobes. In cases where perforation has occurred and/or the appendix is markedly gangrenous, more prolonged therapeutic antibiotic regimens are justified.

BILIARY SURGERY

In the case of biliary surgery, it is widely quoted (perhaps erroneously) that there is a close correlation between 'infected' bile (bactibilia) and post-operative septic complications, especially wound infections. Although bile cultures are more often positive in 'high-risk' patients (e.g. those with acute cholecystitis, jaundice, choledochal stones, diabetes mellitus, a non-functional gall-bladder, or age over 70 years) (Reiss *et al.*, 1982; Lewis *et al.*, 1987, Wells *et al.*, 1989), it is apparent that bile colonization cannot be accurately predicted pre-operatively using such associations (Wells *et al.*, 1989). In addition, a recent study showing the benefits of laparoscopic gall-bladder surgery over open surgery with regard to subsequent surgical site infection rates found no relationship between positive bile cultures and subsequent wound infection (den Hoed *et al.*, 1998). It is therefore unreasonable to limit prophylaxis to supposedly high-risk patients, and most patients undergoing biliary surgery now routinely receive prophylaxis. With acute cholecystitis, antibiotics are used as adjuncts to cholecystectomy to reduce the incidence of post-operative septic complications, although some authors have questioned the need for antibiotic prophylaxis in cases where laparoscopic techniques are employed (den Hoed *et al.*, 1998).

CLEAN PROCEDURES

It is universally accepted that prophylactic antibiotics are beneficial in patients undergoing clean operations with a foreign-body implant (e.g. many vascular, cardiac and orthopaedic operations), or where subsequent sepsis would have horrendous repercussions (e.g. cardiac surgery, transplant surgery). Prophylaxis could also be supported by duration of operation in many of these situations. General agreement suggests that this list should be extended to include patients undergoing a number of clean procedures without foreign implants (e.g breast surgery, hernia repair, central nervous system operations, transurethral prostatectomy). This is because all infections produce physical and psychological disability and stress, while some may result in recurrence of the original problem, requiring that the operation be repeated.

INVASIVE INVESTIGATIONAL PROCEDURES

It is also prudent to use prophylactic antibiotic cover in patients who are undergoing various invasive investigational procedures that are known to be accompanied by a high prevalence of transient bacteraemia, especially in individuals who, because of pre-existing problems (e.g. foreign body implant), are at serious risk from bacteraemic episodes. The more common investigational procedures include endoscopy, percutaneous abscess drainage, abdominal exploration for penetrating abdominal trauma with no observed gastrointestinal leakage, and placement of a chest tube to correct a pneumothorax associated with chest trauma (Nichols, 1994).

ROUTE AND TIMING OF ANTIBIOTIC ADMINISTRATION

It is possible to administer prophylactic antimicrobial agents by a variety of routes – orally, locally and parenterally. All of these routes have been used with varying degrees of success.

ORAL USE

In many areas of the world it is still the custom to use oral pre-operative antibiotic preparations before elective colonic resection. These are often accompanied by mechanical cleansing, whole gut lavage and dietary restrictions. The antibiotics are given during the 24 h before operation in an attempt to obtain significant intraluminal (local) and serum (systemic) levels. A commonly used combination has been oral neomycin and erythromycin base given as 3 doses of each during the 19 h prior to incision. Erythromycin alone is absorbed. Clearly the use of such oral prophylaxis offers cost savings and other benefits. However, equal efficacy with parenteral regimens is not always apparent. In some centres, combinations of cheap oral and parenteral antibiotics have been utilized (see Playforth et al., 1988). Readily absorbed oral antibiotics (e.g. ciprofloxacin) are now being used prophylactically for a number of other surgical procedures (e.g. gastroduodenal surgery, cholecystectomy, breast surgery) (McArdle, 1994; McArdle et al., 1995). Metronidazole can also be given intrarectally or orally, although the timing of the administration of this and other oral antimicrobial agents is critical if maximum tissue levels are to be obtained at the time of surgery.

LOCAL USE

There is a long history of local antibiotic use for the prevention and/or treatment of orthopaedic infections (Wininger and Fass, 1996). Early studies employed the direct instillation of antibiotic powder into wounds which had been carefully debrided and immobilized, while techniques such as closed wound irrigation/suction and isolation perfusion gained favour for the treatment of osteomyelitis as systemic antibiotics became more widely available.

In 1970 it was shown that antibiotics incorporated into bone cement reduced early post-operative arthroplasty infections (Buchholz and Engelbrecht, 1970). The antibiotics appeared to diffuse into the surrounding tissues over a period of months. Extension of this idea resulted in the use of gentamicin-impregnated beads to fill the dead space created after debridement of infected bone, and to treat the associated chronic infection (Klemm, 1979). Absorbable ofloxacin-impregnated beads have been shown experimentally to reduce post-operative *Staphylococcus aureus* osteomyelitis significantly (Nicolau et al., 1998). Nowadays, antibiotic-impregnated cement (often in conjunction with systemic peri-operative antibiotics) is used to prevent infections in arthroplasties, while beads are used to provide temporary high local antibiotic concentrations and to fill the dead space after debridement in patients with chronic osteomyelitis or compound fractures. As with the joint procedures, systemic antibiotics are usually given as well (Wininger and Fass, 1996). While the advantages of antibiotic-impregnated cement and beads in the prophylaxis of orthopaedic infections are supported by some animal and human studies, comparisons of the efficacy of such procedures with systemic antibiotics alone are lacking. While the rates of adverse reactions appear to be low with cements/beads, problems concerning the potential to select resistant microbes are of major concern.

Intra-incisional instillation of nafcillin (with lidocaine) has been suggested as being worthy of consideration as prophylaxis in dermatological surgery. It appears to reduce post-operative wound infection rates (Griego and Zitelli, 1998).

It has also been demonstrated that the surgical-site infection rate in cardiothoracic surgery patients can be reduced following the peri-operative elimination of nasal staphylococcal carriage using mupirocin (Kluytmans et al., 1997).

PARENTERAL USE

For most surgical prophylaxis, antibiotics administered parenterally have found universal acceptance and favour. In an elegant study, Classen et al. (1992) endorsed Burke's pioneering studies showing that the timing of antibiotic administration was critical, with optimum results being achieved when the dose was given within 2 h pre-operatively. Significantly increased infection rates were found when the intravenous antibiotic was administered more than 2 h before incision, or after commencement of the operation. Most regimens now state that the antimicrobial agent(s) should be given at the time of induction of anaesthesia. Maximum levels of antibiotic are required in tissues at the time of incision, and should be maintained until after wound closure. Thus, in cases where the duration of a procedure exceeds 2 h, antibiotics with half-lives in the range 30–90 min require repeat doses 2-hourly intra-operatively (or at intervals of twice the plasma half-life) (see Wittmann and Condon, 1991; Nichols, 1994; 1996). The value of using antibiotics with an extended half-life is obvious. However, unacceptably long intervals between antibiotic doses still seem to be relatively common in some situations (e.g. hip arthroplasty) (Lewis, 1998).

With cardiac surgery, sternal wound and cardiac site infections have serious complications. Cardiac surgeons often continue prophylaxis for 24 to 48 h, or at least while invasive lines and drains remain in place, because of the known propensity for plastic tubes, haematomas and other collections to serve as a nidus for microbial colonization and subsequent infection. However, it has been demonstrated that a single appropriate dose of antibiotic will achieve and maintain serum levels sufficient for prophylaxis, several hours after coronary artery bypass grafting procedures (Vuorisalo et al., 1997). Despite suggestions that a single dose of antibiotic is inadequate prophylaxis for patients undergoing vascular surgery, a survey of relevant literature has endorsed the adequacy in general surgery of single-dose prophylaxis (McDonald et al., 1998).

POSSIBLE PROPHYLACTIC ANTIBIOTIC CHOICES AND REGIMENS

Table 5.1 lists some currently recommended prophylactic regimens. The advantages of one regimen over another are still largely unproven, and many other apparently effective alternatives exist. When deciding which antibiotic to use for surgical prophylaxis, it must be borne in mind that most wound infections involve *Staphylococcus aureus*. It is therefore not surprising that cephalosporins (e.g. cephazolin) have become the mainstay of prophylaxis for most situations, and are often used alone. Unfortunately, their general lack of anaerobic activity has meant that additional specific anaerobic cover must be added in cases where the potential exists for contamination by anaerobes. This applies especially to large bowel surgery. Activity against Gram-negative enteric bacilli increases with progression from first- to third-generation compounds (see p. 80). There is evidence that some cephalosporins (e.g. cefodizime) may help to prevent the postoperative neutrophil dysfunction that is seen following coronary artery bypass grafting (Wenisch et al., 1997). As yet, these findings appear to have had little influence on the antibiotic choice for cardiovascular prophylaxis.

Experienced gastrointestinal surgeons know they must expect the unexpected and plan for worst-case scenarios. It can be justified (and not criticized as being purely defensive) to formulate prophylactic antibiotic policies for gastrointestinal surgery as if all operations were on the colorectum with anticipated minor spill. There are many established formulae suited to this policy, such as cefuroxime plus metronidazole, gentamicin plus metronidazole, co-amoxyclav alone and cefotetan alone. All of these are inexpensive and so flexible that some advocate their use in most abdominal operations.

First-generation cephalosporins (e.g. cephazolin) have minimal but probably adequate activity against Gram-negative enteric bacilli. The replacement of gentamicin or cefuroxime with a third-generation cephalosporin (e.g. cefotaxime, ceftriaxone) is reasonable, but significantly more costly and without proven benefit apart from the fact that ceftriaxone has a long half-life. The major drawback of most cephalosporins in this setting is their short half-life and general lack of anaerobe activity – hence the addition of metronidazole to most cephalosporin-containing regimens. While clindamycin has in the past been favoured as the anti-anaerobe component, emerging problems of resistance have meant that it is now inferior in this respect to metronidazole. With most appendectomies, microbial participation, especially obligate anaerobes, will be advanced, and appropriate anaerobe cover must be included in any regimen; cefoxitin or cefotetan alone are probably inadequate.

Enterococci are often isolated from specimens obtained from intra-abdominal infections, but apparently create few problems unless the patient is severely debilitated. As a general rule, prophylactic cover for elective gastric, biliary tract and intestinal operations (where enterococci are significant normal flora) need not specifically cover these potential pathogens. Cephalosporins such as cephazolin or cefuroxime alone appear to be adequate for simple gastroduodenal operations. A wide range of antibiotics (e.g. ciprofloxacin, piperacillin, cefoxitin, cefotetan, co-amoxyclav) have proved to be equally effective with biliary operations. Only some of these possess enterococcal activity (e.g. co-amoxyclav, piperacillin and possibly ciprofloxacin). In cases where cancer exists in the stomach or biliary tree, or severe blockage of the bile duct has occurred, additional anaerobe cover is advisable if it is not already included by the antibiotic(s) used.

For surgery involving the pelvic regions (e.g. vaginal or abdominal hysterectomy, Caesarean section), the contaminating microbes are likely to be

Table 5.1 Antibiotic prophylaxis in surgery

Type of surgery	Likely pathogens	Suggested regimen(s)	Relative cost/dose[a]
Clean procedures			
Orthopaedic (total joint replacement; open reduction or internal fixation of fractures)	Staphylococcus aureus, coagulase-negative staphylococci. Less commonly enteric Gram-negative bacilli and clostridia	Cephazolin 1 g IV or cefamandole 2 g IV or teicoplanin 400 mg IV[b]	1+ 3+ 4+
Cardiovascular (prosthetic valve, coronary artery, other open heart, abdominal aorta, prosthesis or groin incision)	As above	As above	
Thoracic (non-cardiac)	As above	Prophylaxis of uncertain benefit and does not prevent pneumonia	
Neurosurgery (craniotomy, shunt spinal insertion, spinal surgery)	As above	Prophylaxis of uncertain benefit Regimens as above	
Breast surgery and inguinal herniorrhaphy	As above	As above	
Ocular	Staphylococcus aureus, coagulase-negative staphylococci, streptococci, enteric Gram-negative bacilli	Topical drops[c] plus cephazolin 100 mg subconjunctivally at *end* of operation	
Potentially contaminated procedures			
Gastroduodenal, biliary	Enteric Gram-negative bacilli, Gram-positive cocci, occasionally anaerobes[d]	Cephazolin 1 g IV or co-amoxyclav 1.2 g IV or ceftriaxone 1 g IV	1+ 1+ 4+
Colorectal	Enteric Gram-negative bacilli, anerobes, Staphylococcus aureus, enterococci, Streptococcus anginosus	Cefotetan 1 g IV alone or co-amoxyclav 1.2 g IV alone or cefuroxime 1.5 g IV[e] plus metronidazole 500 g IV[f]	2+ 1+ 2+ 1+
Appendicectomy	As above (colorectal)	Cefotetan 1 g IV or ceftriaxone 1 g IV or ciprofloxacin 500 mg IV each plus metronidazole 500 mg IV[f]	2+ 4+ 4+ 1+
Gynaecological (vaginal or abdominal hysterectomy, Caesarean section, other pelvic surgery, termination of pregnancy)	Enteric Gram-negative bacilli, group B streptococci, enterococci, anaerobes, Staphylococcus aureus	Cefotetan 1 g IV or co-amoxyclav 1.2 g IV	2+ 1+
Urological[g]	Enteric Gram-negative bacilli	Gentamicin at least 3 mg/kg IV or ceftriaxone 1 g IV or ciprofloxacin 500 mg IV	1+ 4+ 4+

Table 5.1 Continued

Type of surgery	Likely pathogens	Suggested regimen(s)	Relative cost/dose[a]
Ear, nose and throat[h] with mucosal incision	*Staphylococcus aureus*, streptococci, oral anaerobes	Co-amoxyclav 1.2 g IV	1+
Amputation of ischaemic lower limb	Clostridia, staphylococci, streptococci	Flucloxacillin 1.5 g IV ± metronidazole 500 mg IV[f] or co-amoxyclav 1.2 g IV alone	1+ 1+ 1+

[a]Costs range from around $US 3 (1+) to $US 30 (4+) per dose.
[b]Where methicillin-resistant staphylococcal infections are common or the patient is allergic to β-lactam antibiotics.
[c]Parenteral antibiotics do not penetrate the aqueous or vitreous humour adequately.
[d]Some consider prophylaxis to be unnecessary in uncomplicated vagotomy and pyloroplasty and cholecystectomy in younger patients without stones or obstruction. Trials do not support the need for routine cover of anaerobes and the enterococci in gastroduodenal or biliary surgery in the absence of malignancy and/or obstruction.
[e]Third-generation cephalosporins such as cefotaxime and especially ceftriaxone are effective, but are more costly and have unproven benefit over early-generation compounds.
[f]Ornidazole or tinidazole may be preferred for oral use because of their longer half-lives.
[g]Prophylaxis is of uncertain benefit when the pre-operative urine is sterile.
[h]Prophylaxis is not usually necessary in tonsillectomy, adenoidectomy or rhinoplasty.
Adapted from Lang, S. (ed.) (1997) *Guide to pathogens and antibiotic treatment 1998*. Auckland: Adis International, with permission.

similar to those 'released' during intestinal operations (e.g. enteric Gram-negative bacilli and obligate anaerobes), although the potential exists for other microbes (e.g. group B streptococci) to become prominent. Prophylactic regimens include cefotetan, co-amoxyclav and the combination of cephazolin and metronidazole. In patients in whom sexually transmitted pathogens may have been the cause of the problem for which surgery was required, the addition of doxycycline or erythromycin may be advantageous. These cover the mycoplasma/ureaplasma group of bacteria on which β-lactams have no effect (they lack a cell wall).

With vascular grafts and also hip and knee arthroplasty, *Staphylococcus aureus* and coagulase-negative staphylococci such as *Staphylococcus epidermidis* are the major bacterial types to be covered, although the close proximity of the groin perineum may result in wounds being contaminated with enteric Gram-negative organisms. Some second-generation cephalosporins (e.g. cefamandole, cefuroxime) are particularly active against both *Staphylococcus aureus* and *Staphylococcus epidermidis*-type species, and are probably superior in this respect to most first- and third-generation compounds (Mauerhan *et al.*, 1994). They also have some activity against Gram-negative enteric bacilli. In cases where the patient is known to harbour methicillin-resistant *Staphylococcus aureus*, or local sensitivity patterns suggest a high incidence of such microbes, vancomycin may be used in place of cephalosporins. However, vancomycin prophylaxis may cause considerable problems (e.g. vasodilation), especially if infused at an accelerated rate (Romanelli *et al.*, 1993). Teicoplanin or, if sensitivity patterns suggest this, netilmicin, may be better alternatives.

While many texts recommend a combination of flucloxacillin plus penicillin for lower limb amputations and open-fracture 'prophylaxis' – presumably to cover staphylococci and streptococci/anaerobes, respectively – it is apparent that levels of flucloxacillin that give satisfactory cover for staphylococci are more than adequate for streptococci and the important anaerobes (clostridia). There is really no need for the double β-lactam combination. In cases where the significance of obligate anaerobes is likely to be increased (e.g. in diabetics), metronidazole can be added to the flucloxacillin regimen, or co-amoxyclav can be used alone.

If pre-operative urine is free of bacteria, prophylaxis for urological operations is of uncertain benefit. The important potential pathogens are enteric Gram-negative bacilli and *Staphylococcus aureus*; prophylaxis with antimicrobial agents such as gentamicin or ciprofloxacin can be considered.

Infections following ear and nose surgery involve streptococci, *Staphylococcus aureus* and penicillin-susceptible anaerobes. A prophylactic agent such as co-amoxyclav may be beneficial in reducing the rate of post-operative infection, notwithstanding the fact that penicillin and metronidazole have been standard therapy.

It is generally accepted that prophylactic principles also apply where infection is established and minimal contamination of surrounding tissues has occurred or will do so during the surgical procedure (e.g. removal of a non-perforated acute appendix, lower limb amputation without ulceration, percutaneous drainage of an abscess). Some surgeons advocate a peri-operative pulse of antibiotic rather than a more extended course of antibiotics for a ruptured appendix. However, until such a 'prophylactic' approach is evaluated further, most surgeons would be unwilling to refrain from extended therapeutic antibiotic cover in this situation. Consensus as to how long antibiotics should be continued in the treatment of established intra-abdominal infections is still far from being reached (see Wittmann *et al.*, 1996). Patients with obstruction of the bowel or biliary tree should have parenteral cover which must be continued until they are afebrile and have recovered from post-operative ileus. A significant proportion of these patients have Gram-negative bacteraemia. If there is supporting clinical or biochemical evidence, 1 week of therapy may be appropriate.

Expensive third-generation cephalosporins such as ceftriaxone should be reserved for serious infectious episodes, or for high-risk patients who are having long operations. However, there is accumulating evidence that, at least in some types of surgery (e.g. biliary), prophylactic use of ceftriaxone may significantly reduce subsequent respiratory infections and be cost-effective (Morris, 1994). Whether this is related to the antibacterial activity of ceftriaxone or is a reflection of some other pharmacological property (e.g. long half-life), is not known. Previously, antibacterial prophylaxis has not been aimed at reducing infection at sites distant from the operational area, although in these days of financial stringency a reduction in the total cost of a procedure by reducing the length of hospital stay is seen as a very desirable goal.

CONCLUSIONS

There is no doubt that surgical antibiotic prophylaxis should be a dynamic and evolving process. Most institutions now have in place a series of proven and acceptable regimens which have a number of features in common. These can be summarized as follows (see also Clunie, 1997):

1. High concentrations of antibiotic should be present in target tissues at the time of incision, and should be maintained until completion of wound closure.
2. The optimal timing for prophylaxis by parenteral administration is at the time of induction of anaesthesia, i.e. about 20 min prior to incision.
3. The timing of any oral regimen is critical. Before commencement of the operation, the antibiotic(s) used must reach maximum levels at the operation site as well as the skin incisional site (oral regimens employing only non-absorbed antibiotics do not fulfil this requirement).
4. For elective procedures lasting 2 h or less, a single dose of prophylactic antibiotic is sufficient. Prophylactic courses continued after wound closure should be restricted because of unproven benefit and the risk of selection of resistant bacteria.
5. For procedures lasting > 2 h, or where there is pronounced blood loss, one or two further doses at intervals of twice the plasma half-life (i.e. around 2-hourly for most antibiotics) may be required. In such cases, a good time for the final dose is just prior to wound closure. The use of antibiotics with longer half-lives is a consideration for operations that are expected to exceed 2 h, as this may negate repeat dosing.
6. Complete cover of all possible microbes that are likely to be encountered is not necessary. The antibiotic(s) chosen must be active against the microbes that are most likely to result in surgical site infection following the operation, and should always possess *Staphylococcus aureus* cover. In complex situations such as colon resection for cancer, cover of skin (e.g. staphylococci) and bowel (e.g. obligate anaerobes, coliforms) flora with parenteral and oral antibiotics may be optimal.
7. In cases where there is an emergency, the likelihood of active infection, an obstruction, unusual blood loss, pronounced trauma or some other very special circumstance, the antibiotic regimen may be prolonged and should involve maximal doses of drugs that are most likely to reach the extravascular spaces. In reality, this is therapy and not prophylaxis.

SELECTIVE DECONTAMINATION OF THE DIGESTIVE TRACT

Many publications have discussed the potential value and/or problems of selective decontamination of the digestive tract (SDD), an approach to infection 'prophylaxis' in critically ill patients using multiple antimicrobial agents, at least some of which are given orally. It has been suggested as a means of preventing – or at least reducing – the occurrence of infections in intensive-care patients, especially those receiving mechanical ventilation, total gastrectomy for gastric cancer (Schardey *et al.*, 1997), and in patients with acute liver disease requiring transplantation (Anon., 1999a). SDD aims to prevent infection by eliminating colonization of mucosal surfaces by pathogenic microbes, and by the early treatment of infections which may be incubating as the patient is admitted (Dever and Johanson, 1993; Tetteroo *et al.*, 1994). Whatever 'prophylactic' regimen is contemplated, oral vancomycin should not be part of the antimicrobial cocktail (Ford et al., 1998).

More carefully constructed and controlled trials (e.g. Gastinne et al., 1992; Hammond et al., 1992; Holzapfel et al., 1993; Saunders et al., 1994; Lingnau et al., 1998), have failed to demonstrate improved survival or a significant decrease in nosocomial sinusitis or pneumonia following SDD in patients receiving mechanical ventilation. In addition, SDD does not appear to influence the formation of biofilms on endotracheal tubes (Adair et al., 1993). Such microbes may be responsible for sinusitis, which can be an unappreciated source of fever in the intubated patient (Dellinger, 1993). Among patients who have been ventilator-assisted for 1 week or longer, the occurrence of bacterial sinusitis has been estimated at < 10% (Westergren et al., 1998). However, there is some debate concerning the role of biofilms on ventilator equipment as a potential source of microbes in ventilatory-associated pneumonia (van Saene et al., 1998; Kollef, 1999).

Although SDD may limit the spread of some problematic microbes within the intensive-care environment, it should not be used to replace conventional infection control measures. It is apparent that mechanisms other than bacterial colonization of the gut and mucosal surfaces contribute to many secondary infections, including wound infections. In addition, selection of resistant strains or a change in the spectrum of microbes may be induced by SDD. For example, ciprofloxacin SDD in critically ill patients has been shown to increase the level of ciprofloxacin-resistant bacteria and to result in a shift towards colonization and/or infection by Gram-positive bacteria, including MRSA (Lingnau et al., 1998). There are also practical problems associated with administering an SDD regimen to patients awaiting cadaver liver transplants (Arnow et al., 1996).

ORGAN TRANSPLANT PATIENTS

Extended 'prophylaxis' is now a feature of transplant surgery, especially during periods of high immunosuppression and/or prolonged neutropenia. In a review of 284 liver transplant patients, 298 episodes of bacterial infection occurred in 159 patients (56%) (Casewell, 1997). The most common sites of infection were blood (bacteraemia, 22%), wounds (21%), the urinary tract (16%) and chest (13%). Around 75% of infections occurred in the first month post-operation. Prolonged pre-surgery hospitalization appeared to be a significant independent risk factor. It was interesting to note that wound infections were not considered to be associated with the duration of surgery or other peri-operative variables, and that pre-transplant antibiotics had a protective effect on subsequent infections. Apart from klebsiellae, the five most common bacteria causing infections were all Gram-positive cocci, indicating that empirical antibiotic regimens in the early post-operative period should have good anti-staphylococcal and anti-enterococcal activity. A rather surprising result was that *Enterococcus faecium* (often vancomycin-resistant) was second only to *Staphylococcus aureus* as a causal agent of infections. Whether or not vancomycin formed part of an SDD or 'prophylactic' regimen was not stated.

chapter 6

ANTIFUNGAL AGENTS

The successful control and treatment of some deep-seated fungal infections is at last a reality. As newer agents are produced and marketed, some of the older and more widely known antifungal agents have become outdated and obsolete. A clear understanding of the potential role of the various agents now available is therefore essential if optimal therapy is to be practised. Fungal infections in surgical patients are increasing, with the mortality associated with invasive disease approaching 50%. Predictors of fungal colonization include broad-spectrum antibiotic and metronidazole use and duration of ventilation, while APACHE II scores, duration of total parenteral nutrition use and sustained fungaemia appear to be associated with mortality (Fraser *et al.*, 1992; D'Amelio *et al.*, 1995).

MODE OF ACTION

Both fungal and mammalian cells are typically eukaryotic. Therefore the formulation of antifungal agents that are free from human toxicity has been a problem. However, fungi do possess a cell wall composed of a polysaccharide matrix in which glucans appear to be structurally important. Although antifungal agents that are active on the cell wall constituents have been developed, these have yet to become commercially available. They are targeted at enzymes associated with chitin, β-glucan or mannoprotein synthesis.

Most of the current systemic antifungal agents act by interfering with the synthesis and function of the cytoplasmic membrane (see Figure 6.2). Unfortunately, the molecular events involved in most cases are still poorly understood. Alterations in membrane porosity and permeability, and impairment of cytochrome P_{450}-dependent enzyme activities seem to be important. Some antifungal agents inhibit the cytochrome P_{450}-dependent demethylation of lanosterol – a precursor to the formation of ergosterol during the biosynthesis of the cell membrane (see Figure 6.1). Ergosterol is the primary sterol in the fungal cell, as opposed to cholesterol in the mammalian cell membrane; most of the currently available systemic antifungal agents have ergosterol as their 'target site' in some way. Depletion of ergosterol in the fungal cell membrane results in altered membrane fluidity, thereby reducing the activity of membrane-associated enzymes. This leads to increased permeability and subsequent inhibition of cell growth and replication (Kauffman and Carver, 1997). Unfortunately, cytochrome P_{450} is reasonably well conserved throughout the biological world, and is important in some human synthetic pathways (e.g. adrenal androgen biosynthesis), which has resulted in a number of significant drug interactions and adverse events with some antifungal agents.

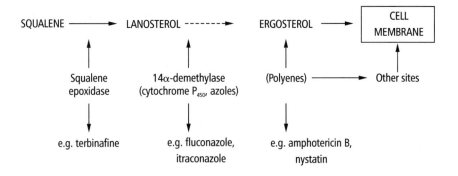

Figure 6.1 Site of action of some membrane-active antifungal agents.

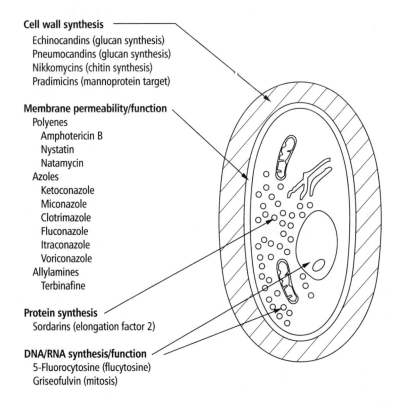

Figure 6.2 Site of action of some common antifungal agents.

Other potential sites of activity are the nucleic acids and protein synthesis. Antifungal agents that inhibit RNA or DNA synthesis and function do so either as antimetabolites (e.g. 5-fluorouracil for uracil in RNA) or, less clearly understood, by blocking mitosis through abnormal microtubule assembly. Protein synthesis is considered to be an attractive site for antimicrobial activity. Unfortunately, fungal and mammalian cells are relatively similar in this respect, although elongation factor 3 is an essential factor for protein synthesis (translation phase) which is unique to yeasts.

The sordarin family of compounds show inhibitory activity against yeasts, and are thought to act on this elongation factor (or possibly elongation factor 2).

RANGE OF ANTIFUNGAL AGENTS

As with antibacterial agents, antifungal drugs are grouped according to their molecular structure. The main systemic antifungal agents that are currently available belong to one of two families, namely polyenes and azoles (see Table 6.1). The other major family of systemic drugs consists of the allylamines, while the remainder comprise a group of unrelated compounds. Important adverse effects of the major systemic agents are listed in Table 6.2. Careful consideration of the benefit to the mother and the risk to the fetus is required when prescribing antifungal therapy during pregnancy (see Table 6.3 and King et al., 1998).

POLYENES

The polyenes include two main products, namely amphotericin B and nystatin. Of these, only amphotericin B can be used parenterally and in the treatment of

Table 6.1 Characteristics of important currently available[a] systemic antifungal agents

Antifungal agent	Formulations (dose)	Patient tolerability	Therapeutic spectrum (other than dermatophytes)
Polyenes			
Amphotericin B	IV (0.3–1.5 mg/kg q 24 h); increasing number of lipid associated formulations[b]; (also available as suspension or lozenges for local use)	Poor	Most fungi (see text for notable exceptions)
Azoles (i) imidazoles			
Ketoconazole	Oral only (200 mg q 24 h); (also as shampoo)	Good	Yeasts
Azoles (ii) triazoles			
Fluconazole	IV and oral (200–1000 mg q 24 h PO)[c]	Good	Yeasts
Itraconazole	Oral only (200–600 mg q 24 h)	Good	Yeasts, some other fungi, including aspergilli
Others			
Flucytosine	IV and oral (150 mg daily divided 6-hourly)	Good	Yeasts

[a] Yet to be released antifungal agents include the cell-wall-active compounds (e.g. echinocandins) and the triazole, voriconazole.
[b] Liposomal and lipid complex formulations (e.g. amphotericin B collidal dispersion, ABCD).
[c] Loading doses (twice normal dose) for 24–48 h are usual with 200 mg regimen.

Table 6.2 Some documented adverse effects associated with systemic antifungal agents

	Antifungal agent				
Characteristic	AmpB[b]	5FC[a]	Ket	Flu[a]	Itr
Toxicity					
Kidney	++				
Liver	+	+	++	+	+
CNS	+				
Gastrointestinal disturbance	++	+	+	+	+
Endocrine effects			++		
Drug interactions	+		++	+	+
Anaemia	+				
Leukopenia	+	++			
Phlebitis/pain at injection site	++				
Skin rash	+	+	+	+	+
Pruritus	+		+		
Potentially teratogenic/fetotoxic		+	+	+	+

[a]Oral formulation.
[b]Lipid formulations are less toxic but expensive.
AmpB, amphotericin B; 5FC, flucytosine; Ket, ketoconazole; Flu, fluconazole; Itr, itraconazole; +, uncommon and/or mild adverse effects; ++, most common and/or serious adverse effects.

systemic mycoses; it remains the most common choice for treatment of infection in seriously ill patients and seems safe during pregnancy, although there are no adequate human studies (FDA category B). Although amphotericin B is absorbed too poorly to be used orally in systemic infections, lozenges and suspensions are available for the topical therapy of oral and gastrointestinal lesions. Local irrigation is suitable for some urinary tract infections. Aerosol preparations ('aero-AmB') for the possible treatment of fungal pneumonias, lipid emulsions and liposomal encapsulated formulations (ampholiposomes) are also under investigation.

Conventional amphotericin B is a deoxycholate preparation which is diluted in a dextrose solution for intravenous use. In an attempt to reduce the toxicity of this formulation (see below), various lipid formulations have been proposed (e.g. intralipid emulsion, lipid complex, colloidal dispersion, liposomal encapsulated) (de Pauw et al., 1998; Anon., 1999d). The simplest (and cheapest) has been an emulsion prepared by vigorously shaking amphotericin B for a considerable period of time (over 12 h) in the intralipid used for parenteral feeding (1 to 2 mg amphotericin/mL intralipid). While it is favoured in some European countries (see Walker et al., 1994), this crude preparation is not licensed in most other countries. Other delivery systems include true liposomes, e.g. AmBisome or AmBi, and lipid complexes, e.g. amphotericin B colloidal dispersion (Amphotec or ABCD) and amphotericin B lipid complex (Abelcet or ABLC) (Richardson, 1997; Boswell et al., 1998; Coukell and Brogden, 1998;

Table 6.3 Relative safety of some antimicrobial agents in pregnancy[a]

Antimicrobial agent	Category[b]		Comment
	Australian	FDA (USA)	
Antifungal agents			
Amphotericin B	B_2	B	In view of the very low systemic exposure, topical agents are generally safe unless applied to extensive, occluded or raw areas. Amphotericin appears to be safe, but there have been no adequate human studies. Fluconazole has been associated with craniofacial, skeletal and cardiac anomalies in three infants born to separate mothers. The similarity of abnormalities to those described in rodent embryos suggests that fluconazole is teratogenic (*Clinical Infectious Diseases* (1996) **22**, 336–40)
Fluconazole	B_3	C	
Flucytosine	B_3	C	
Itraconazole	B_3	C	
Antiparasitic agents			
Albendazole	D	–	Teratogenic in some animal species
Mebendazole	B_3	C	Embryotoxic and teratogenic in animals, but limited human experience has shown no adverse effects
Thiabendazole	B_3	–	Limited experience suggests that thiabendazole is probably safe
Antiviral agents			
Acyclovir	B_3	C	Greatest experience of antiviral use in pregnancy has accrued for acyclovir and zidovudine, and these both appear to be safe. However, acyclovir is still not recommended for the treatment of other than life-threatening disease in pregnancy. Ganciclovir and zalcitabine are teratogenic in animals
Didanosine	B_2	B	
Famciclovir	B_1	B	
Foscarnet	B_3	C	
Ganciclovir	D	C	
Indinavir	–	C	
Interferon α 2a/2b	B_3	C	
Lamivudine	B_3	C	
Ribavirin	X	–	Ribavirin is teratogenic, mutagenic and embryotoxic. Pregnant women should not tend patients receiving ribavirin aerosol treatment
Ritonavir	–	B	
Saquinavir	–	B	
Stavudine	B_3	C	
Valaciclovir	B_3	B	
Zalcitabine	D	C	
Zidovudine	B_3	C	

[a]See also Table 3.6.
[b]See Table 3.5 for category description.
Adapted from Lang, S. (ed.) (1997) *Guide to pathogens and antibiotic treatment 1998*. Auckland: Adis International, with permission.

Wong-Beringer *et al.*, 1998; Anon., 1999d and Table 6.4). The *in vivo* activity of these lipid preparations compared to the conventional formulation is still largely unknown; the lipid preparations generally seem to be less potent on a mg amphotericin/kg basis, but much higher doses (e.g. 5 mg/kg) can be given in order to produce a similar outcome (Stevens, 1997). Whether the higher drug levels delivered to some tissues (e.g. liver, spleen) and cells (e.g. macrophages) are clinically relevant has yet to be determined.

In a recent trial of ABCD (4 mg/kg daily) in 133 renally compromised patients with invasive fungal infections (mainly aspergillosis and candidosis), it did not appear to affect renal function adversely (Anaissie et al., 1998). A complete or partial response to treatment was reported for 50% of the 133 intend-to-treat patients, and 67% of the 58 evaluable patients. In another trial, it was shown that ABLC could be used successfully to treat invasive fungal infections in patients who were intolerant of or refractory to conventional amphotericin B therapy (Walsh et al., 1998). ABLC has also proved successful (when combined with reduced chemotherapy and granulocyte colony-stimulating factor) in controlling renal zygomycosis in an immunosuppressed patient (Weng et al., 1998). In addition, it has been suggested that the enhanced activity of ABLC *in vivo* compared to the conventional formulation may be related to the release of active drug from the lipid

Table 6.4 Comparison of some features of conventional amphotericin B with three commercially available lipid-based formulations[a]

Feature	AmB	Lipid-based formulations		
		LAB	ABCD[c]	ABLC[c]
Shape and size (nm)	Micelle (< 25)	Unilamellar liposomes (< 100)	Lipid disc (115 × 4)	Ribbon-like sheet (1600–11100)
Lipid carrier/composition	Deoxycholate	Cholesterol esters plus phospholipids	Cholesteryl sulphate	Phospholipids
Amphotericin concentration (mol/L)	340	100	50	330
Pharmacokinetics[b]				
C_{max} (mg/L)	1.1	83	3.1	1.7
AUC (mg/L/h)	17.1	555	43	14
Approved indications	US UK	US UK	US UK	US UK
Empirical in febrile neutropenia patients	+ +	+ −	− −	− −
Severe invasive candidosis	+ +	− −	− −	− +
Patients ineligible for AmB	− −	+ +	+ +	+ +
Dosage (mg/kg/day)	0.25–1.5	3–5	1–4	5
Comparative daily cost (approximate $US)	12–17	800–1300	340–680	600

[a]Adapted from Anon. (1999d).
[b]5 mg/kg dose LAB, ABCD, ABLC and 0.6 mg/kg AmB.
[c]Unavailable in some European countries.
AmB, conventional amphotericin B; LAB, liposomal amphotericin B; ABCD, amphotericin B colloidal dispersion; ABLC, amphotericin B lipid complex.

complex at the site of fungal growth by the action of fungal or host-cell-derived phospholipases (Swenson et al., 1998).

In perhaps the first trial comparing deoxycholate (conventional) amphotericin B with the three more accepted lipid formulations, it was concluded that the rank order of efficacy was ABCD ≅ AmBi > ABLC ⩾ deoxycholate, although it seems that this may depend on the fungus involved and the anatomical location of the lesions (Clemons and Stevens, 1998). In this trial, an experimental model of systemic cryptococcosis was used.

It also appears that liposomal amphotericin B has physico-chemical properties and a pharmacokinetic profile that are considerably different to those of other currently available lipid-complexed amphotericin B formulations, with a greatly increased area under the plasma concentration–time curve and much lower clearance of equivalent doses (see Table 6.4 and Boswell et al., 1998). Clearly, lipid formulations are less nephrotoxic than the older conventional preparations, but in most centres they are reserved for situations where renal toxicity has limited the continued use of the more proven deoxycholate formulation. Apart from 'home-made' intralipid emulsions, lipid formulations of amphotericin B are expensive, and they seem to be no more effective than the deoxycholate formulation (Anon., 1999d).

Amphotericin B is regarded as the treatment of choice for most serious fungal infections, including invasive (systemic) candidosis, aspergillosis and zygomycosis (mucormycosis). It has also been the mainstay of therapy for cryptococcal meningitis, usually in combination with flucytosine (5-fluorocytosine, 5-FC) or fluconazole. Fungi which appear to be less susceptible to amphotericin B include *Candida lusitaniae, Scedosporium apiospermum* (the asexual state of *Pseudallescheria boydii*), the zygomycete *Cunninghamella bertholletiae, Paecilomyces lilacinus, Trichosporon cutaneum (Trichosporon beigelii), Fusarium* species, and some agents of chromomycosis. Although most of these would seem to be rare, their potential involvement in human disease cannot be ignored, and amphotericin B may not always be the drug of choice for mycotic infections in the seriously ill patient. The isolation and identification of the causal agents is just as important with the mycoses as it is with other microbial diseases. Unfortunately, antifungal susceptibility testing (see below) is still in its infancy, and drugs are mainly prescribed on an empirical basis. In surgical patients, most yeast infections can be treated with oral azoles (see below), which are clearly more tolerable and acceptable to the patient than amphotericin B. If there is any question regarding the numbers or activity of neutrophils, amphotericin B remains the antifungal agent of choice. Unlike azoles, it is cidal rather than static in its action.

The major disadvantage of the conventional formulation of amphotericin B is the high incidence of adverse reactions (see Table 6.2). The most commonly observed side-effects are chills, fever, vomiting, generalized pain, thrombophlebitis and pain at the infusion site, abnormal renal function (e.g. azotaemia, renal tubular damage, hypokalaemia) and anaemia (Peacock et al., 1993). Most patients who receive the drug experience at least one side-effect. When total doses exceed 4 g, renal tubular damage is predictable. Although it has not been possible to prevent renal side-effects, the infusion of 500–1000 mL of normal saline before the administration of amphotericin B appears to ameliorate renal toxicity. In this respect, lipid formulations appear to be less nephrotoxic. Concurrent use of other antimicrobial agents and drugs with known renal toxicity (e.g. aminoglycosides, the diuretic chlorothiazide, cyclosporin) should be avoided.

Amphotericin B is basically a specialist/hospital-only drug. The details of its complicated pharmacology and procedures for clinical use are well documented (Peacock et al., 1993) and need not be repeated here. A practice that is gaining support is to administer the drug in full dose over 1–2 h with prior intravenous hydrocortisone (100 mg). The latter is rapidly decreased and withdrawn over a few days in the absence of amphotericin B reactions. In most patients, neither the traditional stepwise starting doses nor prolonged infusion appear to minimize adverse reactions, although concern has been expressed about the use (especially in the first few days of therapy) of more rapid infusions (e.g. over 1 h) in patients with underlying renal insufficiency and cardiac disease. As with most other antifungal agents, few guidelines concerning the optimal duration of therapy are available. These are largely governed by considerations such as the causal fungus, the organ(s) involved, the severity of renal dysfunction and the patient's tolerance of prolonged hospitalization or out-patient therapy. Suggested regimens revolve around daily doses in the range 0.3–1.5 mg/kg, bearing in mind that kidney damage will occur after about 6 weeks of therapy (when the total dose exceeds 4 g). Patients on prolonged treatment may benefit from 48-hourly administration (i.e. twice the 'daily dose' given on alternate days). In one report concerning a series of patients with systemic candidosis, the dose was not critically calculated, and they simply used 25 mg (half a 50-mg ampoule) routinely. Lower doses and shorter treatments have been found to be effective in less severely compromised patients (e.g. candidaemia in post-surgical patients), with higher doses and/or prolonged courses being reserved for more serious disease or disease in highly immunocompromised patients. In cases where it is feasible, changing patients to a more tolerable antifungal agent (e.g. fluconazole) following a short 'burst' of amphotericin B is a practice that is rapidly gaining acceptance.

AZOLES

This rapidly expanding family of antifungal agents consists of two groups – the imidazoles and triazoles (see Kauffman and Carver, 1997; Sheehan et al., 1999).

Imidazoles

The imidazoles include a number of compounds intended primarily for topical use (e.g. miconazole, clotrimazole, econazole, isoconazole, ketoconazole, tioconazole), and ketoconazole and intravenous miconazole for systemic use. The latter is basically now obsolete. The activity of the imidazoles is directed mainly against yeasts and dermatophytes (ringworm fungi).

Ketoconazole This water-soluble compound is well absorbed orally under acid conditions; it is poorly absorbed in patients with achlorhydria and in those taking antacids or H_2-receptor antagonists, and in neutropenic patients. Ketoconazole is metabolized mainly in the liver and excreted largely as inactive metabolites – very little active drug appears in the urine. Penetration into the cerebrospinal fluid and other body fluids is poor (see Table 6.5).

Ketoconazole toxicity includes nausea, vomiting, pruritus and headache, although serious side-effects have been rare (see Table 6.2). Apart from mild gastrointestinal disturbances, hepatotoxicity (which has been estimated to be

Table 6.5 Pharmacology of azole antifungal agents[a]

Feature	Ketoconazole	Itraconazole	Fluconazole
Molecular weight	531	706	306
Water solubility	Poor	Poor	Excellent
Protein binding (%)	High (99)	High (> 99)	Low (12)
Cerebrospinal fluid (CSF) penetration (CSF/serum)	Poor (< 10%)	Poor (< 10%)	Excellent (> 80%)
Affinity for mammalian cytochrome P_{450}	High	Low	Low
Elimination half-life (h)	8	21–64	30
Excretion in urine (%)	< 5	< 1	80
Reduction of dose in renal failure	Not necessary	Not necessary	> 50 mL/min: no reduction 20–50 mL/min: ↓ by 50% 10–20 mL/min: ↓ by 75%
Oral bioavailability (%)	75[b]	40–50[b,c]	90
Influence of food on oral bioavailability	Variable	Increase	None
Effect of raised gastric pH on oral bioavailability	Decrease	Decrease	None
Dosage formulations	Oral tablets	Oral[d] capsules	IV/oral tablets, solution, suspension
Usual daily dose (mg)	200–800	100–600	100–1000
Dosing regimen	Once daily	≤ 200 mg: once daily > 200 mg: twice daily	Once daily

[a]Adapted from Kauffman and Carver (1997).
[b]Decreased and unpredictable in some seriously ill patients and in bone-marrow-transplant patients immediately after operation (possibly chemotherapy-induced alteration to gastric mucosa).
[c]Cyclodextrin solution improved bioavailability (possible availability).
[d]Newly developed intravenous formulation under trial.

symptomatic in approximately one in 10 000 to 15 000 patients and not necessarily dose related), gynaecomastia (related to androgen-blocking activity) and anaphylaxis have been reported. The effect of ketoconazole on the liver varies from asymptomatic transient abnormalities of the enzymes to symptomatic and potentially fatal acute hepatic necrosis. Monitoring of liver function tests is recommended. Ketoconazole is mainly employed in dermatology.

Triazoles

The triazoles may revolutionize the treatment of many fungal infections. At present two systemically active triazoles are available, namely fluconazole and itraconazole (see Table 6.5), although others (e.g. voriconazole) are currently under investigation (Kauffman and Carver, 1997; Vora et al., 1998; Sheehan et al., 1999).

Fluconazole Fluconazole has excellent water solubility, permitting both intravenous and oral administration. Clinically it shows excellent penetration into the CSF (> 50% serum concentration) and other sites such as the peritoneum – a feature that is not shown by other systemic antifungal agents apart from flucytosine. It also has a relatively prolonged half-life (> 24 h) and is largely excreted unchanged via the kidneys, making it also potentially useful in urinary tract infections. Fluconazole is an excellent oral alternative for the treatment of infections by many common *Candida* species in surgical patients, for the treatment of cryptococcosis in non-immunocompromised patients, and for long-term suppressive treatment of cryptococcal meningitis and oesophageal candidosis in AIDS patients. The latter regimen has the potential to select yeasts that are less susceptible to fluconazole, and may have been a reason for the emergence of *Candida dubliniensis* in the oral cavity of AIDS patients (see below). Yeasts with inherent reduced susceptibility and/or resistance to fluconazole include *Candida (Torulopsis) glabrata* and *Candida krusei*.

Although fluconazole is generally well tolerated, nausea, abdominal pain, diarrhoea, dry mouth and skin rash have all been recorded, but are uncommon (see Table 6.2). The causal relationship of abnormal hepatic, renal and haematological tests observed in some seriously ill patients receiving multiple doses of fluconazole has yet to be clarified. However, dose-dependent hepatotoxicity has been described in an AIDS patient receiving fluconazole for persistent oesophageal candidosis.

With most yeast infections it is usual to start with higher oral doses of fluconazole (e.g. 400 mg daily for the first 2 days). This is then reduced to the more normal 100–200 mg daily level. As fluconazole has a relatively long half-life, consideration can be given to more extended dosing periods (e.g. every 2 or 3 days), but maintaining the same total dose. One feature of fluconazole therapy is the rapid clinical response – remission of chronic mucocutaneous disease has been noted within 5 days. Vaginal candidosis has been reported to respond favourably to a single oral dose of 150 mg fluconazole, and this would seem to be highly acceptable from the patient's point of view.

Apart from localized skin and/or mucosal *Candida* lesions, fluconazole has been shown to be effective and well tolerated in patients with a variety of surgery-related invasive *Candida* problems (e.g. intra-abdominal infections, catheter-related candidaemia with or without visceral involvement) (Kauffman and Carver,

1997). The doses used have been in the range 400–800 mg daily, with clinical responses comparable to those achieved with amphotericin B (0.5–0.6 mg/kg daily). Chronic invasive or hepatosplenic candidosis has also been successfully treated with fluconazole, although most patients receive prior amphotericin B or concurrent flucytosine. The efficacy of fluconazole is also proven in prosthetic valve endocarditis attributable to *Candida*, although wherever possible surgical removal and replacement of involved valves is of prime importance. For many invasive yeast infections, a short (e.g. 7–10 days) pulse of amphotericin B followed by high-dose oral fluconazole (e.g. 400–800 mg) to complete a total course of 6–12 weeks has been recommended. Long-term, high-dose (e.g. > 800 mg daily) oral fluconazole appears to be a regimen with relatively few problems (Stevens *et al.*, 1997) apart from possible drug interactions (see below).

Reports supporting the use of fluconazole (400 mg daily) in prophylaxis (e.g. bone-marrow-transplant patients, repeated gastrointestinal surgery) are convincing (Goodman *et al.*, 1992; Eggimann *et al.*, 1999; Rotstein *et al.*, 1999), although the potential to select more resistant fungi (e.g. *Candida krusei*) and strains of *Candida albicans* is a concern. Fluconazole (400 mg daily) prophylaxis (in reality presumptive therapy) also appears effective in limiting intra-abdominal candidosis in surgical patients with recurrent gastrointestinal perforations or anastomotic leakages (Eggimann *et al.*, 1999).

Itraconazole Although on paper it does not possess the excellent pharmacokinetic properties of fluconazole (i.e. itraconazole is insoluble in aqueous solvents, less well absorbed, strongly protein bound and has poor CSF penetration; see Table 6.5), itraconazole does appear to have a wider spectrum of activity and has been shown to be effective in the treatment of a variety of fungal infections, including those by *Candida*, cryptococci, aspergilli and dermatophytes, as well as a variety of more unusual fungi that are apparently unresponsive to amphotericin B (e.g. *Sporothrix schenckii*) and some agents of cutaneous chromomycosis (e.g. *Fonsecaea pedrosoi, Cladosporium carrioni*). In an attempt to improve the irregular and often unpredictable absorption of capsule formulations of itraconazole in seriously ill patients, oral-solution cyclodextrin formulations have been developed. These appear to have greatly improved absorption (Barone *et al.*, 1998; Zhou *et al.*, 1998). However, the use of itraconazole in seriously ill patients requires constant monitoring of blood levels to confirm that therapeutic concentrations are being obtained and maintained (> 0.5 mg/L by HPLC). Giving the daily dose of itraconazole in two or three divided doses (rather than as a single dose) may also result in more reliable and consistent blood levels (Poirier *et al.*, 1997). An intravenous formulation is currently under investigation.

Perhaps the main role of itraconazole is as an alternative to amphotericin B in invasive aspergillosis, and in the treatment of a variety of more chronic cutaneous infections (e.g. sporotrichosis, mucocutaneous candidosis and onychomycosis). Side-effects appear to be similar to those seen with ketoconazole, but are probably less common (e.g. elevation of liver function tests in < 5% of patients). Regimens (50–600 mg daily) have been continued satisfactorily for many months. It seems that itraconazole and amphotericin B may be antagonistic when given concomitantly or sequentially (Sugar and Liu, 1998).

Drug interactions with azoles

Drug interactions seen with azoles can generally be placed into one of two broad categories: first, decreases in azole bioavailability due to chelation or secondary to increases in gastric pH, and secondly, interactions with other cytochrome P_{450} metabolized drugs (Kauffman and Carver, 1997). The latter may result in increases or decreases in the concentration of the azole or of the interacting drug, or of both compounds (see Table 6.6).

Table 6.6 Effects of azole antifungal drugs on serum concentrations of concomitantly administered drugs

	Azole antifungal drug		
Drug affected by azole	Ketoconazole	Itraconazole	Fluconazole
Alterations in cytochrome P_{450}			
Warfarin	↑[a]	↑	↑[a]
Cyclosporin	↑[a]	↑[a]	↑[a]
Phenytoin	↑[a]	None known[c]	↑[a]
Triazolam, alprazolam, midazolam	↑	↑	↑
Diltiazem	↑	None known	None known
Lovastatin	None known	↑	None known
Zidovudine	None known	None known	↑
Carbamazepine	None known	None known	↑[a]
Terfenadine	↑[b]	↑[b]	No effect
Astemizole	↑[b]	↑[b]	None known
Loratidine	↑	None known	None known
Cisapride	↑[b]	↑[b]	↑
Prednisone/methylprednisolone	↑	None known	None known
Sulphamethoxazole	None known	None known	↑
Indinavir	↑	None known	↓
Saquinavir	None known	None known	↑
Ritonavir	↑	None known	None known
Oral hypoglycaemic agents	↑	None known	↑
Isoniazid	↑[a]	None known	None known
Rifampin	↑[a]	None known	None known
Rifabutin	None known	None known	↑
FK 506	None known	None known	↑[a]
Quinidine	↑	None known	None known
Unknown mechanisms			
Digoxin	None known	↑[a]	None known

[a] Clinically significant interaction; serum concentrations of drug and/or clinical status of patient should be monitored.
[b] Life-threatening interaction causing arrhythmias; avoid use of combination.
[c] Interaction has not been studied in human subjects; however, caution should be exercised when using this combination until further information is available.

Reproduced from Kauffman and Carver (1997).

Some important drug interactions are shown in Table 6.6. In most cases, the azole interferes with the metabolism of the other cytochrome P_{450}-metabolized drug. More noteworthy are the interactions of azoles and cisapride, terfenadine, astemizole and loratidine (Kauffman and Carver, 1997). The first three cardiotoxic drugs may accumulate because of inhibition of their metabolism by concurrent azole use. Fatal arrhythmias may result. Like the azoles, these drugs are metabolized almost entirely via the cytochrome P_{450} IIIA4 subfamily. As increasingly higher doses of fluconazole are employed to treat fungal infections, the number of clinically relevant interactions is likely to increase.

Predictably, drugs such as rifampicin, isoniazid and phenytoin, which are known to induce the activity of cytochrome P_{450}, result in increased metabolism of azoles and potential therapeutic failures.

OTHER SYSTEMIC ANTIFUNGAL AGENTS

Flucytosine (5-fluorocytosine)

Flucytosine can be given orally or intravenously and has the advantage over most other systemic antifungal agents (apart from fluconazole) of good penetration into body cavities and fluids such as the CSF and urine. Unfortunately, its widespread use has been limited by a number of factors, including a poor antifungal spectrum (confined to yeasts), rapid emergence of resistant forms, and toxicity in the form of bone-marrow suppression at serum levels above 100 mg/L. The use of flucytosine is now restricted to combination treatment with amphotericin B in serious yeast infections. Its only role as a sole agent is in the short-term treatment of urinary candidosis, but even in this setting fluconazole may now be more appropriate. The necessity for routine serum level assays seriously curtails the prolonged use of flucytosine.

During therapy, patients should be monitored regularly (with blood counts, creatinine levels and liver function tests) and the serum level of flucytosine should be maintained between 50 and 100 mg/L. The maximum daily dose must be lowered in the presence of impaired renal function, in order to prevent excessive accumulation.

Terbinafine

This highly lipophilic fungicidal drug belongs to a new class of antifungal agents known as the allylamines, and it is active both orally and topically. It is readily taken up into body fat and the stratum corneum, and transported by sebum. Not surprisingly, it is also now the drug of choice for chronic dermatophyte infections including those involving the nails. Although early experimental animal trials suggested that, despite promising *in vitro* sensitivity data, terbinafine was ineffective in diseases such as aspergillosis, it now appears that this was a defect of the animal model involved, and that terbinafine may be effective in treating invasive aspergillosis and other mould infections in humans. Further clinical trials are required to confirm this observation.

TREATMENT OF SELECTED FUNGAL DISEASE

The available treatment options for the more significant systemic mycoses are summarized in Table 6.7.

INVASIVE ASPERGILLOSIS

Invasive aspergillosis is the leading cause of infectious death after allogenic bone-marrow transplant (Wald *et al.*, 1997; Denning, 1998; Latgé, 1999). It is an exogenous mycosis with many unknown factors concerning its pathogenesis. Aspergillosis appears to develop after inhalation of spores in the setting of retained secretions (e.g. chronic obstructive respiratory disease) and underlying lung pathology. Depending on the existing lung insult and the patient's immunological capability, no disease, allergic disease, chronic localized disease or rapidly spreading invasive disease may result. Dissemination from the primary lung focus occurs predominantly in association with neutropenia.

In a review of 2500 bone-marrow-transplant (BMT) patients over a 6.5-year period in the USA, 158 patients (6.3%) had proven or probable aspergillosis. A further 56 patients (2.2%) had aspergilli isolated from situations regarded as representing colonization or contamination rather than invasive disease (Wald *et al.*, 1997). The incidence of invasive disease increased from 5.7% to 11.2% of patients during the study period (January 1987 to June 1993). Infection was most common at around 16 days and 96 days post-transplant (i.e. bimodal). Risk factors associated with early infection (i.e. within 40 days of transplant) included

Table 6.7 Treatment recommendations for major systemic mycoses in surgical patients

Disease[a]		Recommended agent(s) (dose)[e]	Possible alternatives
Candidosis	Oral/vaginal thrush	Topical imidazole	Flu, Itr, Ket
	Chronic/recurrent mucocutaneous	Flu (200 mg q 24 h)	Itr, Ket
	Candidaemia	Flu (400 mg q 24 h)	AmpB
	Invasive/systemic	Flu (400–800 mg q 24 h)	AmpB (0.6–1.0 mg/kg q 24 h) ± 5FC (150 mg daily divided 6-hourly)
	Cystitis	Flu (150 mg stat)	5FC, AmpB[b]
Crytococcosis (non-immunosuppressed)	Meningeal	Flu (800 mg q 24 h)	AmpB + 5FC
	Non-meningeal	Flu (400 mg q 24 h)	AmpB + 5FC
Aspergillosis[d]	Invasive	Itr (600 mg q 24 h), or AmpB (up to 1.5 mg/kg q 24 h)	? Ter[c]
Zygomycosis[d] (mucormycosis)	Invasive	AmpB (up to 1.5 mg/kg q 24 h)	

[a]Azoles (e.g. itraconazole) may be superior to amphotericin B with diseases (mainly cutaneous) such as sporotrichosis, chromoblastomycosis, scedosporiosis, paecilomycosis, trichosporonosis, fusariosis and some forms of zygomycosis (see text).
[b]Local irrigation.
[c]Terbinafine may be of use (given orally) for some systemic mycoses, including aspergillosis.
[d]Surgical removal of involved tissue where possible.
[e]These may vary compared to those recommended for immunocompromised patients.
AmpB, amphotericin B; 5FC, flucytosine; Ket, ketoconazole; Flu, fluconazole; Itr, itraconazole; Ter, terbinafine.

underlying disease (haematological malignancy), donor type (autologous), season (summer) and transplant outside a laminar-flow room. For patients who developed disease more than 40 days after transplant, increasing age, underlying disease, unrelated donor, neutropenia, the development of graft vs. host disease and the use of corticosteroids were associated with increased risk. The 1-year survival of patients with invasive aspergillosis was 7%, compared to 54% for control patients. The importance of graft vs. host disease and the level of prednisone used have also been highlighted by Ribaud et al. (1999) with regard to survival after aspergillosis in BMT patients.

In view of the high mortality in patients who do become infected, the prevention of aspergillosis in immunosuppressed and/or neutropenic patients using some form of prophylaxis and/or high-efficiency particulate air (HEPA) filtration has been investigated. The results of prophylaxis trials (either systemically or intranasally) have not been encouraging, although HEPA filtration has proved effective in reducing invasive disease, and has allowed increasingly aggressive treatment regimens to be introduced (Bretagne et al., 1997; Wald et al., 1997; Denning, 1998; Withington et al., 1998). Improved antifungal prophylaxis is clearly needed in potential bone-marrow-transplant recipients with a history of probable or proven previous invasive aspergillosis (Offner et al., 1998).

The mortality rate in BMT patients with invasive disease is very high, particularly in cases where the brain is involved. While therapy has in the past consisted of high-dose amphotericin B (e.g. 1.0–1.5 mg/kg daily), it is now apparent that this is largely unsuccessful and that other drugs and/or additional (surgical) procedures are required (Denning, 1998). It seems that aspergilli may reveal *in vivo* resistance to amphotericin B that is not detected by *in vitro* susceptibility tests (Verweij et al., 1998). As neutropenia has been reported as a major risk factor for invasive disease, the use of granulocyte colony-stimulating factor and possibly granulocyte transfusions can be considered as adjunctive therapy. Itraconazole (600 mg daily) and possibly terbinafine (dose unknown) and voriconazole are additional drugs worthy of consideration. The former is considered to be superior to amphotericin B in the treatment of aspergillosis by some authorities. It is considered to be a reasonable first choice in patients who are eating and are not on drugs that induce the metabolism of cytochrome P_{450} (e.g. rifampicin). Pulmonary lesions of aspergillosis are said to have a distinctive radiological (computerized tomography) appearance with characteristic pleural involvement. In neutropenic patients, early aspergillus lesions present as small nodules and/or small pleural-based lesions with straight edges and surrounding low attenuation (halo sign). As disease progresses, the nodules may cavitate (often as the neutrophil count recovers), resulting in the air-crescent sign. Both the halo and air-crescent signs are highly characteristic for invasive fungal disease of the lung (Denning, 1998). This may allow early diagnosis in susceptible individuals and the surgical removal of involved tissues. Surgery is now considered to be a significant therapeutic adjunct, despite the inherent problems of bleeding in the type of patient that is likely to be involved. Other indications for surgery include persisting lung shadows prior to a bone-marrow transplant or more aggressive chemotherapy, significant haemoptysis, and lesions impinging on the great vessels or major arteries (Denning, 1998). The role of additional cytokine therapy in this disease is largely unknown.

As already stated, pulmonary aspergillosis reflects the underlying host abnormality. In otherwise normal patients with pre-existing lung cavities (e.g. old tuberculous cavities, necrotic cancerous areas), inhalation of spores may result in the development of compact, localized fungal growth. These 'fungus balls' or aspergillomas may slowly spread into the surrounding tissue, and require surgical resection with peri-operative itraconazole or amphotericin B cover. Unfortunately, there is a high surgical complication rate with such procedures. It has therefore been suggested that aspergillomas should be left and monitored, with surgical removal being a last resort in cases where progressive tissue invasion and the risk of massive haemoptysis become apparent.

Apart from the lower respiratory tract, aspergilli are important causal agents of infection involving the paranasal tissues and orbit, especially in hot dry geographical areas such as the Sudan. In such countries, *Aspergillus flavus* lesions involving the orbit (paranasal granuloma) are a significant cause of unilateral proptosis. Aspergilli have also been recovered from brain abscesses, heart lesions following cardiac surgery, osteomyelitis following penetrating injury, and skin lesions either as a reflection of haematogenous dissemination or as a result of traumatic injury. In all cases, surgical intervention in conjunction with antifungal therapy is the mainstay of therapy.

The diagnosis of invasive aspergillosis in BMT patients relies on a high degree of suspicion, particularly in patients with increasing lung infiltrates that are apparently unresponsive to antibacterial drugs (Latgé, 1999). The recognition that pulmonary lesions have a characteristic appearance on computerized tomography has been a major recent advance. It seems that antifungal therapy should be initiated immediately aspergillosis is contemplated, and while confirmatory tests are being undertaken. Any delay in starting therapy could have disastrous consequences. Clearly consideration should be given to the early surgical resection of involved tissue.

The value of sputum cultures and/or microscopy has been debated for years. To distinguish between colonization, contamination and invasive disease is difficult if not impossible; patients with invasive pulmonary disease may be culture negative. Current diagnostic methods lack sensitivity and have not really been improved by serum antigen and/or metabolite detection techniques. Computerized tomography is perhaps the one recent advance, and is clearly essential in cases where there is a high level of suspicion of invasive pulmonary disease.

SYSTEMIC CANDIDOSIS

Candida albicans and related species (e.g. *Candida parapsilosis*) are emerging as significant systemic pathogens in abdominal surgery, particularly in patients who are debilitated and require intravenous nutrition and/or who have received broad-spectrum antimicrobial agents (Solomkin *et al.*, 1982a,b; Smith and Payne, 1994; Hazen, 1995; Casewell, 1997; Flanagan and Barnes, 1998; Levy *et al.*, 1998; Pfaller *et al.*, 1998a; Stratov *et al.*, 1998). Post-surgical infections have been associated with an extrinsically contaminated intravenous anaesthetic agent (McNeil *et al.*, 1999). *Candida* infection should be a major consideration in patients with possible TPN-related central-line sepsis, and in endophthalmitis and endocarditis in intravenous drug abusers. Surgery plays a major role in the treatment protocol of patients with eye and heart lesions (Martinez-Vázquez *et al.*, 1998).

Uncontrolled infection is a potent source of organ failure (Edwards, 1991; Casewell, 1997). As well as direct seeding of the abdominal cavity during surgery, colonization of long-term intravascular catheters is a significant factor in the establishment of many infections. These may present as local sepsis or more commonly as a catheter-related fungaemia with the potential to seed distant organs and tissues (e.g. eye, kidneys, muscle).

The incidence of nosocomial fungal infections, with *Candida albicans* the most frequently isolated fungus, increased from 2.0 to 3.8 per 1000 discharges in the USA during the 1980s (Beck-Sagué, Jarvis and the National Nosocomial Infections Surveillance System, 1993). The highest overall infection rate found in 1990 was in the surgical services (5.6 per 1000 discharges), having risen by 124% since 1980. Within individual surgical subspecialities, the highest rate occurred in burns/trauma, followed by cardiac and general surgery. In another survey conducted in the USA in the late 1980s (Fraser *et al.*, 1992), parenteral hyperalimentation was an associated risk factor in 61% of 106 patients with candidaemia, while 45 patients (42%) had undergone surgery (15 cases of intra-abdominal surgery). Fungaemia associated with intravascular catheterization is now clearly a major nosocomial consideration, and the cross-infection potential of yeasts via hands and inanimate objects should also not be underestimated (Flanagan and Barnes, 1998; Huang *et al.*, 1998).

In a more recent multi-country review of invasive candidosis, 306 episodes of bloodstream infection were analysed (Pfaller *et al.*, 1998a). Around 50% of cases occurred in intensive-care patients, with *Candida albicans* the dominant isolate overall (53% isolates), followed by *Candida parapsilosis* (16%), *Candida glabrata* (15%), *Candida tropicalis* (8%), *Candida krusei* (2%) and *Candida guilliermondii* (1%). On a geographical basis, 44% of infections in the USA were non-*albicans* (*Candida glabrata* was most common), 48% of fungaemias in Canada were non-*albicans* (*Candida parapsilosis* was most common) and 59% of infections in South America were non-*albicans* (mainly *Candida parapsilosis*). Apart from around 10% of *Candida glabrata* and all the *Candida krusei* isolates, all other species were 99–100% susceptible to fluconazole (MIC $\leq$ 32 mg/L).

It is also apparent that *Candida* species are a significant cause of infection in post-organ-transplant patients, with return to surgery being a significant risk factor (in addition to the more widely quoted risk factors such as antibacterial use, neutropenia and vascular catheterization).

Chemotherapy

Amphotericin B is still considered by many to be the mainstay of therapy for systemic candidosis; daily doses range from 0.3–1.0 mg/kg given over 1–4 h as an infusion of 5% dextrose. Side-effects and toxicity are common, and are often cited as reasons for withholding therapy. However, amphotericin B therapy, is safe (see p. 126). Lipid formulations of amphotericin B (e.g. liposomal capsules, colloidal dispersions, lipid emulsions) are slowly being marketed. These expensive drugs are still largely of unproven mycological advantage over the conventional formulation, although they are clearly less toxic, probably less compromising to the immune system, and can be given in higher doses.

For candidaemia in post-surgical patients or where infected vascular lines are encountered, 25 mg/day (half an ampoule) for 7–10 days is usually sufficient. Higher doses and/or protracted schedules are reserved for more serious deep-seated invasive disease. The added value of concurrent flucytosine in catheter-related candidaemia is unknown. While most clinicians recommend that infected catheters be removed and the patient be treated with amphotericin B, there is some disagreement as to whether removal of the catheter is always mandatory, especially if the patient is not neutropenic. Treatment through the catheter may be feasible in many surgical settings where the patient has a mild subacute illness and for whom no convenient place for another catheter is evident.

Fluconazole has shown promise as an alternative to amphotericin B. It can be taken orally once a day (and is also available as an intravenous fluid), has excellent absorption and body distribution (including the urine, peritoneum and cerebrospinal fluid), is relatively free of serious adverse effects, and is certainly more patient tolerable than amphotericin B (see p. 138). In surgical patients with candidaemia, loading doses of 400 mg/day for 2–3 days followed by 200 mg daily for 4 weeks can be considered as an alternative to amphotericin B, although in seriously ill intensive-care patients, intravenous doses of 10 mg/kg body weight per day have been recommended (Graninger *et al.*, 1993).

As with many other infections, surgical debridement and removal of any necrotic foci are important adjuncts to be considered when attempting treatment of deep-seated fungal infections. If *Candida* poses a potential nosocomial problem (e.g. in an intensive-care unit), selective decontamination of the oral cavity and gastrointestinal tract using nystatin should be performed (Damjanovic *et al.*, 1993). Other forms of 'prophylaxis' are not recommended in general surgical patients. Following bone-marrow transplantation, gastrointestinal surgery and in patients with cancer complicated by neutropenia, relatively high-dose fluconazole (e.g. 400 mg daily) prophylaxis would appear to be of benefit in reducing the occurrence of invasive candidosis (Goodman *et al.*, 1992), although some authorities would argue that there is no survival benefit of antifungal agents given prophylactically.

Laboratory aspects

Confirmation of the diagnosis of systemic candidiosis requires the demonstration by culture or histology of *Candida* in otherwise sterile sites, or on positive blood cultures where transient catheter-related fungaemias can be excluded. Blood cultures are positive in only about 50% of patients with invasive disease, while culture of biopsy material from hepatosplenic lesions is invariably negative. The value of serological tests (for antigen or antibody) in surgical patients is questionable, especially in view of the availability of oral fluconazole. In many patients, particularly those in intensive care, the diagnosis is often made using inferential evidence of yeast infection, (e.g. a persistent fever despite apparently appropriate antibiotics) (Flanagan and Barnes, 1998).

Antifungal susceptibility testing of fungi is still in its infancy, although standardized methods (especially for yeasts) are slowly being developed and utilized. National Committee for Clinical Laboratory Standards (NCCLS) guidelines for *Candida* and a number of antifungal agents (e.g. fluconazole, amphotericin B) have recently been published (Rex *et al*., 1997). The emerging importance of azole resistance in yeasts has been attributed to the increasing use of fluconazole in the therapy, prophylaxis and suppression of *Candida* infections, although some surveys have found no link with high use in at-risk patients (e.g. those with cancer). Whether or not fluconazole use is responsible for the apparently increasing significance of inherently less susceptible species such as *Candida glabrata* and *Candida krusei* is another debatable point (see p. 147).

ZYGOMYCOSIS (MUCORMYCOSIS)

Present-day interest in zygomycosis in humans stems from a paper by Gregory *et al*. (1943), who in a comprehensive review of the literature accompanying their report of four cases of orbital and central nervous system (CNS) infection, noted 11 papers dealing with superficial and cutaneous mucormycosis, 10 reports of pulmonary infection, five descriptions of gastric ulceration and a single case of CNS involvement.

Since this report, zygomycosis has been described in association with a variety of clinical syndromes (see Smith, 1989). While generalizations may be misleading, it has been apparent since the late 1950s that the anatomical distribution of lesions does correlate to a certain degree with defined predisposing conditions (e.g. craniofacial involvement in individuals with diabetic acidosis, pulmonary and disseminated infection in patients with acute leukaemia, and gastrointestinal and cutaneous lesions following local trauma). Indeed, cutaneous infections have possibly replaced craniofacial and pulmonary disease as the prevalent clinical manifestation of zygomycosis, a change that is mirrored by the emergence of *Rhizopus microsporus* var. *rhizopodiformis* (as *Rhizopus rhizopodiformis* in most cases) as a significant pathogen. In addition to the predisposing factors listed above, the use of deferoxamine (a heavy-metal chelator) to remove metal overload has been shown to favour the development of zygomycosis.

Unlike most other mycoses, the infectious propagule responsible for initiation of lesions in zygomycosis is unproven. While most investigators implicate spores, others have suggested that small pieces of viable hyphae are responsible. As spores of mucoraceous fungi are both uncommon in air and readily contained by the phagocytic cells that are normally found in pulmonary tissues, it is possible that in some cases, especially where trauma is the associated factor, hyphal fragments represent the infectious propagule. The association of lesions with contaminated elastoplast dressings (on which the fungus was possibly growing), and the ease with which starved animals can be infected following ingestion of hyphae but not spores, lend support to this hypothesis.

Mucoraceous fungi rank third after *Candida* and cryptococci as causal agents of cerebral mycoses. Nervous system disease can occur by direct or indirect extension from paranasal lesions (e.g. hard palate, orbit), and by haematogenous spread from primary lesions in the lungs or other tissues. A growing number of cases have been associated with parenteral drug abuse (probably due to contaminated fluids/syringes).

In the few but increasing number of rhinocerebral cases that have been cured, successful treatment was dependent on a combination of early diagnosis (clinical associations, direct smears and culture of seropurulent biopsy material), control of the underlying disease and cessation of predisposing chemotherapy, surgical removal of involved tissues, and early administration of amphotericin B (1.0–1.5 mg/kg daily). An increasing survival rate (up to 70%) has been noted over the last decade. Patients with hemiplegia, facial necrosis or nasal deformity have a poor prognosis.

While most cases of rhinocerebral infection have been sporadic and devoid of any epidemiological considerations (apart from the association with drug abuse), it has been suggested by Bottone *et al.* (1979) that the nasal and pulmonary lesions seen in half of their six patients who developed *Rhizopus microsporus* infections over a 9-month period resulted from inhalation of spores from contaminated adhesive bandages. A common source and air-borne route of transmission (from an air-conditioner filter) has since been suggested in a number of other cases.

The great majority of patients with zygomycosis of the pulmonary tissues have leukaemia or lymphoma as the underlying disease. Not surprisingly, pulmonary disease is often accompanied by widespread dissemination of the offending fungus. Mucoraceous fungi have a marked affinity for vascular invasion and growth in blood-vessel walls. Infarction and necrosis secondary to this are the hallmark of pulmonary zygomycosis, and pulmonary arteriography has been suggested as a possible diagnostic aid. A persistent fever accompanied by progressive cavitary pneumonia that is refractory to antibacterial or steroid therapy in patients with poorly controlled leukaemia or lymphoma and with chemical diabetes suggests the need for aggressive diagnostic procedures to eliminate or confirm zygomycosis. Early diagnosis (e.g. by the use of fine-needle aspiration) may lead to successful treatment with amphotericin B with or without surgery. The importance of control of the underlying cancer and the rarity of positive sputum cultures have been stressed on numerous occasions.

Gastrointestinal zygomycosis is perhaps underestimated. Areas of the alimentary tract that have been found to be affected include the oesophagus, stomach and intestines – the stomach being the area most commonly involved. Widespread dissemination of the fungus from primary gastric lesions is reputedly uncommon, which is rather surprising in view of the remarkable predilection of the fungi for growth within blood vessels. It may be that in many cases infection was acquired shortly before death, giving the fungus little time to spread. Clinical, radiographic and endoscopic features may suggest gastric cancer. An operation may allow a diagnosis to be made during life, as stomach lesions are characteristically much larger than those found with the usual gastric ulcer, and they contain large amounts of black necrotic material. The surrounding tissue tends to be extremely hard.

Lesions involving cutaneous tissues are now perhaps the most frequently encountered presentation of zygomycosis. These usually follow some form of skin trauma. In 1978, 23 cases of cutaneous zygomycosis associated with the use of elastoplast dressing were described by the Center for Disease Control in Atlanta. Most of the patients (who came from 11 hospitals in the USA) had been subjected to traumatic/invasive investigative procedures, following which elastoplast bandages or dressings had been placed over sterile gauze pads covering the involved skin area. In one case, the dressing covered a bite wound. Lesions were usually present on removal of the dressing, and ranged from vesiculo-pustular eruptions to ulceration with eschar formation. In some patients, skin necrosis was present and required debridement. One of the patients had diabetes mellitus, three had cancer and at least one was on steroids. In cases where only raised plaques and pustules were present, spontaneous recovery occurred. However, where deep abscesses occurred (e.g. at the iliac crest bone-marrow biopsy site of two patients with lymphocytic leukaemia), amphotericin B therapy accompanied by surgical excision was necessary.

Mucoraceous fungi can also colonize or invade burn wounds. In one survey (Nash et al., 1971) of fungal infection of burn wounds, 22 of the 30 cases studied were attributable to mucoraceous-like fungi. These 30 patients represented just under 1% of all burn cases treated, and the high incidence of fungal invasion appeared to be related to the topical use of mafenide acetate (sulphamylon) cream. Dissemination occurred in 27% of cases and was always fatal. Early diagnosis by direct smear or biopsy followed by an aggressive surgical approach gave the best results. Isolation of the aetiological agent by culture was only possible in about one-third of the cases.

An unusual case of cutaneous zygomycosis has been described involving a breast prosthesis in an actress engaged to be married to a young surgeon (Symmers, 1968). The prosthesis was of an experimental type, designed with the intention that its effect could be finally determined by injecting or withdrawing saline solution once the patient had recovered completely from the implantation operation. The unfortunate woman developed a *Rhizopus* infection (complicated by pseudomonads) at the site through which the left-sided prosthesis was usually needled. Prosthetectomy was successful in curing the infection.

Other complications of zygomycosis have included osteomyelitis, usually in association with a contiguous tissue infection (e.g. the sinuses and frontal bones of diabetics with rhinocerebral zygomycosis), and myocarditis and endocarditis (e.g. following cardiac surgery). Fungal lesions in the myocardium and on the valve cusps may be a prominent feature of some disseminated infections, invariably in association with multiple abscesses and occlusion of vessels in other organs.

OTHER SYSTEMIC MYCOSES

In 1983, a fungus identified as *'Scedosporium inflatum'* was isolated in pure culture from a bone biopsy in a boy with osteomyelitis, and was given pathological significance (Malloch and Salkin, 1984). Since this report, numerous other isolates of this fungus now known as *Scedosporium prolificans*, have been recovered from human lesions – mainly involving bone. This fungus appears to be an emerging pathogen with a predilection for bone. Apart from the need for surgical removal of infected tissue, *Scedosporium prolificans* and other *Scedosporium* species tend to be resistant to amphotericin B; optimal chemotherapy involves the use of triazoles such as itraconazole (in the past, the now obsolete miconazole was the drug of choice). It remains to be seen how effective new cell-wall-active antifungal agents, terbinafine or the new triazole voriconazole will be against these fungi. *In vitro* results with the latter suggest a possible role (Espinèl-Ingroff, 1998).

An increasing problem in cancer patients concerns infection by *Fusarium* species, which are often unresponsive to currently available antifungal agents (Krcmery *et al.*, 1997).

Entomophthoramycosis is primarily a disease of the subcutaneous tissues and the nasal submucosa which occurs in otherwise normal individuals. Lesions may spread locally to involve adjacent tissues (e.g. the palate). Causative fungi belong to the genera *Conidiobolus* (syn. *Entomphthora*) and *Basidiobolus*. The disease is clearly more common in hotter, tropical regions. Surgical removal of affected tissues is the primary objective of the clinician.

A variety of 'black moulds' (dematiaceous fungi) have been recovered from infections requiring surgical intervention. The pathogenesis appears to involve some form of cutaneous trauma, with neurotropism being a feature of some species (e.g. *Xylohypha bantiana*). The optimal antifungal therapy is not known, although local excision should be accompanied by azole therapy (e.g. with itraconazole). In a review of dematiaceous fungal infections in organ-transplant recipients (see Singh *et al.*, 1997a), two distinct patterns of infections were identified. Around 80% of infections involved skin and/or soft tissues or joints, and were predominantly caused by *Exophiala* species. The mortality rate was low (7%). On the other hand, the remaining infections were invasive, often involving the brain, and were attributed to *Ochroconis gallopavum* (formerly *Dactylaria* species). In this group, the mortality rate was about 57%.

It is now obvious that the 'sleeping giant' of the mycological world, namely cryptococcosis, has emerged to become the commonest systemic mycosis worldwide. Infection is acquired by inhalation of one of the three varieties of *Cryptococcus neoformans* from the environment. In cases where the patient has impaired cell-mediated immune activity, dissemination to the central nervous system (e.g. meningitis) and other organs may occur from the primary lung focus. Cryptococcal meningitis is now seen in up to 30% of HIV-infected individuals in certain geographical areas (e.g. Africa). The significance of the disease for surgeons occurs when they are asked to remove localized pulmonary or cerebral tumour-like cryptococcal lesions (cryptococcomas). Such lesions have in the past been mistaken for solid tumours. High-dose fluconazole (e.g. 800 mg daily) provides satisfactory antifungal cover for such surgery.

RESISTANCE TO ANTIFUNGAL AGENTS

The mid-1990s have seen the development of standardized laboratory tests for the detection of yeast sensitivity to a number of the important systemic antifungal agents (Rex et al., 1997). A reasonable correlation appears to exist between the suggested interpretive guidelines and clinical outcome. As well as the usual susceptible (S) and resistant (R) categories, a susceptible–dose dependent (S-DD) grouping has been formulated. With many fungi displaying borderline or intermediate susceptibility, effective treatment can often be achieved by increasing the drug dose (hence the S-DD rating).

Although a number of fungi are intrinsically less susceptible or resistant to defined antifungal agents (see Table 6.8), acquired resistance to antifungal compounds such as fluconazole is a rapidly emerging problem. This appears to be especially true in AIDS patients receiving long-term fluconazole therapy. Biochemical mechanisms that have been found to be associated with the development of resistance include altered drug target site (usually an enzyme), overproduction of the target, and decreased intracellular levels of drug brought about by either impaired penetration or active drug efflux (see Figure 6.3). Mechanisms analogous to the β-lactamase story in bacteria (i.e. enzymatic modification of the drug) have yet to be discovered.

Flucytosine resistance in *Candida albicans* develops readily in yeasts that are exposed to the drug, and is now common in many countries. It appears to be a feature of the B-serotype of this yeast. Resistance is associated with changes to one or more of the enzymes required for flucytosine metabolism, or the loss of feedback control of pyrimidine biosynthesis.

The protracted and widespread therapeutic, suppressive and prophylactic use of fluconazole in certain patient groups (e.g. those with AIDS, and neutropenic patients) has been accompanied by the emergence of strains of *Candida albicans* with decreased susceptibility and even resistance to this agent. In most cases this is associated with decreased intracellular drug levels (possibly due to active efflux),

Table 6.8 Naturally occurring decreased antifungal drug susceptibility or resistance in fungi

Antifungal agent	Decreased susceptibility/resistance seen in:
Amphotericin B	*Candida lusitaniae, Scedosporium* species, *Cunninghamella bertholletiae, Paecilomyces lilacinus, Trichosporon cutaneum (beigelii),* fusaria, some agents of chromomycosis (e.g. *Cladosporium carrionii), Sporothrix schenckii, Candida rugosa* (nystatin)
Fluconazole	*Candida glabrata, Candida krusei, Candida norvegensis, Candida dubliniensis*
Flucytosine	*Candida krusei, Candida tropicalis*

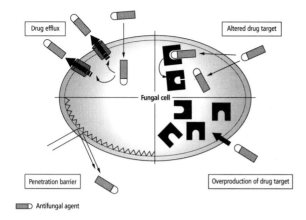

Figure 6.3 Biochemical mechanisms associated with resistance to antifungal agents. Adapted from Cannon (1997).

which in some cases can be countered by using higher doses of fluconazole. Mutational changes to the 14-α-sterol demethylase target site are a recorded but rare event.

In addition to the selection of resistant mutants of a particular strain, prolonged fluconazole therapy or prophylaxis may enhance the selection of inherently less susceptible strains of *Candida albicans,* or of *Candida* species that are normally less susceptible. The apparently increasing significance of yeasts such as *Candida krusei* in clinical medicine has been blamed on such a phenomenon by some authorities, but this has been disputed by others who consider that fluconazole use has not influenced the prevalence of non-*albicans* species (see Kauffman and Carver, 1997). A significant proportion of isolates of the recently described species *Candida dubliniensis* show reduced susceptibility to fluconazole, and susceptible strains of this species have been shown to develop fluconazole resistance rapidly *in vitro* (Sullivan and Coleman, 1998). Whether the apparently increasing incidence of this '*Candida albicans* look-alike' in the oral cavity of AIDS patients is associated with fluconazole use is still a matter of speculation.

While some strains of *Candida albicans* that are resistant to fluconazole may show cross-resistance to other azoles, most do not. This suggests that there are differences in cell permeability and/or efflux mechanisms between the azoles. Clearly there is still a great deal to be learned about the occurrence and clinical relevance of azole resistance in *Candida albicans* and related species (White, 1998).

Mutant strains of *Candida albicans* and several other fungi that are resistant to polyenes such as amphotericin B can be readily obtained in the laboratory. However, clinical resistance to amphotericin B in *Candida albicans* is a very uncommon event; the interaction of amphotericin B with the cytoplasmic membrane is clearly complex, and it seems that multiple changes are required to prevent a disruptive interaction. In addition, alterations in the membrane that lead to resistance also reduce virulence and slow the growth rate. Prophylactic topical nystatin has been incriminated in the emergence of *Candida rugosa* (resistant to nystatin) – an important invasive pathogen in seriously burned patients.

chapter 7

OTHER ANTIMICROBIAL AGENTS

The other antimicrobial agents of interest to surgeons include a variety of antiviral agents and the antihelminthic drugs that are used as adjuncts to surgery with hydatid disease. Blood-borne viruses present potential problems to the surgeon (Fry, 1996; Telford, 1996).

ANTIVIRAL AGENTS

Virus replication is so intimately associated with the host cell that it has been difficult to develop agents which act selectively against viruses without being unduly toxic to mammalian cells. Potential antiviral targets are associated with the entry, reproduction and assembly and release of new virus from the infected cell (Figure 7.1). Of the antiviral agents that are currently in use, most are nucleoside analogues (Balfour, 1999). These are phosphorylated within the host cell and, after incorporation into the elongating nucleic acid chain, result in chain termination (i.e. they inhibit nucleic acid synthesis). In recent years, no doubt fuelled by the global HIV epidemic, novel antiviral agents which inhibit the activity of viral proteases (enzymes which cleave precursor polyproteins to generate essential viral structural proteins and enzymes) have been developed. These are the so-called protease inhibitors.

ANTI-HERPES VIRUS AGENTS

Acyclovir and related compounds
Acyclovir (now often spelt aciclovir) or acycloguanosine is an analogue of the purine nucleoside, guanosine, in which the ribose moiety has lost its cyclic configuration. Acyclovir itself has no antiviral activity. It is phosphorylated to the active acyclovir triphosphate within infected cells by a series of enzymes, the first of which is a viral thymidine kinase produced by certain herpes viruses. This viral enzyme is much more efficient in producing the initial acyclovir monophosphate than is the cellular form of the enzyme. Thus the active form of the drug is produced only in virus-infected cells, which then take up more acyclovir from the surrounding milieu – a truly unique form of selective antimicrobial specificity.

Acycloguanosine triphosphate inhibits viral DNA replication by two mechanisms. First, as an analogue of guanosine triphosphate it is incorporated into the elongating DNA chain and causes chain termination. Secondly, acyclovir triphosphate is a direct inhibitor of viral DNA polymerase.

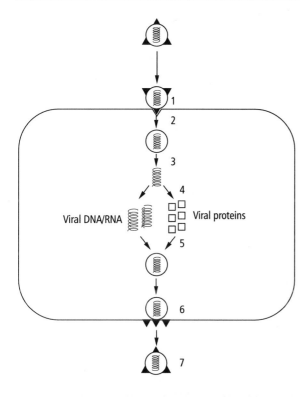

Figure 7.1 Events in the virus replication cycle which are possible target sites for antiviral agents. 1, attachment; 2, entry; 3, uncoating; 4, macromolecular synthesis; 5, assembly; 6, release/budding; 7, maturation.

Acyclovir has potent antiviral activity, and is virtually free of toxic side-effects. As it is poorly absorbed when given orally, it must be given intravenously in life-threatening infections. It is recommended for the treatment or prophylaxis of infections by herpes simplex viruses (e.g. genital herpes, herpes encephalitis, herpes simplex infection in immunocompromised patients) and for varicella zoster (shingles) (see Table 7.1).

A number of structural analogues of acyclovir are available (e.g. valaciclovir – an oral prodrug with improved oral bioavailability). In addition, penciclovir (a guanine analogue) and its prodrug famciclovir exhibit antiviral activity similar to that of acyclovir. Penciclovir triphosphate has a considerably longer half-life within infected cells than does acyclovir triphosphate, which may allow fewer doses per day (Robson, 1997; Balfour, 1999). It seems that none of these drugs displays superior efficacy or safety to acyclovir, although they can be administered two or three times daily, compared to five times daily for acyclovir.

Viruses with altered thymidine kinases or, clinically more significant, DNA polymerases, have emerged during acyclovir therapy, especially in immunocompromised patients who require prolonged treatment. A monophosphate nucleotide analogue, cidofovir, that undergoes intracellular phosphorylation to its active diphosphate form has been developed. This drug inhibits herpes virus DNA polymerase but, unlike acyclovir, does not require thymidine kinase for phosphorylation. It is therefore active against viruses where resistance to acyclovir is related to thymidine kinase mutations. It may be a viable alternative for some acyclovir-resistant herpes infections.

Table 7.1 Some anti-herpes virus agents and their use

Antiviral agent	Virus	Approved indications	Possible regimen
Acyclovir	Herpes simplex	Keratoconjunctivitis	3% eye ointment 5 times daily during waking hours until at least 3 days after healing
		Mucocutaneous disease including herpes labialis	200–400 mg orally 5 times daily (topically of no benefit)
			400 mg orally 12-hourly for 6–12 months as prophylaxis
		Progressive mucocutaneous disease in immunocompromised patients	5–10 mg/kg IV 8-hourly
			200–400 mg orally 5 times daily
		Encephalitis, neonatal and visceral herpes	10 mg/kg (500 mg/m^2) IV 8-hourly for 10–21 days
		Zoster (shingles)	800 mg orally 5 times daily for 7 days
Ganciclovir		CMV retinitis, oesophagitis and radiculopathy (uncertain benefit in other situations)	5 mg/kg as IV infusion over 1 h 12-hourly for 2–3 weeks, followed by 5 mg/kg IV daily as maintenance[a]
Foscarnet		As above for ganciclovir	60 mg/kg as an IV infusion over 2 h 8-hourly for 2–3 weeks, followed by 90 mg/kg infused daily as maintenance

[a]Similar regimens are used for the treatment and prophylaxis of CMV infection in bone-marrow and solid-organ transplant recipients.

Ganciclovir is a derivative of acyclovir which exhibits therapeutic activity against the herpes virus, cytomegalovirus (CMV). Acyclovir is inactive against CMV, as this virus does not possess a thymidine kinase. Like acyclovir, ganciclovir must be activated by phosphorylation, but clearly the viral enzyme involved is not thymidine kinase. Unfortunately, cellular enzymes can phosphorylate ganciclovir, so active drug is also generated in uninfected cells. Not surprisingly, ganciclovir is considerably more toxic than acyclovir; it impairs bone-marrow function and is nephrotoxic. It must be given by intravenous injection. Clinically it is used for the treatment of CMV retinitis and other severe acute CMV infections in immuno-compromised post-transplant patients (Balfour, 1999).

Foscarnet

Foscarnet is not a nucleoside analogue, and it does not require phosphorylation for activation. It works directly as a DNA and RNA polymerase inhibitor, with some selective toxicity for viral rather than host-cell enzymes. It has activity against all of the herpes viruses, but is usually reserved for the treatment of serious CMV infection or for acyclovir-resistant herpes simplex virus infections. It has a number of toxic side-effects (e.g. nephrotoxicity, renal failure, hypocalcaemia, penile ulceration) and, because of its poor oral bioavailability, should be given by intravenous infusion.

CMV infection

The development and use of sensitive diagnostic assays such as the polymerase chain reaction (PCR) has helped to clarify our understanding of the incidence and course of CMV infection after haematopoietic stem-cell and solid-organ allograft transplantation. This has allowed improved strategies for the prophylaxis and treatment of CMV disease to be developed and used, something that is still to a large degree lacking in the 'low-risk' transplant patient (e.g. kidney allograft recipient) (Hebart et al., 1998).

After solid-organ transplantation, intravenous ganciclovir alone results in resolution of the clinical signs of CMV disease in most patients. In high-risk patients after transplantation, especially seronegative individuals receiving organs from seropositive donors and patients requiring severe immunosuppression to treat allograft rejection, prophylactic or short pre-emptive courses of ganciclovir are beneficial. In these situations, acyclovir is better tolerated but clearly less effective.

Despite the use of intravenous ganciclovir in combination with immune globulin (passive immunization), the outcome remains poor in patients with established CMV infection after allogenic bone-marrow transplantation. Prevention of such infections using ganciclovir thus becomes a priority. It seems that following autologous stem-cell transplantation, neither antiviral prophylaxis nor pre-emptive therapy can currently be recommended (Hebart et al., 1998).

ANTI-HUMAN IMMUNODEFICIENCY VIRUS (HIV) AGENTS

HIV exists in two forms, namely HIV-1, which causes most HIV disease worldwide, and HIV-2, which is mainly confined to West Africa. Until recently, treatment for HIV-1 infections was limited to the use of nucleoside inhibitors of the viral enzyme reverse transcriptase (RT). This enzyme uses the RNA of the virus as a template to generate DNA, which is then incorporated into the host-cell DNA and in turn transcribed to produce viral RNA. This RNA serves to provide the genetic material and polyproteins for new virus particles. While RT inhibitors initially appeared promising, they have only modest antiviral activity, and the benefits of treatment are limited by the emergence of drug-resistant strains and dose-limiting toxic effects (McDonald and Kuritzkes, 1997).

Reverse transcriptase (RT) inhibitors

The commonly available RT nucleoside analogues which have been used to treat HIV infection include zidovudine (azidothymidine or AZT), didanosine (dideoxyinosine or ddI), zalcitabine (dideoxycytidine or ddC), stavudine (didehydrodeoxythymidine or d4T) and lamivudine (3TC).

Like other nucleoside analogues, zidovudine is activated by phosphorylation to the triphosphate form; these steps are catalysed by cellular enzymes. Incorporation of the zidovudine triphosphate into a growing DNA chain results in chain termination. Although it was originally used as an anticancer agent, zidovudine inhibits RT activity of retroviruses at concentrations considerably lower than those needed to interfere with host-cell DNA. Bone-marrow cell toxicity is dose related, and temporary cessation of therapy usually results in restoration of bone-marrow function. Early common non-specific side-effects which often abate include headache, anorexia and nausea.

Unfortunately, prolonged use of zidovudine (i.e. for 6 months or longer) is associated with the emergence of less susceptible strains of HIV. Apparently these arise from mutations in the gene coding for reverse transcriptase leading to reduced binding of zidovudine triphosphate. This phenomenon has significant implications concerning the possible long-term use of AZT in asymptomatic AIDS patients.

The other later anti-HIV deoxynucleosides act in a similar way to zidovudine, but have varying toxicity profiles (see Table 7.2). Resistance may arise, although cross resistance between these drugs and zidovudine is uncommon. The differing properties of these drugs have led to the use of treatment regimens formulated to reduce the emergence of resistant strains or intolerable side-effects. These have met with some success.

There are also a number of non-nucleoside reverse transcriptase inhibitors (NNRTIs) which have limited availability but have been used as part of combination therapy in clinical trials. These include efavirenz, delavirdine, loviride and nevirapine, which are all taken by mouth.

Protease inhibitors

During the drive to develop more potent drugs that target other stages in the HIV virus life cycle, a new class of compounds that inhibit HIV-1 protease activity has been discovered. These protease inhibitors appear to suppress HIV-1 replication to a far greater extent than was previously considered possible. Four such agents, namely saquinavir mesylate, ritonavir, nelfinavir and indinavir sulphate, are currently available and undergoing clinical investigation, while others appear to be imminent (Anon., 1997a; McDonald and Kuritzkes, 1997; Flexner, 1998; Moyle *et al.*, 1998).

Protease inhibitors inhibit cytochrome P_{450} enzyme systems and can potentially interact with other similarly handled drugs (Anon., 1997a; McDonald and Kuritzkes, 1997; Moyle *et al.*, 1998). This is especially true of sanquinavir, but may be used to advantage by combining it with other protease inhibitors such as indinavir. The sanquinavir tends to reduce the metabolism of indinavir, resulting in more consistent blood levels of the latter. In contrast to the RT inhibitors, protease inhibitors appear to be capable of reducing the production of infectious virus particles from chronically infected cells. As these may constitute a major reservoir of virus within the body, this is an important advance. Combinations of various protease inhibitors with nucleoside and NNRTIs have been studied extensively in the laboratory. In all cases, the combinations have been at least additive and usually synergistic (Anon., 1997a; McDonald and Kuritzkes, 1997; Moyle *et al.*, 1998). It is this laboratory data which has prompted clinical studies into the use of combined therapy for the treatment and prophylaxis of HIV infections.

Treatment strategies

While North American and some European authorities favour a triple combination of two nucleoside analogues (e.g. AZT and 3Tc) plus a protease inhibitor (e.g. indinavir)

Table 7.2 Some characteristics of anti-HIV drugs[a]

Drug	Dosing schedule	Bone-marrow toxicity	Pancreatitis	Peripheral neuropathy	Hypo-uricaemia	Gastrointestinal disturbance	Paraesthesiae (mouth, hands, feet)	Elevated liver enzymes	Kidney stones	Rash	Lipodystrophy	Comments
Nucleoside reverse transcriptase inhibitors (NRTIs)												
Zidovudine (AZT)	2–5 times daily	++[b]										Best in cells actively producing virus
Didanosine (ddI)	Twice daily		+	+	+	++						–
Zalcitabine (ddC)	Three times daily		+	++		++						Active on 'resting' infected cells
Stavudine (d4T)	Twice daily	+	+	++				+				Best in cells actively producing virus
Lamivudine (3TC)	Twice daily		+/–	+/–		+						Headache, fatigue
Non-nucleoside reverse transcriptase inhibitors (NNRTIs)												
Nevirapine	Once daily initially, then twice daily									+[c]		In combination with NRTIs
Delavirdine	Three times daily									+[c]		Raises levels of indinavir and sanquinavir in blood
Protease inhibitors (PIs)												
Ritonavir	Twice daily with food					++	++	+			+	First PI to show increased survival in advanced disease
Indinavir	Three times daily around a meal					++		+	++		++	Good control of HIV when combined with NRTIs
Sanquinavir	Three times daily					++					+	Weakest PI
Nelfinavir	Three times daily with food					++					+	Combined with NRTIs

[a]Adapted from Anon. (1997a).
[b]Dose dependent.
[c]Self-limiting.

for the treatment of HIV infection, many European authorities favour a combination of one or two nucleoside analogues (e.g. AZT and ddI) plus an NNRTI (e.g. nevirapine) as initial therapy when the viral load is low (less than 50 000 copies), saving protease inhibitors for those patients with more aggressive and pronounced disease. Unfortunately, the development of a rash and the rapid emergence of resistance (single mutation) make NNRTIs less robust than protease inhibitors. However, whatever the combined therapy, the aim is to reduce the viral load to undetectable levels, with a parallel rise in CD_4 numbers.

There is no doubt that all protease inhibitors are extremely potent inhibitors of viral replication, and in combination with two nucleoside analogues produce profound and sustained inhibition of viral replication. The major disadvantage of the present protease inhibitors, which are all peptide mimetics, is their individual and cross-class toxicities. Indinavir is associated with renal stones (of indinavir crystals) in at least 4% of patients; ritonavir is associated with considerable gastrointestinal side-effects, particularly diarrhoea and an unpleasant taste in the mouth; sanquinavir is linked with poor bioavailability (possibly improved with new soft gelatine capsule formulations), and its potent inhibition of a number of cytochrome P_{450} enzymes, resulting in significant drug interactions (e.g. with NNRTIs and azoles); and nelfinavir is associated with diarrhoea (see Table 7.2).

The present protease inhibitors also have a complex effect on intermediary metabolism, with a small incidence of diabetes and an almost invariable rise in the mean cholesterol and triglyceride levels, which may be serious in a small number of patients. In addition, some individuals have developed an increase in visceral fat, with a loss of fat from the face and limbs. This lipodystrophy may be associated with other more serious metabolic disturbances. Whether or not lipodystrophy is a consequence of protease inhibition *per se*, or whether it is an effect of any effective antiretroviral agent, is not known. Another problem with combination therapeutic regimens is the sheer number of tablets that have to be taken daily, and the complexity of the dosing regimen. In this respect, the use of NNRTIs has some advantages.

It is not the purpose of this text to elaborate more fully on the rationale behind, and use of, combined anti-HIV treatment. This can be found in a number of publications (e.g. Feinberg *et al.*, 1998). As far as surgeons are concerned, it is the prophylaxis of accidental needle-stick or similar injuries (e.g. lacerations, mucosal or conjunctival splash, contamination of damaged skin) that is of most interest. The risk of HIV transmission following needle-stick injury involving contaminated blood is about 0.4%, and it is around 0.1% following mucosal exposure. Transmission via intact skin is not documented. Factors which increase the risk include deep injury, visible blood on the injuring device, a procedure involving a device being placed directly into a blood vessel, and a terminally ill source patient (Anon., 1997a). In the past, zidovudine prophylaxis (1000 mg daily for 3 to 4 weeks following exposure) has been shown to reduce the transmission rate by about 80%. Prophylactic guidelines in the USA and elsewhere now recommend combination prophylaxis based on zidovudine (200 mg three times daily) plus lamivudine (150 mg twice daily) and possibly indinavir (800 mg three times daily), all for 4 weeks (i.e. two reverse transcriptase inhibitors and one protease inhibitor) (Anon., 1997a; Cloeren and Perl, 1998; Masterton, 1999). Ideally, treatment should commence within 1–2 h of exposure. In cases where the source patient's HIV status is not known, decisions as to whether or not to start prophylaxis depend on estimation of the likely HIV-exposure risk.

Although treatment of HIV infection is essentially medical, some patients require surgical procedures, especially for gastrointestinal problems. Surgeons may be required for the diagnosis of disease (e.g. tissue biopsy) or its complications (e.g. drainage of abscesses), in addition to normal surgical procedures and vascular access. Operative morbidity and mortality following emergency surgery are significantly higher in AIDS patients than in those not infected with HIV (Francis, 1997).

Opportunistic infections in HIV-infected patients

Infection of the oral cavity and oesophagus with *Candida*, herpes simplex virus (HSV) and cytomegalovirus (CMV) is common in HIV patients. Such infection can be controlled, but not cured in most patients, by the use of appropriate long-term suppressive antimicrobial therapy (e.g. fluconazole for *Candida*; see p. 134). Problems concerning the treatment of refractory mucosal candidosis in patients with HIV infection have recently been reviewed (Fichtenbaum and Powderly, 1998). Bleeding from the gastrointestinal tract in HIV patients may be due to Kaposi's lesions, in addition to those causes that are seen in non-HIV-infected patients. Kaposi's lesions may also be responsible for gastric and intestinal obstruction. Chronic diarrhoea and abdominal pain are frequent in HIV patients with or without AIDS. A variety of microbes that are reliant on an adequate cell-mediated immune response for their control may be involved (e.g. viruses such as CMV, bacteria such as salmonellae or *Mycobacterium avium-intracellulare* complex and protozoa such as *Cryptosporidium*). Appendicitis may be difficult to diagnose and is more common in HIV patients. Anorectal problems are common among HIV-infected homosexuals and AIDS patients, and include perianal warts which may recur after diathermy, anorectal ulcers associated with sexually transmitted pathogens (e.g. *Treponema pallidum*, gonococci), perirectal abscesses and malignancies. Operations in the anorectal region must be undertaken judiciously because of the risk of subsequent infection, and may require a diverting colostomy.

Pulmonary disorders are common in AIDS patients (e.g. *Pneumocystis carinii* pneumonia). Invasive procedures (e.g. bronchoscopy, lavage, transbronchial biopsy) carry with them a significant risk from bleeding and transmission of the virus. Tracheostomy is only rarely indicated because of the risk of overwhelming sepsis and difficulties in preventing HIV contamination of the immediate environment (Francis, 1997).

HEPATITIS B AND HEPATITIS C VIRUS INFECTIONS

Chronic hepatitis B and C infections may benefit from treatment with interferon. This should be given under specialist supervision. One suggested regimen has been 3 million units of interferon α-2b given three times a week for 18 months (see Poynard *et al.*, 1995).

All surgeons and allied health-care workers should be protected from hepatitis B infection by vaccination. In the non-immune individual, the risk of seroconversion is high following percutaneous or mucosal exposure (Cloeren and Perl, 1998). Up to 23% of individuals who are exposed to blood from a hepatitis B early antigen (HBeAg) positive source may seroconvert following a hollow-needle injury. Prophylaxis is not required if the exposed person is surface antigen positive (HBsAg positive) or immune (antibody to HBsAg positive). Immunity to hepatitis B should not be assumed – vaccination does not guarantee immunity. If the exposed person is surface antigen–antibody negative (anti-HBs negative) and surface antigen negative (HBsAg negative), hepatitis B vaccine should be given (i.e. initiate a course, complete the course or give a booster). If the donor (source) blood is available for testing and is infectious (i.e. HBsAg positive), hepatitis B immune globulin should be given immediately or at least within 7 days. If the source blood is of unknown status, the use of immune globulin is controversial and depends on the perceived risk of the exposure. Follow-up serology should be obtained after 6, 16, 26 and 52 weeks in exposed individuals at risk of infection.

Blood containing antibodies to hepatitis C virus is assumed to be infectious. The risk of seroconversion following a hollow needle-stick injury with blood from an antibody-positive source is up to 10% (Cloeren and Perl, 1998). While immunoglobulin is not of established value, some authorities advocate its use at 0.2 mg/kg intramuscularly following exposure and repeated after 1 month. Follow-up serology should be obtained after 6, 12, 26 and 52 weeks.

ANTIPARASITIC AGENTS

TREATMENT OF HYDATID DISEASE

Of the parasitic infections of significance to surgeons, hydatid disease is perhaps the most important. There are two classical hydatid tapeworms, namely *Echinococcus granulosus* and the closely related *Echinococcus multilocularis*. A third parasite, *Echinococcus vogeli*, has been reported from northern areas of South America (Wen *et al.*, 1993). These parasites are unusual in that humans are the intermediate host, harbouring the larval form in hydatid cysts which occur in viscera, usually the liver. Cysts develop following ingestion (or possibly inhalation) of eggs (ova) shed by the adult tapeworm. These live in the intestines of the definitive (final) host, which is always a carnivore. Fragments (proglottids) of the tapeworm containing mature eggs are passed out in the faeces of the infected animal.

Hydatid disease occurs in many countries, and mainly involves a sheep and dog cycle. However, these hosts may vary geographically (e.g. a caribou deer and fox cycle is found in regions of Canada). It appears to be a zoonosis of increasing and largely unrecognized importance (Jenkins, 1998). Humans can only become infected from eggs shed by the definitive host. The definitive carnivore hosts acquire infections by eating liver and/or lungs (offal) of infected intermediate hosts. These organs contain cysts, and protoscolices within the cysts develop into adult tapeworms in the carnivore's intestines.

It has generally been accepted that there is really no effective chemotherapy for hydatid disease, although claims have been made for success with benzimidazoles, including mebendazole, albendazole and oxfendazole (e.g. albendazole 5 mg/kg 12-hourly for 3 months with non-calcified liver cysts). Reported cure rates have been in the range 50–70% (Gil-Grande *et al.*, 1993; Wen *et al.*, 1993; Salama *et al.*, 1995). More recent studies in experimental animals have suggested that oxfendazole is at least as effective and easier to administer than albendazole (Blanton *et al.*, 1998). Combinations of a benzimidazole with cimetidine have been suggested as meriting further clinical evaluation (Wen *et al.*, 1993). Praziquantel may also be of benefit (Wen *et al.*, 1993). Such observations led to the opinion in the mid-1990s that chemotherapy remained only an adjunct to surgical removal, which was not without risk from the spillage of viable protoscolices into the peritoneal cavity or other tissues, or as an alternative in the initial treatment of uncomplicated liver cysts (Gil-Grande *et al.*, 1993; Wen *et al.*, 1993). However, in 1999 it was suggested by two surgeons in New Zealand that surgery (with albendazole pretreatment) should be reserved for patients with either curable disease or complications, and that all others should initially, at least, be managed by albendazole alone (see Lynch and Stubbs, 1999 and Figure 7.2). Medical treatment with albendazole would now appear to be the treatment of choice in patients with uncomplicated disseminated disease. Once complications develop, surgery is usually required – this may necessitate multiple operations with an increasing morbidity risk.

Echo-guided percutaneous cyst puncture has been used successfully in the non-operative treatment (and diagnosis) of hepatic cysts. Injection of a scolicidal agent

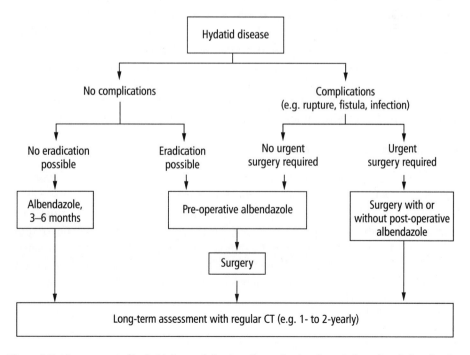

Figure 7.2 Management of hydatid disease following diagnosis, showing relative roles of albendazole chemotherapy and surgery. Adapted from Lynch and Stubbs (1999).

(e.g. hypertonic saline) before withdrawal of the needle used for aspiration has reputedly given excellent results with no complications over a 3-year follow-up (Salama et al., 1995). Others consider such practice to be ill-advised (Lynch and Stubbs, 1999), although it can be considered for patients who are thought to be inoperable or who decline surgery.

Human infestation with *Echinococcus multilocularis* is much more aggressive than that seen with *Echinococcus granulosus*, and until recently was almost uniformly fatal. It seems to be limited geographically to Alaska, Central Europe and Japan (possibly due to the habitat of intermediate hosts). Radiologically it can be confused with hepatoma. Wherever possible, treatment is by liver resection with albendazole cover. The latter drug appears to be at least parasitostatic, and should probably be continued indefinitely to prevent recurrence.

TREATMENT OF *TAENIA* INFECTION (NEUROCYSTICERCOSIS)

Infestation of the human central nervous system with tissue cysts of the pig tapeworm, *Taenia solium*, is probably the most common parasitic disease (neurocysticercosis) of the human nervous system. It appears to be common not only in developing countries, but also in the USA, especially in areas with large immigrant populations from Latin America. Estimates suggest that there are at least 1000 new cases per year in the USA. Although most of these occur in patients infected in developing countries, the number of locally acquired infections is increasing.

Reports on the use of antiparasitic agents in neurocysticercosis first appeared in the 1970s. Praziquantel was one of the first drugs reported to be active against cysticercosis, and has been used at doses of up to 100 mg/kg daily. In recent trials, albendazole has been the favoured drug, and it was approved by the US Food and Drug Administration in 1996 for the treatment of parenchymal cysticercosis in patients found to have non-enhancing cysts. In patients with cysticercosis the usual dosage of albendazole is 15 mg/kg daily in two or three divided doses for 8 to 30 days. Unlike praziquantel, albendazole can be given in combination with steroids and anticonvulsants without affecting its serum level.

MISCELLANEOUS PARASITIC INFECTIONS

Other parasites of some significance to surgeons are the protozoans *Entamoeba histolytica* and *Toxoplasma gondii*, the fungus/protozoan *Pneumocystis carinii*, and the liver flukes *Fasciola hepatica* and *Clonorchis sinensis*. In most cases treatment is with antiparasitic drugs (see Table 7.3), although flukes may require operation to clear the biliary tree. Complications such as cholangitis and pancreatitis which may accompany *Clonorchis* infection are treated by standard surgical and endoscopic techniques. Percutaneous drainage is seldom required for amoebic liver abscesses (Morris, 1997).

Table 7.3 Chemotherapeutic options for some parasitic infestations

Parasite	Treatment
Echinococcus granulosus (hydatid)	Albendazole 5 mg/kg 12-hourly for 3 months is an alternative to surgery in the case of non-calcified cysts
Taenia solium (neurocysticercosis)	Albendazole 5 mg/kg 8-hourly for 2–4 weeks *or* praziquantel 15 mg/kg 8-hourly for 2 or more weeks
Entamoeba histolytica (protozoan)	Metronidazole 400–800 mg orally for 10 days (children 35–50 mg/kg/day)[b]
Pneumocystis carinii[a]	Cotrimoxazole 4 tablets or 4 ampoules 6-hourly (15–20 mg trimethoprim/75–100 mg sulphamethoxazole/kg/day) for 21 days, then suppressive/prophylaxis doses (2 tablets 3 times a week)
Toxoplasma gondii (protozoan)	Pyrimethamine 100-200 mg (2 mg/kg) orally daily for 2 days, then 25-50 mg orally daily *plus* sulphadiazine 1–2 g (35 mg/kg) orally 6-hourly *plus* folinic acid 10–15 mg/day, all for 3–4 weeks
Clonorchis sinensis (liver fluke)	Praziquantel 25 mg/kg orally 8-hourly for 1 day
Fasciola hepatica (liver fluke)	Bithionol 20 mg/kg orally 12-hourly on alternate days for 10 to 15 doses

[a]Considered by most to be more closely related to fungi than to the protozoa.
[b]Tinidazole 2 g orally once daily for 3 days or 1 g twice daily for 3 days (children 50–60 mg/kg/day), and ornidazole 1.5 g orally once daily for 3 days or 1 g twice daily for 3 days (children 40 mg/kg/day) are alternatives.

chapter 8

LABORATORY ASPECTS OF ANTIMICROBIAL USE

The microbiology laboratory is available to assist clinicians in a variety of ways, including the isolation and identification of microbes associated with infectious diseases, sensitivity testing of significant isolates against an appropriate range of antimicrobial agents and, where appropriate, the monitoring of antibiotic blood (serum) levels to ensure that these are in the therapeutic range and/or that levels known to be associated with toxicity are not occurring. The latter is especially significant in the case of patients receiving prolonged courses of aminoglycosides. Two of these laboratory aspects, namely sensitivity testing and blood level monitoring, will be discussed in more detail.

ANTIMICROBIAL SENSITIVITY TESTING

Determining the antibiotic sensitivity of an isolate is important if the incidence of potentially resistant strains is high or increasing, if potentially life-threatening illnesses are involved, or if it is important to know accurately the susceptibility (expressed in mg/L of antibiotic) of an isolate to particular antibiotics. Although the laboratory conditions employed for sensitivity testing are far removed from those which occur in the body, there does seem to be a reasonable correlation between the results generated and subsequent clinical outcome.

DISC TESTS

Sensitivity of an isolate to a range of antibiotics (usually 4–6) is generally determined by placing discs containing set concentrations of the antibiotic on plates containing a 'lawn' of the isolate to be tested. Strict criteria apply to the composition of the agar medium, the incubation conditions and the concentration of the bacterial inoculum. After incubation, the diameter of the inhibition zone produced around the disc is measured and compared with international interpretive standards, or with zones produced with a known susceptible control bacterium. This allows designation of the isolate as susceptible, intermediate (moderately susceptible) or resistant to each of the antibiotics being tested (see below). Such tests are strictly non-quantitative and fail to record the level of susceptibility or resistance of isolates.

DILUTION TECHNIQUES

If sensitivities need to be quantified, tubes or wells containing a series of doubling dilutions (e.g. 128 mg/L down to 0.03 mg/L) of the antibiotic in an appropriate liquid medium are inoculated with a standard suspension of the isolate to be tested. After incubation, the presence or absence of growth is determined by examining for cloudiness, which provides an indication of growth and resistance to that concentration of antibiotic. The minimum inhibitory concentration (MIC) of the antibiotic being tested is the level of drug in the last tube or well that shows no visible growth. If required, the minimum bactericidal concentration (MBC) can then be determined by inoculating a small amount of liquid from each of the clear tubes or wells on to an agar plate. If the bacteria have been inhibited but not killed, growth will appear on these agar plates after suitable incubation, while cidal activity is reflected by no growth. Nowadays, most therapeutic regimens are based on MIC data.

Once the MIC in mg/L (μg/mL) is known, consultation of or reference to internationally available and accepted tables permits the isolate to be recorded as susceptible, intermediate or resistant to the antibiotic being tested (see below).

E-TESTS

A simplified quantitative MIC test has recently become available for routine laboratory use. This is the E-test, which is performed in a similar manner to the disc test, but utilizes a plastic strip containing a linear gradient of antibiotic, rather than a disc containing a single antibiotic concentration. By observing where bacterial growth bisects the strip, an accurate estimation of the MIC can be obtained (see Figure 8.1). This technique is much quicker and simpler than the dilution method, and the results appear to correlate well with those obtained using more labour-intensive dilution techniques.

OTHER PROCEDURES

Sensitivities can also be determined by incorporating set concentrations of an antibiotic into an appropriate agar medium and inoculating (usually with a multi-point inoculating device) bacterial suspensions on to the agar surface. After incubation, susceptibility or resistance is measured by the absence or presence of growth, respectively. This is a 'break-point' technique and allows the determination of susceptibility to a given concentration of antibiotic.

Automated techniques, where the presence or absence of bacterial growth is determined by techniques such as light scattering or changes in electrical impedance in liquid media containing known concentrations of the antibiotic, are now being employed for the rapid determination (i.e. within 4–6 h) of bacterial sensitivities. These are sophisticated modifications of the break-point technique mentioned above.

INTERPRETATION OF LABORATORY REPORTS

In laboratory reports that are prepared for clinicians, sensitivity results for isolates regarded by the laboratory staff as being potentially significant are simply listed as susceptible, intermediate or resistant to the drug in question. Only in exceptional circumstances will actual MIC figures (in mg/L) be quoted. 'Susceptible' implies

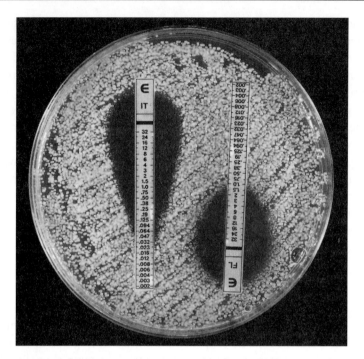

Figure 8.1 E-test sensitivity tests on the yeast *Candida albicans*. The MIC for fluconazole (FL) is 1.5 mg/L, and for itraconazole (IT) it is 0.64 mg/L.

that the infection should respond to normal doses/regimens of that particular antibiotic. 'Intermediate' means that doses above those normally given are probably required for a satisfactory outcome, or that a favourable response can only be expected if infection occurs in an area of the body where natural concentration of the drug occurs (e.g. in the bladder/urine). 'Resistant' implies that the microbe will be unaffected by any safely achievable drug levels. In practice, many laboratories only report susceptible and resistant categories. Antibiotics that have been demonstrated to be ineffective (resistant) should be withdrawn as soon as possible, and replaced by those listed as potentially effective (susceptible), in order to reduce the risks of morbidity and mortality.

In general, sensitivity testing presents few problems for laboratory staff in terms of procedure and interpretation, although detection of potential resistance in some microbes (e.g. methicillin resistance in *Staphylococcus aureus*) relies heavily on the use of appropriate techniques. Important implications may be drawn from sensitivity reports (e.g. methicillin resistance in *Staphylococcus aureus* implies that the microbe is resistant to all β-lactams).

While 100% of cells of some methicillin-resistant *Staphylococcus aureus* reveal the resistance phenotype under normal sensitivity-testing conditions (e.g. 35°C for 18 h on plain Mueller-Hinton agar), some strains appear to be heterogenous in that only a very low percentage of cells display the resistance phenotype under these conditions. With such strains, incubation at lower temperatures (e.g. ≤ 30°C) and/or for extended time periods (e.g. 24–48 h) on media containing additional salt

(2% NaCl), and the use of a heavier than normal inoculum density, is often required to detect resistance reliably. Demonstration of the *mec*A gene using recently developed molecular techniques is clearly now the most appropriate way of confirming the methicillin-resistance potential of isolates, but is possibly beyond the scope of many laboratories.

ESBL-elaborating Gram-negative bacilli may be incorrectly recorded as susceptible to third-generation cephalosporins by some automated machines. Their detection relies on the demonstration of a significant reduction in cephalosporin MIC following the addition of a β-lactamase inhibitor such as clavulanic acid or, less conclusively, on the demonstration of synergy between discs containing a third-generation cephalosporin and a β-lactamase inhibitor such as clavulanic acid (see Figure 8.2). Whether or not it is clinically necessary or cost-effective to test all Gram-negative bacilli routinely for possible ESBL production, is open to debate (Emery and Weymouth, 1997).

In addition to the tests mentioned above, rapid tests are available for detecting β-lactamase activity in bacteria (e.g. the chromogenic cephalosporin assay). These are performed by inoculating a heavy suspension of the isolate on to moistened filter paper containing the β-lactam drug and an indicator which changes colour under acidic conditions. If the bacterium being studied produces the enzyme, it will hydrolyse the β-lactam ring to form acid residues which will change the colour

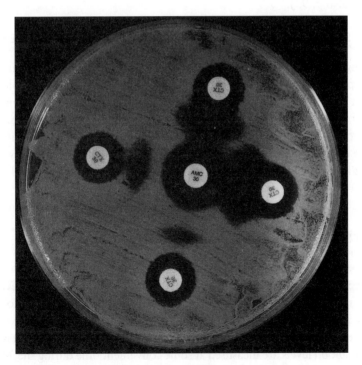

Figure 8.2 Enhancement of ceftazidime (CTX) susceptibility by the clavulanic-acid component of a co-amoxyclav (AMC) disc. The strain of *Klebsiella pneumoniae* under test is producing a clavulanate-susceptible ESBL.

of the indicator within seconds or minutes. The clinical purpose of these assays is to provide a rapid indication of β-lactamase activity, which may not always be apparent with more conventional sensitivity tests.

SERUM ASSAYS ON PATIENTS DURING ANTIBIOTIC THERAPY

In seriously ill patients who are not responding to an antibiotic dosing regimen that would be expected to be adequate, it is often necessary to monitor blood levels to show that the therapy being used is achieving a suitably 'high' antibiotic blood level. With bacterial endocarditis (one of the few diseases that has been thoroughly studied), a satisfactory clinical response is known to be associated with maintaining serum antibiotic levels of at least 8 times the MBC at all times. Some debate currently exists concerning the need and/or usefulness of serum assays, especially in patients who respond well to therapy. Tests known as SBAs (serum bactericidal assays) are usually carried out on blood specimens obtained immediately before the next antibiotic dose (i.e. when levels will be at their lowest).

All patients who are receiving aminoglycosides *must* have their blood levels determined at regular intervals. The methodology and interpretation of the results vary depending on the dosing regimen being used. A brief summary is presented below (see also p. 95).

PATIENTS ON 8-HOURLY GENTAMICIN DOSING SCHEDULES

Monitoring should commence 24–36 h after initiation of treatment, and should then be repeated every 3 to 4 days, or at shorter time intervals in unstable patients who are at risk of deteriorating renal function. In practice, blood is taken immediately before and 60 min after administration of the next dose. These represent the pre-dose or trough, and post-dose or peak samples. The former is used to determine tissue accumulation and potential nephrotoxicity (associated with levels ≥ 2.0 mg/L), and the latter is used to give some idea of whether the therapeutic range (5–10 mg/L) is being achieved. Aminoglycoside failure is often associated with inadequate peak serum concentrations (i.e. less than 5 mg/L for gentamicin).

PATIENTS ON ONCE DAILY (24-H) GENTAMICIN DOSING SCHEDULES

Patients receive about 5 mg/kg of gentamicin as a single 30-min infusion every 24 h. Peak serum levels are clearly well above minimum therapeutic levels and do not need to be measured. However, to minimize potential nephro- and ototoxicity, it is important to maintain the 24-h trough levels below 0.5 mg/L. As this level is difficult to measure, blood levels are measured at some convenient time during the 24-h cycle, and by use of a computer or other methodologies, the blood level that is likely to be present after 24 h is calculated. Alternatively, blood levels at 18 h can be measured. These blood levels should be below 1 mg/L. In cases where trough levels are seen to be exceeding the safety level, increasing the time interval between doses, rather than lowering the dose, is recommended. Regular monitoring, especially in cases where renal function is poor or deteriorating, is mandatory. Although these dosing schedules are clearly less toxic than multiple daily dosing schedules, unacceptably high levels of drug accumulation and related toxicity can occur with once daily routines.

Note that the peak and trough concentrations cited above apply to gentamicin, tobramycin and netilmicin, but not to amikacin (see Table 4.8).

PART THREE

SURGICAL INFECTIONS

chapter 9

NEUROSURGICAL INFECTIONS

BRAIN ABSCESS

GENERAL

Descriptions of focal suppurative lesions within the brain date back to the Edwin Smith Papyrus (c. 1500 BC). While trephining may have been of value, it was not until 1893 that MacEwen reported encouraging results with surgical intervention. He documented a remarkable 8 out of 10 survival rate after neurosurgical drainage of temporal lobe abscesses (see Wispelwey and Scheld, 1995). With improvements and advances in surgical and imaging techniques and the availability of new antimicrobial agents over the last decade or so, a significant improvement in mortality and morbidity from brain abscesses has occurred. It has been estimated that brain abscesses account for approximately 1 in 10 000 general hospital admissions, although the global AIDS epidemic has led to increased numbers of patients with focal intracranial infections.

Significant epidemiological features which are apparent include the dominance of the male sex, and the importance of otitis media, mastoiditis, paranasal sinusitis and dental extractions as sources of the causal microbes. About 25% of all cases occur in children, with a peak around 4–7 years of age (see Wispelwey and Scheld, 1995). Chronic rather than acute middle ear infections are more likely to lead to intracranial extension. A biomodal age distribution (paediatric and those over 40 years old) is often reported.

About 50% of brain abscesses appear to arise directly from contiguous infectious foci, with haematogenous spread from a more distant focus (e.g. chest) accounting for around 25% of cases. Other reported associations include trauma, although in about 15% of cases no predisposing factor can be identified. Not surprisingly (in view of the significance of anatomically related infectious foci), abscesses most often involve the frontal and temporal regions, followed in order of frequency by the frontoparietal, parietal, cerebellar and occipital regions. Intrasellar abscesses, often in association with sphenoid sinusitis, and brainstem abscesses arising by blood-borne spread from a distant focus, are other sites. While early surveys suggested that multiple brain abscesses were uncommon, computerized tomography (CT) scanning suggests that these occur in up to 50% of patients (Wispelwey and Scheld, 1995).

In the pre-antibiotic era, culture of intracranial pus revealed *Staphylococcus aureus* in about 30% of cases, streptococci in around 30% and coliforms in 12%.

Microbes were only recovered from about 50% of the cases studied (see Wispelwey and Scheld, 1995). However, with the introduction of anaerobic technology in the 1960s, the significant and dominant role of obligate anaerobes, particularly those of otitic origin, soon became apparent. A general trend in recent years has been a decrease in the frequency of staphylococci as causal agents and an increase in the significance of Gram-negative coliforms.

MICROBIAL AETIOLOGY

Pyogenic brain abscesses are often (30–60% of cases) polymicrobial. The more commonly involved microbes are listed in Table 9.1, and more complete lists are available in the excellent review by Wispelwey and Scheld (1995). If the usual microbiological culture techniques fail to reveal a causative microbe, molecular procedures can be contemplated (Logan et al., 1999). *Staphylococcus aureus* can be recovered from up to 15% of cases (usually in pure culture), and is the most common pathogen associated with trauma. Common Gram-negative bacilli include *Proteus* species, *E. coli* and pseudomonads. Streptococci have been implicated in over 50% of brain abscesses, and usually belong to the *Streptococcus anginosus* ('*Streptococcus milleri*') group. Of the obligate anaerobes, *Bacteroides* species, including *Bacteroides fragilis*, and *Prevotella* species dominate. They arise from both haematogenous spread from distant tissues (e.g. intestines) and local spread from the oral cavity, and they are usually present in mixed culture. Apart from bacteria, yeast and dimorphic fungi have assumed an increasingly important role (up to 17% of cases in some surveys). Various protozoa, roundworms (helminths) and tapeworms may also cause brain abscesses.

Of growing concern world-wide is the increasing incidence of focal CNS lesions in patients with AIDS. Single or multiple abscesses may be seen with toxoplasmosis (*Toxoplasma gondii*), and are difficult to distinguish from pyogenic lesions by CT.

Table 9.1 Microbes that are more commonly isolated from brain abscesses[a]

Microbe	Relative frequency	Comment
Streptococci (e.g. *Streptococcus anginosus* (*milleri*))	++++	Frontal lobe (possibly sinusitis); often monomicrobial
Obligate anaerobes (e.g. *Bacteroides*, *Prevotella* species)	+++	From oral cavity and intestines; usually polymicrobial
Gram-negative bacilli (e.g. *E. coli*, *Proteus*, pseudomonads)	++	Chronic otitic infections; polymicrobial
Staphylococcus aureus	+	Common following trauma
Fungi (e.g. aspergilli, mucoraceous fungi, dematiaceous fungi)	+	Immunocompromised; trauma

[a]Adapted from Wispelwey and Scheld (1995).

Other less common infectious causal agents in this patient population include various fungi, protozoa, and bacteria such as mycobacteria, *Nocardia asteroides*, *Listeria monocytogenes*, salmonellae and pneumococci.

Location of lesions within the brain, or arising in association with particular underlying pathology, may give some clue as to the most likely causation. For example, post-traumatic abscesses are usually staphylococcal, frontal lobe abscess in association with sinusitis is often caused by the *Streptococcus anginosus* group of microbes (usually in pure culture) and abscesses arising from chronic otitis media are virtually always polymicrobial, with the more dominant isolates being combinations of streptococci, *Bacteroides* species and Gram-negative enterics such as *Proteus* (see Wispelwey and Scheld, 1995).

TREATMENT STRATEGIES

Although some patients with brain abscesses respond to prolonged medical therapy alone, most require surgery for optimal management. This includes aspiration after burr-hole placement, or complete excision after craniotomy. With regard to antimicrobial regimens, these are commonly empirical and reflect the known aetiologies mentioned above. Some antibiotics enter and remain active in brain abscesses more readily than others. These include fusidic acid, rifampicin, cotrimoxazole, chloramphenicol, metronidazole and possibly vancomycin. High doses of others (e.g. penicillin, clindamycin) may result in significant CNS levels, although whether they are retained for any length of time is debatable. Little information is available about the third-generation cephalosporins (e.g. ceftriaxone), although they do achieve excellent levels in the CSF. High-dose cefotaxime (3 g q 8 h) has been shown to be effective in a small trial, suggesting its suitability for infections involving a number of Gram-positive and Gram-negative microbes. Imipenem is contraindicated because of the increased risk of seizures, although the related meropenem may be more acceptable. Ampicillin/sulbactam has been used with excellent results, which suggests a potential role for this and the related amoxycillin/clavulanic acid combination (co-amoxyclav).

Because of the important role of streptococci such as *Streptococcus anginosus* and penicillin-sensitive oral anaerobes, it has been suggested that penicillin should be included in all regimens where brain abscesses appear to have arisen from contiguous foci and pyogenic lung lesions. However, a better alternative may be a cephalosporin such as cefuroxime or ceftriaxone. These have excellent streptococcal activity as well as broad-spectrum Gram-negative bacillus (especially ceftriaxone) and *Staphylococcus aureus* (especially cefuroxime) activity.

For decades chloramphenicol has been included in regimens because of its significant anaerobe activity and lipid solubility, resulting in high concentrations of the drug in brain tissue. However, it seems that metronidazole is more ideally suited as an anti-anaerobe drug for brain lesions. It shows excellent and sustained concentrations in brain abscess pus, even with concomitant steroid use. An anti-anaerobe agent such as metronidazole should be included in the treatment of abscesses complicating otitis media, mastoiditis or pyogenic lung disease.

If staphylococci are suspected (e.g. in abscesses following head trauma) or are recovered in culture, a penicillinase-stable penicillin or cephalosporin is indicated. The possibilities include flucloxacillin, nafcillin and cefuroxime. Vancomycin, or possibly fusidic acid, depending on local sensitivity patterns, is substituted when MRSA are involved. Fusidic acid is probably underestimated as an anti-staphylococcal agent in the treatment of brain abscesses.

To cover Gram-negative enterics, a third-generation cephalosporin or the lipid-soluble cotrimoxazole is added to empirical regimens pending laboratory results. As pseudomonads are often found in abscesses complicating chronic middle ear disease, the choice of the antipseudomonal ceftazidime as the third-generation cephalosporin is logical. As yet, the role of ciprofloxacin in the chemotherapy of brain abscesses is largely unknown.

Taking all of the above considerations into account, empirical antimicrobial therapy for brain abscesses should centre around combinations such as metronidazole (anaerobes), ceftriaxone (streptococci, Gram-negative bacilli excluding pseudomonads) and fusidic acid (staphylococci), although several other suitable combinations clearly exist. More specific chemotherapy should be introduced following receipt of the laboratory results.

EPIDURAL ABSCESS

Epidural abscess represents localized infection between the outermost layers of the meninges, the dura mater, and the overlying skull or vertebral column. Intracranial epidural abscesses form by stripping periosteum from bone, most often adjacent to frontal sinuses. These sharply confined lesions are invariably accompanied by focal osteomyelitis. Abscess may also follow traumatic head injury. Causative microbes include *Staphylococcus aureus*, streptococci, Gram-negative bacilli and obligate anaerobes. In most cases therapy consists of appropriate antimicrobial agents, and emergent surgical drainage to stop the development of subdural empyema. Where appropriate, additional surgical therapy of any sinusitis, otitis or orbital infection or osteomyelitis may also be required. Considerations for empirical antimicrobial therapy are similar to those for brain abscesses (see p. 171).

Spinal epidural abscesses often spread to occupy several (e.g. 2–9) vertebral segments, and within the narrow confines of the vertebral canal they may cause extensive cord compression and necrosis. Cervical and thoracic regions are commonly involved. These abscesses arise by a variety of means (e.g. by direct extension from vertebral osteomyelitis, by haematogenous spread from some other focus, or secondary to surgery or epidural anaesthesia). Occasionally, epidural abscesses arise as a complication of abdominal surgery. In up to 30% of cases there is a history of back trauma, and often a history of diabetes mellitus, intravenous drug abuse or pregnancy. *Staphylococcus aureus* is the dominant causal microbe (60–90% of cases), and may be the only microbe recovered. Of the other microbes which have been isolated, streptococci and Gram-negative bacilli such as *E. coli* and *Pseudomonas aeruginosa* are most common. Atypical microbes may be associated with intravenous drug abuse. Tuberculosis cannot be ignored. The potential danger of spinal cord necrosis requires surgical drainage (usually by laminectomy) as soon as possible, with prior empirical antimicrobial cover. In view of the dominance of *Staphylococcus aureus* as a causal agent, a logical regimen is flucloxacillin or some other penicillinase-stable penicillin. If other microbes are involved, alternatives include ceftriaxone for streptococci and Gram-negative bacilli (excluding pseudomonads), and metronidazole for obligative anaerobes. It is possible that antimicrobial agents with superior pharmacokinetics to some of these should now be considered, e.g. fusidic acid or rifampicin (both plus flucloxacillin) or cefuroxime for staphylococci, and ciprofloxacin for Gram-negative bacilli, including pseudomonads. Treatment may have to be extended for up to 8 weeks if osteomyelitis is present.

INTRAMEDULLARY ABSCESS OF THE SPINAL CORD

Intramedullary abscess of the spinal cord (IASC) is a rare infection of the CNS, but has been associated with significant mortality and neurological morbidity (Chan and Gold, 1998). This suppurative infection has features remarkably similar to those seen in cases of pyogenic brain abscess, and the inflammatory process may extend to the meninges and into the CSF space.

In the pre-antibiotic era, about 50% of cases resulted from haematogenous spread of infection from an extraspinal focus (e.g. the lung), but nowadays this figure is less than 10%. Most of the recent cases appear to be cryptogenic in origin and to involve patients with structural abnormalities of the spinal cord and/or vertebral bodies. Important causative microbes are *Listeria monocytogenes*, viridans streptococci, *Haemophilus* species, Enterobacteriaceae and obligate anaerobes. Figures provided in a review covering the period 1977–1997 suggested that about 25% of cases resulted from contiguous spread of infection through a dermal sinus tract (Chan and Gold, 1998). In addition, lesions may result from post-surgical complications. Significant microbes associated with contiguous spread and post-surgical infections include staphylococci, enteric Gram-negative bacilli, *Pseudomonas aeruginosa* and obligate anaerobes such as *Bacteroides fragilis*.

Empirical antimicrobial therapy should be based on the presumed mechanism of infection and, where possible, the results of the Gram stain of an aspirate. Ampicillin (or amoxycillin) should be included in all apparently cryptogenic cases to cover *Listeria monocytogenes*. Other possible antimicrobial considerations include meropenem, piperacillin/tazobactam, or a combination of a cephalosporin (e.g. cefuroxime, ceftriaxone) or ciprofloxacin with metronidazole. The role of monotherapy with newer quinolones (e.g. clinafloxacin) in therapy is as yet unknown. A minimum of 4–6 weeks of parenteral therapy followed by 2–3 months of oral therapy should be considered (Chan and Gold, 1998).

In addition to the use of antimicrobial agents, surgical therapy is recommended for the treatment of IASC. Decompression/evacuation of the lesion should be urgently performed in an attempt to limit the neurological injury. Diagnostic and therapeutic drainage of the abscess may be accomplished through an open surgical approach (myelotomy), or by stereotatic needle aspiration of the lesions under CT guidance (see Chan and Gold, 1998).

SURGICAL SITE INFECTION FOLLOWING SPINAL INFUSION

Post-operative spine infection can cause significant morbidity, and may compromise the outcome of spinal surgery to correct scoliosis. Surgical management – including drainage and debridement – in conjunction with parenteral antimicrobial agents is the mainstay of therapy. Significant causative microbes include *E. coli* and related coliforms (e.g. *Proteus*, *Klebsiella*), *Staphylococcus aureus*, enterococci, and obligate anaerobes such as the *Bacteroides fragilis* group, peptostreptococci, propionibacteria, clostridia and veillonellae (see Brook and Frazier, 1999). Mixed aerobes/anaerobes occur in about 50% of lesions.

Treatment is therefore similar to that employed for intra-abdominal infections (e.g. monotherapy with meropenem or piperacillin/tazobactam, or a combination of ciprofloxacin plus metronidazole). Newer fluoroquinolones (e.g. clinafloxacin) may have a role in the treatment of this type of infection, although this possibility will be dependent on appropriate clinical trials.

chapter 10

INFECTIONS OF THE HEAD AND NECK

SINUSITIS

Acute sinusitis involving one or more of the paranasal sinuses may progress to more complicated intracranial infections such as meningitis, epidural and subdural abscess, and brain abscess. Although viruses may play an initiating role, the disease should be regarded as bacterial and treated accordingly. Common microbes are listed in Table 10.1. Antimicrobial therapy should be directed at the more significant pathogens (e.g. *Streptococcus pneumoniae, Haemophilus influenzae* and *Moraxella catarrhalis*). Possible regimens include one of co-amoxyclav, cefuroxime axetil, cefaclor or cotrimoxazole. With severe frontal or sphenoid sinus infections, or if there is a likelihood of intracranial or orbital complications, surgical drainage may be required.

Once sinusitis has reached a chronic state, the microbial aetiology becomes more complicated. When it is hospital-acquired, there is increased participation of anaerobes, resistant Gram-negative bacilli and staphylococci. Infection may follow nasotracheal intubation, and in this setting Gram-negative bacilli predominate. Obstruction of the ostiomeatal area in the nose seems to be a major contributing factor in the development of chronic disease; correction of this defect by traditional or functional endoscopic sinus surgery is rapidly replacing earlier surgical procedures for promoting sinus drainage (Gwaltney, 1995). Antimicrobial agents which can be considered for chronic and/or hospital-acquired sinusitis include ciprofloxacin (for Gram-negative bacilli) and co-amoxyclav (for anaerobes and staphylococci).

FACE

After major maxillofacial trauma, particular attention should be paid to fractures that may traverse the sinus cavities and tooth-bearing areas of the maxilla or mandible, as secondary infection rates in these areas are high and osteomyelitis may result. The dominant pathogen is *Staphlyococcus aureus*; treatment is with co-amoxyclav or a flucloxacillin-type drug. Therapy is essential if there is persistent otorrhoea or rhinorrhoea, which may suggest a cerebrospinal fluid leak.

Infectious complications (e.g. osteonecrosis of the mandible) following irradiation, surgical resection and the use of chemotherapeutic agents to treat malignancies involving the head and neck are often associated with *Staphylococcus aureus* and

Table 10.1 The microbial aetiology of acute community-acquired antral sinusitis[a]

Microbe	Relative incidence	
	Adults	Children
Bacteria		
Streptococcus pneumoniae[b]	+++	++++
Haemophilus influenzae (non-capsulated)[b]	++	++
Moraxella (Branhamella) catarrhalis	+	++
Gram-negative bacilli[c]	++	+/−
Staphylococcus aureus	+	−
Streptococcus pyogenes	+	+
Obligate anaerobes[d]	+	−
Viruses		
Rhinoviruses	++	−
Influenza viruses	+	−
Parainfluenza viruses	+	+
Adenoviruses	−	+

[a]Adapted from Gwaltney (1995).
[b]Often mixed together.
[c]Include *E. coli*, *Klebsiella pneumoniae* and *Pseudomonas aeruginosa*; more prevalent following previous antibiotic therapy.
[d]Peptostreptococci, prevotellae and fusobacteria are most common.

Pseudomonas aeruginosa. A combination of cefuroxime and ciprofloxacin is a logical therapeutic regimen, although the former could be replaced with antimicrobial agents such as flucloxacillin, cefazolin or co-amoxyclav.

ORAL CAVITY

Infections of the oral cavity are most commonly odontogenic in origin and include dental caries, pulpitis, peri-apical abscess, gingivitis, and periodontal and deep fascial space infections (Chow, 1995). Microbes from such infections can spread locally to involve intracranial or lung structures, or they may spread via the blood to a variety of locations (e.g. heart valves, prostheses). Suppurative orofacial infections can also arise from the teeth, middle ear, nasopharynx, mastoids and paranasal sinuses. Infections of the head and neck region in adults may result from animal (including human) bites, trauma, irradiation or surgical procedures (Chow, 1995).

Microbes associated with odontogenic infections reflect the normal flora of the oral cavity. Although this flora is complex (see Table 10.2), it does appear that certain microbes occur in defined localities and are commonly associated with specific disease processes. Examples include *Streptococcus mutans* and dental caries, the anaerobes *Prevotella intermedia* (previously *Bacteroides intermedius*), *Prevotella melaninogenica* and *Porphyromonas gingivalis* (formerly *Bacteroides gingivalis*) with periodontitis, *Actinobacillus actinomycetemcomitans* in juvenile periodontitis, and a polymicrobial flora including *Fusobacterium nucleatum*, *Prevotella* species, peptostreptococci, actinomycetes and streptococci with peri-apical abscess or deep fascial space infections (Chow, 1995). Gram-negative enteric bacilli and *Staphylococcus aureus* are seldom recovered from odontogenic infections.

In addition, it is apparent that apart from disease and/or anatomical considerations, a variety of other factors influence the composition of the oral flora

Table 10.2 More common microbes that may be recovered from the oral cavity[a]

Microbe	Relative frequency[b]
Obligate anaerobes	
e.g. Peptostreptococci	++
Prevotellae	++
Porphyromonas species	+++
Bacteroides species	+
Veillonellae	+
Fusobacteria	+
Actinomycetes	++
Gram-negative bacilli (non-obligate anaerobes)	
Eikenella corrodens	+
Coliforms	+
Actinobacillus species	+
Pasteurella species	+
Gram-positive cocci (non-obligate anaerobes)	
Viridans streptococci	++++
Gram-positive bacilli (non-obligate anaerobes)	++
e.g. Lactobacilli	+
Coryneforms	+
Gram-negative cocci (non-obligate anaerobes)	
Moraxella species	+
Neisseria species	+
Spirochaetes	+
Mycoplasmas	+
Yeasts	
Candida albicans	++

[a]Adapted from Chow (1995).
[b]Tends to vary slightly depending on the site involved (e.g. gingival crevice vs. saliva).

(e.g. age, diet and nutrition, oral hygiene, smoking habits, previous antimicrobial therapy, hospitalization, and possibly racial or genetic factors). The anaerobic flora increase dramatically in the presence of mucosal disease and poor oral hygiene.

In view of the relatively common occurrence of odontogenic infections, it is somewhat surprising that osteomyelitis of the mandible or maxilla is not more prevalent. When it does occur, it is usually associated with compound fracture, previous irradiation, osteopetrosis, diabetes mellitus, steroid therapy or Paget's disease (Chow, 1995). Other rare complications of odontogenic infections include suppurative jugular thrombophlebitis, carotid artery erosion, septic cavernous sinus thrombosis and maxillary sinusitis (see Chow, 1995).

Antibiotics which can be considered for periodontitis and necrotizing gingivitis include tetracyclines (contraindicated in children), co-amoxyclav or metronidazole. With suppurative (pyogenic) odontogenic infections, the most important therapeutic modality is surgical drainage and removal of necrotic foci, including the infected tooth. Antimicrobial agents are important for halting the spread of disease. As most of the causative microbes (both aerobes and anaerobes) are penicillin-susceptible, regimens often consist of either penicillin, amoxicillin or co-amoxyclav. Some oral anaerobes (e.g. *Prevotella melaninogenica*) are becoming increasingly resistant to penicillin and amoxicillin. Alternatives to the penicillins include clindamycin, cefoxitin or, as a last resort, imipenem. Metronidazole should not be used on its own.

chapter 11

CARDIOVASCULAR INFECTIONS

VASCULAR GRAFT INFECTION

The development of prosthetic arterial substitutes in the 1940s and 1950s revolutionized the management of patients with aneurysmal and occlusive vascular disease (Bandyk and Esses, 1994). However, despite advances in surgical technology and refinements in conduit fabrication, infections of prostheses still remain a problem. The true incidence of graft infection is unknown. Even with the implementation of antibiotic prophylaxis, the infection rate does not appear to have decreased in recent years, and may be as high as 5%. Although infection may occur early after operation, onset may be delayed, only becoming apparent months to years after implantation. This is especially true where coagulase-negative staphylococci are involved (Bandyk and Esses, 1994).

MICROBIAL AETIOLOGY

Early graft infection (i.e. within the first month) is less common when contemporary vascular prostheses and grafting techniques are employed. Staphylococci are now the dominant causal microbes, with many infections being associated with identifiable risk factors (e.g. emergency procedure, re-operation for thrombosis or haematoma). The ability of microbes to adhere to different graft materials varies, being related to the graft's physical properties (e.g. roughness) and chemical properties (e.g. surface structure). For example, adherence to dacron is 10 to 100 times greater than adherence to polytetrafluoroethylene-containing grafts (Bandyk and Esses, 1994). Once attached to the prosthesis, staphylococci elaborate an extracellular glycocalyx and reside within a biofilm protected to some degree from antibodies, antibiotics and phagocytic cells. In such an environment, the bacteria also show phenotypically decreased susceptibility to antibiotics. Once a prosthetic graft pseudointima is established, the biomaterial becomes less susceptible to colonization following a transient bacteraemia.

Apart from coagulase-negative staphylococci and *Staphylococcus aureus*, other more common and/or significant pathogens are Gram-negative bacilli, including *Pseudomonas aeruginosa*. A high incidence of anastomotic disruption and vessel wall necrosis is typical of Gram-negative infections, especially with pseudomonads. Coagulase-positive staphylococci (e.g. *Staphylococcus aureus*) are also known for their extracellular lysin production, while infection with the less aggressive coagulase-negative species (e.g. *Staphylococcus epidermidis*) relies on their continued presence in biofilms, and adherence to foreign material, grafts and sutures. Although not featuring on many lists of causal microbes, the role of obligate anaerobes may be underestimated, especially in cases where grafts are intraperitoneal.

Infections that occur within 4 months of implantation are most commonly attributable to *Staphylococcus aureus*, with complications of initial wound healing often being noted following surgery. When haemorrhage is a presenting sign of graft infection, the presence of Gram-negative bacteria should be suspected. Late-appearing infections (i.e. occurring after 4 months post-implantation) frequently involve coagulase-negative staphylococci. Other less common causal agents include the yeast *Candida albicans* (Bandyk and Esses, 1994).

Most infections seem to be associated with contamination at the time of surgery and insertion of the graft. Commonly a short course of peri-operative anti-staphylococcal antibiotic prophylaxis is given. Arterial wall colonization and/or contiguous spread as a mode of infection is uncommon. Bacteraemic episodes following colonization of intravascular catheters would appear to be the likely means of graft colonization in cases where coagulase-negative staphylococci are involved.

TREATMENT STRATEGIES

Most prosthetic graft infections are fatal if they are not treated with aggressive surgical and antibiotic therapy. This includes wide debridement of infected tissues (including the artery adjacent to the graft), reconstruction of vital arteries through uninfected tissue, and prolonged post-operative antibiotic administration (Bandyk and Esses, 1994). Although antibiotic irrigation and the placement of closed suction drains are also recommended, their role is subject to some debate. Because of the unpredictable microbial flora involved, graft culture is recommended.

Parenteral antibiotics that are effective against the microbes recovered from blood, wound or graft cultures should be administered for 2 to 4 weeks. Anaerobic as well as the usual aerobic cultures should be requested if grafts have been inserted in the abdominal cavity and/or involve an aorto-enteric fistula, even if this is as proximal as the duodenum. If *Staphylococcus aureus* is involved, considerations include a penicillinase-stable penicillin (e.g. flucloxacillin, nafcillin), various cephalosporins (e.g. cefazolin, cefuroxime), a combination of fusidic acid or rifampicin with flucloxacillin, clindamycin, or vancomycin (for methicillin-resistant *Staphylococcus aureus*). A number of alternatives are available for infection by Gram-negative bacilli (e.g. ceftriaxone, ciprofloxacin, piperacillin/tazobactam).

Coagulase-negative staphylococci present more of a problem. Many strains are resistant to flucloxacillin-type drugs, and may require the use of fusidic acid plus ciprofloxacin (which have the added benefit of better diffusion into biofilms), vancomycin or perhaps a second-generation cephalosporin such as cefamandole. Metronidazole should be included if the infected graft involves the abdominal cavity, until appropriate culture results eliminate anaerobes as a contributing factor. The role of antibiotic (e.g. rifampicin)-impregnated grafts in minimizing post-operative infections is largely unproven (see Bandyk and Esses, 1994).

INTRAVENOUS AND CENTRAL CATHETER-RELATED INFECTIONS

GENERAL POINTS

The overall incidence of infections related to the use of intravascular catheters is around 1%, although with central venous catheters that are used for total parenteral nutrition, this figure may be as high as 4–8%. Central venous line infections carry with them a mortality rate as high as 3% in high-risk patients, and an estimated cost of US$ 6000 per infectious episode (Garrison and Wilson, 1994). Most catheter-related infections are caused by coagulase-negative staphylococci, *Staphylococcus aureus* and yeasts, although infections with Gram-negative bacilli are being increasingly reported (Garrison and Wilson, 1994). Colonization of the catheter most commonly occurs by transcutaneous migration down the catheter tract from the skin insertion site, although other recorded routes include contamination of the infusate solution or the catheter itself, and haematogenous spread and colonization of the luminal end of the catheter.

Apart from possessing adhesins (e.g. slime in coagulase-negative staphylococci) which allow the microbes to adhere to the catheter itself or the fibrin sheath which may occur around the catheter, subsequent microbial colonization and/or infections are related to factors such as the insertion site, the duration of catheterization, insertion and subsequent care techniques, the use of multilumen devices, the degree of hydration of the surrounding cutaneous site, and the type of material used in the construction of the catheter. A number of patient-related factors have also been implicated (e.g. age, underlying disease (such as AIDS) and the presence of concurrent infection) (see Garrison and Wilson, 1994). Catheter-related infections are uncommon in the first 3–4 days after insertion, but increase on a daily basis thereafter. Metallic scalp needles carry the lowest incidence of infection, followed (in order of increasing rate of colonization) by silastic catheters (11–14% colonization rate) and polyethylene catheters (up to 27%). The latter may induce an inflammatory tissue response even in the absence of significant colonization. Randomized and controlled trials have clearly shown a clinical benefit linked to the use of central venous catheters impregnated with antimicrobial agents (e.g. chlorhexidine and silver sulphadiazine, minocycline and rifampicin, colloidial silver – see MacGowan, 1998; Grubbauer, 1999; Mark et al., 1999; Masterton, 1999).

TREATMENT STRATEGIES

In some cases, removal of the catheter without antibiotic therapy may result in resolution of the associated infection. If this is not followed promptly by a positive clinical response (e.g. temperature and vital signs returning to normal), appropriate antibiotics should be given and maintained for 7 to 14 days depending on the physical state of the patient. Removal alone (i.e. without concurrent antibiotics) is not sufficient when either *Staphylococcus aureus* or *Candida albicans* is responsible. The persistent nature of these microbes, and their tendency to colonize heart valves and other tissues following a bacteraemic or fungaemic episode, necessitate continuation of antimicrobial treatment for at least 4 to 6 weeks. If catheters cannot easily be removed or exchanged (e.g. long-term implanted devices, no other access site), antibiotic treatment alone can be considered, although this carries a greater risk of failure compared to antibiotics plus removal. Using a small intraluminal brush, it is now possible to culture indwelling catheters while they remain in place, with results comparable to those obtained by techniques in which the catheter is removed and the tip forwarded to the laboratory for study (Markus and Buday, 1989; Tighe *et al.*, 1996a,b). Such *in situ* techniques have the potential to reduce the removal of sterile catheters that are wrongly suspected of harbouring infection.

Empirical therapy should be directed against staphylococci, especially coagulase-negative types (e.g. *Staphylococcus epidermidis*). Possible alternatives include a penicillinase-resistant penicillin (e.g. flucloxacillin), a first- or second-generation cephalosporin (e.g. cefuroxime), a quinolone, netilmicin, or vancomycin for methicillin-resistant *Staphylococcus aureus*. Sensitivity results are important in guiding therapy against coagulase-negative staphylococcal strains. If there are indications that a *Staphylococcus aureus* catheter-related bacteraemia may have seeded other organs, inclusion of fusidic acid or rifampicin with the antistaphylococcal β-lactam regimen may be warranted. Rifampicin appears to be one of the drugs for which penetration is least affected by bacterial extracellular slime (Souli and Giamarellou, 1998). Therapy for Gram-negative bacilli includes aminoglycosides, third-generation cephalosporins and ciprofloxacin. The latter may be especially useful in cases where catheters are left in place. Amphotericin B or fluconazole are alternatives for *Candida* infections (see p. 141).

chapter 12

CHEST INFECTIONS

PULMONARY CAVITY (ABSCESS)

GENERAL POINTS

Suppurative pulmonary infection may result in the destruction of lung parenchyma and the formation of one or multiple lung cavities or empyema. Although often referred to as abscesses, the term 'cavity' is preferred, as lesions do not show an enhanced outline on radiological viewing. However, they do in fact fall within the definition of an abscess – i.e. a focus of suppuration within a tissue, organ or region of the body. If multiple small (< 2 cm in diameter) cavities exist, the term 'necrotizing pneumonia' has been used (Finegold, 1995c).

Most lung cavities contain obligate anaerobes either alone (around 60% of cases) or mixed with more aerotolerant microbes (see Table 12.1). The predominant obligate anaerobes are Gram-negative rods (e.g. fusobacteria, *Prevotella, Porphyromonas* and the *Bacteroides fragilis* group), and Gram-positive cocci. Non-anaerobes include micro-aerophilic streptococci (e.g. *Streptococcus anginosus*), other streptococci, *Staphylococcus aureus* and Gram-negative enteric bacilli (see Finegold, 1995c, for a more complete list). *Pseudomonas aeruginosa* may be a significant causal microbe in debilitated patients. Although clostridia, including *Clostridium perfringens*, may be recovered from patients with necrotizing pneumonia and empyema, there is usually nothing distinctive about the clinical picture in such cases. Infections are classically polymicrobial, with around four anaerobes and two aerobes or facultative anaerobes per specimen.

Most lung cavities result from aspiration, and there appear to be some differences in the causal bacteria depending on whether this occurs in the community or hospital setting. The former appears to be primarily an anaerobic process with obligate anaerobes from the oral cavity often the only type of microbes found. What role bacteria such as *Staphylococcus aureus* have in initiating such lesions is not known. In comparison, hospital-acquired infections yield pure anaerobes in only about 20% of cases, with important nosocomial pathogens such as *Staphylococcus aureus*, klebsiellae, pseudomonads and *Proteus* species being regularly recovered as part of the infecting flora. Gram-negative bacilli are clearly more common in the oral cavity of hospitalized patients, especially in those who have received antimicrobial therapy. If patients have an increased microbial flora in the stomach (e.g. due to malignancy, or H_2-antagonists for ulcers), then aspiration pneumonia is likely to reflect this change.

In necrotizing pneumonia, obligate anaerobes as well as several major aerobic or facultative anaerobic species have been incriminated (e.g. *Staphylococcus*

Table 12.1 More common microbes recovered from pleuropulmonary infections[a]

Obligate anaerobes
Peptostreptococci, especially *Peptostreptococcus micros, Peptostreptococcus anaerobius, Peptostreptococcus magnus*
Pigmented *Prevotella* (e.g. *Prevotella melaninogenica, Prevotella intermedia*)
Non-pigmented *Prevotella* (e.g. *Prevotella oris-buccae, Prevotella oralis*)
Porphyromonas asaccharolyticus
Fusobacterium nucleatum, Fusobacterium necrophorum
Bacteroides fragilis group
Non-sporing Gram-positive bacilli (e.g. *Actinomyces, Eubacterium*)
Veillonella species

Non-obligate anaerobes
Staphylococcus aureus
Streptococcus anginosus (*milleri*)
Gram-negative bacilli (e.g. *E. coli, Proteus* species (hospitalized patients))
Pseudomonas aeruginosa – debilitated patients
Viridans streptococci

[a]Adapted from Goldstein (1996), Finegold (1995c) and Finegold and Wexler (1996).

aureus, Klebsiella pneumonia and *Pseudomonas aeruginosa*. All of these elaborate powerful exotoxins that are capable of causing lung necrosis. In addition, mycobacteria and various fungi (e.g. *Aspergillus fumigatus*) may also be recovered from necrotizing pneumonia.

Apart from aspiration, haematogenous spread of microbes or septic emboli from some other focus may result in multiple cavity formation. More notable microbes that fall into this category include *Staphylococcus aureus*, Gram-negative enteric bacilli and anaerobes. Lung cavities are seen following staphylococcal bacteraemia, right-sided endocarditis, and pelvic or deep vein thrombophlebitis, spread of Gram-negative bacilli from the urinary tract or bowel, and secondary to intra-abdominal or pelvic infections, where anaerobic or micro-aerophilic streptococci (e.g. *Streptococcus anginosus*) and Gram-negative anaerobic bacilli are frequently involved.

TREATMENT STRATEGIES

Administration of appropriate antimicrobial agents and postural drainage, both of which may need to be continued for 2 to 4 months, are the primary mode of treatment. In view of the importance of anaerobes, metronidazole or clindamycin supplemented with large doses of penicillin G have been recommended by some authorities (see Finegold, 1995c). Some anaerobic cocci and most micro-aerophilic streptococci and actinomycetes are resistant to metronidazole, while resistance to clindamycin is increasing. Other alternatives include combination β-lactam drugs (e.g. piperacillin plus tazobactam), and imipenem. Chloramphenicol is active against essentially all anaerobes, and may be an option in seriously ill patients.

The significant causal role of *Staphylococcus aureus* in lung cavities is often incorrectly obliterated by the emphasis placed on the potential role of obligate anaerobes. If *Staphylococcus aureus* is implicated, possible drugs include a penicillinase-stable penicillin (e.g. flucloxacillin), co-amoxyclav, a parenteral cephalosporin (e.g. cephazolin, cefuroxime), fusidic acid or rifampicin (usually in combination with a β-lactam), ciprofloxacin or a newer fluoroquinolone such as clinafloxacin (again in combination with another agent such as fusidic acid), clindamycin, and vancomycin in the case of methicillin-resistant *Staphylococcus aureus*.

For Gram-negative bacilli, several alternatives are available, including third-generation cephalosporins (e.g. ceftriaxone), several β-lactam combinations (e.g. piperacillin plus tazobactam), meropenem, quinolones (e.g. ciprofloxacin) and cotrimoxazole. Of these, ciprofloxacin, with the potential to switch to oral administration, is possibly the best alternative. Aminoglycoside pharmacokinetics would appear to be unsuited to abscess cavity-type lesions.

For patients who develop infection while in hospital, a logical empirical regimen is clindamycin plus ciprofloxacin, both of which can be given parenterally or orally.

Empyema should be treated surgically.

chapter 13

INTRA-ABDOMINAL INFECTIONS

Intra-abdominal infections requiring either operative or percutaneous intervention are common, and result in substantial morbidity, mortality and cost. These infections may take several forms; they can be diffuse or localized into one or more abscesses, and can be within the peritoneal cavity itself or in the retroperitoneal space. In addition, infection may occur within the intra-abdominal viscera (e.g. liver, pancreas, spleen, kidney, gut) or the mesentery. Abscess or collections may form around the involved organs by walling off around small bowel or omentum, by travelling along paracolic gutters by gravity to the pelvis, to a subphrenic space by negative pressure and lymphatic flow, or by operative manipulation to virtually any location. Intra-abdominal infection is an inflammatory response of the peritoneum to microbes and their toxins, which results in purulent exudate in the abdominal cavity (Wittmann *et al.*, 1996).

PERITONITIS

GENERAL POINTS

There are two main types of peritonitis, namely primary (or spontaneous) peritonitis, in which no primary source of infection is apparent, and secondary peritonitis, in which leakage of microbes from an injured or ruptured viscus (e.g. the appendix) is clearly responsible for the infectious process. Spontaneous bacterial peritonitis is probably the best characterized infectious complication that develops in patients with cirrhosis and ascites (Wittmann *et al.*, 1996; Such and Runyon, 1998). A third form, tertiary peritonitis, refers to persistent peritonitis in seriously ill patients (see Table 13.1). In this situation, the microbes (often resistant Gram-negative bacilli or yeasts) have spread from the initial abdominal focus, resulting in more generalized sepsis with shock. Intra-abdominal sepsis ranks with pneumonia and septicaemia in terms of importance in determining the outcome of many hospitalized patients. From a historical point of view, it seems that the bacteriology of this disease only became clear following the studies of Weinberg around 1922, with the importance of synergy between causal microbes being further enunciated by Meleney *et al.* in 1932 (see Hau, 1998).

The essential therapeutic principles in the management of peritonitis (e.g. immediate operation where possible, elimination of the source of infection, removal of exudate, no medicines administered into the peritoneal cavity, drains not normally

Table 13.1 Classification of intra-abdominal infections

Clinical entity	Examples
Primary peritonitis	
Bacterial infection in which no primary gastrointestinal source of infection is apparent; no disruption of intra-abdominal hollow viscera; monomicrobial	Spontaneous peritonitis in children and adults; peritonitis in patients with continued ambulatory peritoneal dialysis
Secondary peritonitis	
Diffuse or localized (abscess) peritonitis originating from a defect in abdominal viscera; usually polymicrobial	Infection resulting from perforation of the gastrointestinal tract, intestinal ischaemia, peri-operative peritoneal contamination, an anastomotic leak, following abdominal trauma, or leakage from the female genital tract
Tertiary peritonitis	
Persistent peritonitis with systemic spread in debilitated patients – frequently attributable to Gram-negative bacilli or yeasts	

used) were initially summarized by Martin Kirschner in 1926, and following the implementation of these principles the overall mortality rate of peritonitis dropped from above 80% to about 30% (Hau, 1998). However, around 1970, the death rate of septic patients with organ failure was still around 30%, despite the introduction of delayed wound closure, which helped to reduce the magnitude of sepsis and other complications. After the introduction of anti-anaerobe therapy, the mortality started to decline to current levels, although the introduction by Wittmann in the early 1990s of staged abdominal repair (STAR) treatment resulted in a further significant drop in mortality in patients with APACHE-II scores of 25 or less (see Condon, 1996a). In general, there is a correlation between the APACHE-II score and morbidity with intra-abdominal infections (see Figure 13.1 and Levison and Zeigler, 1991; Wittmann *et al.*, 1996). In addition, the incidence of major abdominal infection is linked to the severity of abdominal trauma (see Figure 13.2), including such factors as injury to the colon, repeated blood transfusions and the number of ruptured organs (Nichols and Smith, 1993; Wilson, 1996). Over the last decade, innovations in peri-operative management and technology, image-guided therapy for the diagnosis and management of abscesses, and more effective antimicrobial therapy have continued to lower mortality rates (Montgomery and Wilson, 1996). The classical surgical management of peritonitis is operative control of contamination/pathology and antibiotic therapy with intensive-care support.

The anatomical relationships within the abdomen are an important factor in determining possible sources of infection, as well as the route of any spread and loculation. Fluid that is introduced into the right subhepatic space gravitates

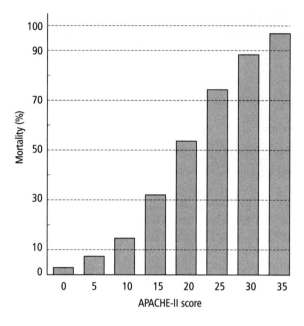

Figure 13.1 Correlation of APACHE-II score with mortality from intra-abdominal infection. Adapted from Wittmann *et al.* (1996).

towards Morison's pouch and then into the right subphrenic space and subsequently via the right paracolic gutter into the pelvic recess. Fluid that is introduced into the false pelvic peritoneal cavity first gravitates to the true pelvic recess and then ascends, regardless of body position, via the right paracolic gutter into the right subhepatic and right subphrenic spaces. Presumably inflammatory exudates will follow similar courses along the right paracolic gutter. Both gravity and normal caudal intraperitoneal fluid flow thus influence the pooling of fluid in peritoneal recesses. Pooling is exacerbated with intra-abdominal lymphangitis and obliteration of lymphatic channels which pass through the diaphragm.

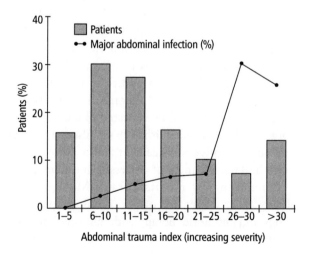

Figure 13.2 The incidence of major abdominal infection is linked to severity of abdominal trauma. Reproduced from Fabian (1996).

Secondary peritonitis is the most common form of peritonitis encountered by the surgeon, and may present either as generalized disease of varying severity, or as a localized abscess. The terms 'intra-abdominal abscess' and 'peritonitis' are clearly not synonyms – abscesses develop from effective host defences and represent a relatively successful outcome of peritonitis (Wittmann et al., 1996).

MICROBIOLOGY

Primary peritonitis was classically a spontaneous disease of children attributable to streptococci (e.g. *Streptococcus pneumoniae*). It is now most often seen in a spontaneous form in cirrhotic patients with ascites and end-stage liver disease, and involves enteric Gram-negative bacilli (e.g. *E. coli*, *Klebsiella pneumoniae*), pneumococci, other streptococci and enterococci. In patients with erosion of an incisional or umbilical hernia, *Staphylococcus aureus* may predominate. Anaerobes are infrequently reported because of the intrinsic bacteriostatic activity of ascites fluid against *Bacteroides* species, the relatively high oxygen levels in ascites, or difficulties with anaerobe isolation and identification in the laboratory.

Peritonitis in patients undergoing continuous ambulatory peritoneal dialysis (CAPD) has been reported in around 45% of cases within the first 6 months of starting treatment and in up to 70% of patients in the first year (Barnish, 1994). Infection is primarily associated with contamination of the catheter or its connections during manipulative procedures, or the dialysis fluid itself. Other lesser sources of infection include the skin insertion site, which should face inferiorly, bacterial migration from the intact gastrointestinal tract or vagina, and haematogenous spread. 'Chemical' peritonitis has to be distinguished from infection. Most infections are bacterial, although fungi (e.g. *Candida*) appear to be an increasingly common cause. Of the bacteria, skin microbes (e.g. coagulase-negative staphylococci and resistant coryneforms) are most common (40% of cases), followed by *Staphylococcus aureus* and other Gram-positive cocci and Gram-negative bacilli, including environmental pseudomonads and *Acinetobacter* species (the latter being cutaneous flora of the foot). Serious streptococcal infection may cause lymphangitis of the peritoneal lymphatics and/or alter the permeability of the peritoneal membrane. Tuberculous peritonitis in CAPD patients should be suspected when routine bacterial cultures are persistently negative and the patient fails to respond to empirical antibiotic therapy. Presumably *Mycobacterium tuberculosis* gains entry by a haematogenous route, although in many reported cases pulmonary disease and systemic illness are absent!

Most cases of secondary peritonitis are endogenous in origin, and involve a mixture of obligate anaerobes in symbiosis or synergy with more oxygen-tolerant aerobic or facultative anaerobic microbes. Monogeneric infections can occur, especially in cases where lesions involve muscles or fascia.

The spectrum of microbes contaminating the peritoneal cavity after perforation of or trauma to the gastrointestinal tract depends on the level of injury. In addition, the risk of serious infection is proportional to the degree of abdominal trauma – major infections increase dramatically with an abdominal trauma index of > 20 (see Fabian, 1996). Leakage from the terminal ileum and colon results in massive bacterial inoculation of some 500 or so different species, with obligate anaerobes clearly dominating. In contrast, the stomach and upper small intestine are practically free of microbes under normal conditions (the numbers increase with the use of antacids and H_2-receptor antagonists, and in the presence of malignancy), with most being Gram-negative enterics, enterococci and yeasts. Patients requiring multiple operations for repair or recurrent gastrointestinal perforations or anastomotic leakage, or acute pancreatitis, seem especially prone to significant intra-abdominal infection with *Candida albicans* (Eggimann *et al.*, 1999).

Despite the enormous variety of anaerobic and aerobic microbial species that inhabit the intestines, only a limited number are consistently recovered from gut-associated infections (see Tables 13.2 and 13.3). The most common anaerobes are *Bacteroides fragilis, Bacteroides thetaiotaomicron* and other *Bacteroides fragilis* group species, peptostreptococci, fusobacteria and clostridia. The latter, especially *Clostridium perfringens*, are particularly significant in infections associated with small-bowel ischaemia (Walker *et al.*, 1994). A recently recognized pathogen is *Bilophila wadsworthia*, one of the more common anaerobes recoverable from abscesses or peritoneal fluid from patients with appendicitis, and apparently the only organism apart from *Bacteroides fragilis* that produces abscesses on its own in a mouse model. Another anaerobe that is apparently commonly recovered from peritonitis secondary to acute appendicitis with perforation, and from female genital tract-associated infections, is *Prevotella melaninogenica*. Surprisingly, the anaerobes that appear to be most common in the normal colonic flora (e.g. veillonellae, *Eubacterium*, bifidobacteria and lactobacilli), are not regularly associated with intra-abdominal infections (Brook, 1989; Wittmann *et al.*, 1991; Gorbach, 1993a,b; McClean *et al.*, 1994; Nathens and Rotstein, 1994; Finegold and Johnson, 1995; Edmiston and Walker, 1996; Solomkin *et al.*, 1996; Wittmann *et al.*, 1996).

The significant gut aerobes/facultative anaerobes are *E. coli, Klebsiella* species and other Enterobacteriaceae, streptococci (e.g. *Streptococcus anginosus*), staphylococci, enterococci, *Pseudomonas aeruginosa* and the yeast *Candida albicans* (Brook, 1989; Barie *et al.*, 1990; Cooper *et al.*, 1993; Solomkin *et al.*, 1996). The latter three microbes or groups of microbes are clearly emerging as significant colonizers and/or pathogens in gut-related abdominal sepsis (Montravers *et al.*, 1996). The potential for *Pseudomonas aeruginosa* to be present in initial cultures from patients (e.g. those with perforated and gangrenous appendicitis) (Yellin *et al.*, 1985; Gorbach, 1993a,b), colon-derived infections (Solomkin *et al.*, 1996) or traumatic wounds (e.g. gunshot) has clearly been under-rated by many clinicians. The influence that the increasing use of cephalosporins in prophylaxis and therapy has had on the emergence of enterococci (e.g. *Enterococcus faecalis*) and yeasts is still largely speculative. However, it is clear that enterococci are an important cause of intra-abdominal infection, particularly with gall-bladder or pancreatic sources (Cooper *et al.*, 1993).

It is apparent that intra-abdominal infections are caused by endogenous microbes from the gastrointestinal tract, and to a lesser extent from the female genital tract, and that the predominant flora that is found reflects the microbes which are likely to

Table 13.2 Microbes that are more commonly recovered from intra-abdominal infections (secondary peritonitis)[a]

Microbe	Relative frequency
Obligate anaerobes[b]	
Bacteroides fragilis	++++
Other Bacteroides species	++
Peptostreptococci	++
Fusobacteria	+
Prevotellae	+
Porphyromonas species	+
Clostridia	+
Bilophila wadsworthia	+
Gram-negative aerobic and facultative anaerobic bacilli	
E. coli	+++
Pseudomonas aeruginosa	++
Klebsiella species	++
Enterobacters	+
Proteus species	+
Citrobacter species	+
Gram-positive cocci (excluding obligate anaerobes)	
Streptococci	+++
Enterococci	++
Staphylococci	++
Gemella species	+
Yeasts	
Candida albicans	+

[a]Adapted from Wittmann *et al.* (1996) and Nathens and Rotstein (1994).
[b]Under-represented in many earlier surveys.

occur at the site of leakage (Walker *et al.*, 1994). However, where problems such as ischaemia, cancer or obstruction occur, increased participation of anaerobes such as *Clostridium perfringens* and *Clostridium septicum* is likely. Not surprisingly, the significance of anaerobes such as *Bacteroides fragilis* (often in association with *E. coli*) increases dramatically with infections involving the lower intestinal tract (i.e. appendix, colon or rectum).

The bacteriology of intra-abdominal infections that arise as complications of female genital tract infections is similar (see Table 13.4) to that of secondary peritonitis due to a gastrointestinal source, except for the occurrence of *Neisseria gonorrhoeae* in cul-de-sac aspirates. Although obligate and facultative anaerobes other than *Bacteroides fragilis* and clostridia tend to dominate in the normal vaginal flora (e.g. *Prevotella*, lactobacilli, *Gardnerella*), the former appear to become more prevalent after surgery and in the puerperium.

Table 13.3 More common bacteria that are recovered from gangrenous or perforated appendicitis[a]

Bacterial species	Approximate frequency[b]
Anaerobes	
Bacteroides fragilis[c]	++++
Bacteroides thetaiotaomicron[c]	+++
Bilophila wadsworthia	+++
Peptostreptococcus micros	+++
Eubacterium species	++
Bacteroides intermedius	++
Bacteroides vulgatus[c]	++
Bacteroides splanchnicus	+
Fusobacterium species	+
Bacteroides ovatus[c]	+
Micro-aerophilic streptococci	+
Peptostreptococcus species	+
Lactobacilli	+
Bacteroides uniformis[c]	+
Bacteroides distasonis[c]	+
Clostridium clostridioforme	+
Other clostridia	+
Unidentified Gram-negative rod	++
Unidentified Gram-positive rod	+
Other identified species (e.g. *Porphyromonas*, actinomycetes, *Bacteroides*)	+
Aerobes and facultative anaerobes	
E. coli	++++
Viridans streptococci	++
Enterococci/group D streptococci	+
Staphylococci	+/−
Pseudomonas aeruginosa	+
Others (mainly Gram-negative bacilli)	+

[a] See Finegold and Johnson (1995) for further details.
[b] ++++, > 75%; +++, 45–60%; ++, 30–40%; +, 20–30%; +/−, < 20%.
[c] Form part of *Bacteroides fragilis* species group.

Although the therapeutic role of antibiotics in acute necrotizing pancreatitis is subject to some debate, it is clear that extended 'prophylaxis' with drugs such as imipenem does reduce the associated sepsis rate (Pederzoli *et al.*, 1993).

As a general rule, antibiotic-resistant strains, especially of *Serratia*, *Acinetobacter* and *Pseudomonas*, are more frequent in patients who develop infections in hospital after having received broad-spectrum antibiotics. Polymicrobial infection occurs in most patients with around 3 to 4 isolates per

Table 13.4 Microbes that are recovered from intra-abdominal infections arising from the female genital tract[a]

Obligate anaerobes
Peptostreptococci
Bacteroides fragilis group
Prevotella species – especially *Prevotella bivia, Prevotella disiens* and pigmented species
Clostridia – especially *Clostridium perfringens*
Fusobacteria
Actinomyces, Eubacterium – associated with intrauterine contraceptive device infections
Non-obligate anaerobes
Streptococci – groups A and B and others
E. coli
Klebsiellae
Gonococci – in sexually promiscuous females
Chlamydia trachomatis – in sexually promiscuous females
Mycoplasma hominis – postpartum
Ureaplasma urealyticum

[a]Adapted from Finegold (1995a) and Finegold and Wexler (1996).

patient at the time of initial laparotomy. Anaerobes or aerobes are seldom found alone.

THERAPY OF INTRA-ABDOMINAL INFECTIONS

Antimicrobial therapy (see Table 13.5) for intra-abdominal infections is intended to eradicate the causal microbes, reduce surgical wound complications and control any bacteraemia (Wittmann *et al.*, 1996). In the majority of cases, therapy is empirical and needs to be commenced well before most laboratory results become available. Whether or not delivering antimicrobials to the peritoneal cavity by way of liposomal capsules (which are possibly taken up by phagocytic cells containing bacteria) will result in improved therapeutic results is open to speculation – experimental results look promising (Martineau and Shek, 1999).

Anaerobes

Appropriate empirical antimicrobial chemotherapy consists in most cases of a combination of drugs covering both anaerobes and aerobes. Laboratory tests (Appelbaum, 1993; Goldstein *et al.*, 1993; Stark *et al.*, 1993; Wiseman *et al.*, 1995; Snydman *et al.*, 1996; Solomkin, 1996) have shown that metronidazole, imipenem and meropenem have excellent broad-spectrum anaerobe activity, although the former is less consistently active against more oxygen-tolerant clostridia, fusobacteria and actinomycetes. Resistance to these three compounds still appears to be rare among *Bacteroides* (Snydman *et al.*, 1996) (see also p. 45). Cefoxitin, cefotetan and especially clindamycin appear to be generally less effective than metronidazole

Table 13.5 Some antimicrobial agents with potentially useful therapeutic activity against microbes that are more commonly associated with intra-abdominal sepsis

Antimicrobial agent	Microbe							
	Bacteroides fragilis group	Non-Bacteroides fragilis anaerobes	Enterobacteriaceae	Pseudomonas aeruginosa	Enterococci	Streptococci[a]	Staphylococcus aureus[b]	Candida albicans
β-Lactams								
Penicillin		S3			GEN	S1		
Amoxycillin		S	V		S1	S2		
Co-amoxyclav	S2	S2	V		S	S	S2	
Flucloxacillin						S	S1	
Piperacillin		V	V	S1	V	S		
Piperacillin/tazobactam	S	S	S2	S1	V	S	S	
Ticarcillin/clavulanic acid	S	S	S2	S1	V	S	S	
Cefuroxime			V			S2	S2	
Cefoxitin, cefotetan	S2	S2	V			S	S	
Cefotaxime		V	S1			S	S	
Ceftizoxime		V	S1			S	V	
Ceftriaxone		V	S1			S	V	
Ceftazidime		V	S	S1		V	V	
Aztreonam			S2	V				
Imipenem, meropenem	S2	S2	S2	S2	V/S	S	S	

Table 13.5 Continued

Antimicrobial agent	Microbe							
	Bacteroides fragilis group	Non-Bacteroides fragilis anaerobes	Enterobacteriaceae	Pseudomonas aeruginosa	Enterococci	Streptococci[a]	Staphylococcus aureus[b]	Candida albicans
Aminoglycosides								
Gentamicin			S1	S2			S	
Tobramycin			S1	S1			S	
Quinolones								
Ciprofloxacin			S1	S1			S	
Clinafloxacin	S	S	S2	S	S2	S	S2	
Nitroimidazoles								
Metronidazole	S1	S1						
Others								
Clindamycin	S	V				S	S2	
Vancomycin		S4			S2	S2	S2	
Antifungal agents								
Amphotericin B								S2
Fluconazole								S1

S, usually susceptible; V, variable or intermediate susceptibility (consult local data); GEN, susceptible in combination with gentamicin; 1, agent of first choice; 2, second-line agent; 3, first-choice agent for *Clostridium perfringens*; 4, Gram-positive anaerobes such as clostridia are susceptible.
[a]Increasing levels of resistance to β-lactams world-wide.
[b]Methicillin-resistant strains resistant to all β-lactams.

and imipenem against *Bacteroides* species, fusobacteria and non-*perfringens* clostridia – a situation that is mirrored to some extent by the decreasing susceptibility noted in clinical isolates over the last few years. *Bacteroides thetaiotaomicron* tends to be less susceptible to cefoxitin and clindamycin than *Bacteroides fragilis*, while the majority of strains of *Clostridium innocuum* are resistant to cefoxitin (Stark *et al.*, 1993). β-lactam combinations – for example, amoxycillin plus clavulanic acid (co-amoxyclav or 'Augmentin'), ampicillin plus sulbactam, ticarcillin plus clavulanic acid, or piperacillin plus tazobactam – have excellent *in vitro* activity against *Bacteroides* species and clostridia, although increasing levels of resistance are being encountered in non-*fragilis* species of *Bacteroides* (although 95–98% of strains are still susceptible; see Syndman *et al.*, 1996). Metronidazole remains the drug of choice for infections involving the *Bacteroides fragilis* group. Meropenem is an acceptable alternative to imipenem. Ceftriaxone has poor *in vivo* activity against *Bacteroides fragilis*, and appears to be rapidly inactivated in mixed communities involving this microbe (Dezfulian *et al.*, 1993). This also applies to the related cephalosporin, ceftizoxime, which was widely touted in the USA in the early 1990s as a useful single agent for intra-abdominal sepsis, including appendicitis (Condon, 1996b; Edmiston and Walker, 1996; Wilson, 1996).

Aerobes/facultative anaerobes

Most Gram-negative Enterobacteriaceae and pseudomonads are still susceptible to a wide variety of antimicrobial agents, including aminoglycosides, β-lactams and quinolones (see Table 13.5). However, extended broad-spectrum β-lactamase (ESBL)-producing strains of *Klebsiella pneumoniae*, *E. coli* and other Enterobacteriaceae have been isolated world-wide. Such strains exhibit resistance to newer cephalosporins (e.g. cefotaxime) and aztreonam, and clearly pose a therapeutic problem should they become more widespread and/or common. Resistance is plasmid mediated, and may be accompanied by resistance to unrelated antibiotics (e.g. gentamicin).

Problems of emerging antimicrobial resistance

Resistance of streptococci and enterococci to the currently available and useful therapeutic options may be a problem in the future, as enterococci that are resistant to gentamicin and/or vancomycin, and enterococci and pneumococci with reduced susceptibility and/or marked resistance to penicillins have emerged in many countries. Further development and/or spread of such resistance obviously has serious therapeutic implications.

The presence of resistant microbes in patients with peritonitis has been associated with a poor outcome. In one study (Montravers *et al.*, 1996) of 100 consecutive patients (over 15 years of age) with diffuse peritonitis occurring after elective abdominal surgery, extensive microbiological studies undertaken at the time of re-operation revealed that resistant microbes were present in the peritoneal fluid or blood of 70 patients, 45% of whom died. By comparison, the mortality rate among those from whom only susceptible microbes were isolated was only 16%. Almost 50% of the patients did not receive adequate antimicrobial therapy within 48 h of re-operation, because of the presence of the resistant pathogens.

Samples from 37 of the 70 patients from whom resistant microbes were obtained contained multi-resistant bacteria. The duration of antimicrobial therapy in these patients before re-operation was significantly longer than it was for other patients, as was the time before re-operation. Fungal infections, all caused by *Candida*, were identified in 23 of the patients, and gastroduodenal and biliary surgery was a common association in those with fungal infection, as was the use of co-amoxyclav after the initial surgery. *The main pathogens to emerge that were apparently not covered by the initial antimicrobial regimen were enterococci (both* Enterococcus faecalis *and* Enterococcus faecium*), staphylococci, pseudomonads and* Candida.

MODEL BASIS FOR THERAPY

Early septicaemia and rapid death from untreated abdominal sepsis is associated with Gram-negative bacilli (e.g. *E. coli*). Survivors develop polymicrobial abscesses in which anaerobes such as *Bacteroides fragilis*, and in particular its polysaccharide capsule, play a key role. Experiments using a mouse model of abdominal sepsis involving *E. coli* and either *Bacteroides fragilis* or *Bacteroides thetaiotaomicron* have provided the rationale for these findings, and have highlighted the importance of combined cover against both aerobic and anaerobic bacteria in mixed abdominal infections (Brook and Gillmore, 1993). Antimicrobial agents effective only against *E. coli* prevented mortality, but did not prevent abscess formation or reduce the number of *Bacteroides*. In all cases, abscesses contained both *E. coli* and the *Bacteroides* species, although the numbers of the former were significantly reduced in animals treated with one of the active anti-coliform agents. Gentamicin in conventional doses appeared to be inferior to the cephalosporins, fluoroquinolones and aztreonam in this respect (Thys *et al.*, 1988). On the other hand, agents that were effective against both the *Bacteroides* species and *E. coli* prevented abscess formation as well as mortality (due to *E. coli*).

In the mouse model, ampicillin/sulbactam prevented abscess formation but was accompanied by a 27% mortality rate, presumably due to a lack of sufficient early coliform activity. The most active single agent (out of cefoxitin, cefotetan, cefmetazole, ceftizoxime, imipenem, ampicillin plus sulbactam and clindamycin) used in this investigation was imipenem. Cefoxitin was superior to the other two 7-methoxy-cephalosporins (cefotetan, cefmetazole) in reducing abscess formation, although the use of all three was associated with early mortality ($\leq 10\%$) and, in general, second-generation cephalosporins have only a limited role in therapy. Unfortunately, the proven anti-anaerobe drug metronidazole, either alone or in combination with other antimicrobial agents, was not included in these experimental studies. However, its effectiveness against the anaerobe component of mixed infections was clearly demonstrated experimentally and clinically in the early 1990s (see Brook and Ledney, 1994; Gorbach 1994).

CLINICAL RESULTS

These experimental results, coupled with the results of clinical trials (Heseltine *et al.*, 1986a,b; Solomkin *et al.*, 1990; Hooker *et al.*, 1991; Wittmann *et al.*,

1991; de Groot *et al.*, 1993; Eklund *et al.*, 1993; Gorbach, 1993a,b; Greenberg *et al.*, 1994; Pefanis *et al.*, 1994; Berne *et al.*, 1996; Solomkin *et al.*, 1996; Wittmann *et al.*, 1996) and the *in vitro* susceptibility data presented by Goldstein *et al.* (1993), highlight the necessity of cover for anaerobes and aerobes in abdominal sepsis. With more serious infections, meropenem or imipenem alone, or a combination of metronidazole plus a parenteral antibiotic covering enteric Gram-negative bacteria and pseudomonads (e.g. ciprofloxacin, ceftazidime, gentamicin) can be considered. Several clinical trials have demonstrated the usefulness of imipenem or meropenem in the treatment of intra-abdominal infections, including those associated with a perforated and gangrenous appendix (Gill *et al.*, 1986a; Heseltine *et al.*, 1986a; Solomkin *et al.*, 1990; Wiseman *et al.*, 1995; Berne *et al.*, 1996; Wilson, 1997). In some comparative trials, the poor performance of aminoglycoside-containing regimens was attributable to a failure to achieve early therapeutic drug levels (Gill *et al.*, 1986b). Some evidence is available that metronidazole and fluoroquinolones may act synergistically in mixed aerobe/anaerobe cultures, and that gentamicin is less efficient than fluoroquinolones and newer β-lactams (e.g. ceftazidime, cefepime) in eliminating coliforms from abdominal lesions (see Burrick *et al.*, 1990). The role of piperacillin/tazobactam and ticarcillin/clavulanic acid combinations as monotherapy for intra-abdominal infections is largely unknown, although early results appear to be promising (Bryson and Brogden, 1994; Schoonover *et al.*, 1995). The latter combination has also been employed successfully as short-term (e.g. 24-h) 'therapy' following penetrating abdominal trauma (Fabian and Boldreghini, 1985). With intra-abdominal infections, the length of hospital stay is an important economic consideration when choosing antibiotic therapy. In this respect, imipenem and meropenem have been shown to be superior to piperacillin/tazobactam (Berne *et al.*, 1996).

Aminoglycosides

Aminoglycoside pharmacokinetic parameters are highly variable in critically ill surgical patients, due to very large and unpredictable volumes of distribution (Cornwell *et al.*, 1997). Constant monitoring and dose adjustment, especially in multiple-dose regimens, are necessary in order to maintain serum levels in the therapeutic range and to avoid renal toxicity and ototoxicity due to excessive drug accumulation. Aminoglycoside activity is also reduced in acidic peritoneal fluid, which may be found in one-third of patients with peritonitis. Increased abdominal infection rates have been associated with failure to achieve sufficiently high early aminoglycoside levels following abdominal trauma and changes in the volume of distribution associated with fluid resuscitation. A potential solution to this problem is to use once daily aminoglycoside dosing schedules that achieve immediate high levels of antimicrobial agent. This appears to be safe and effective in the case of abdominal infections (de Vries *et al.*, 1990; Maller *et al.*, 1993), as well as in other settings. It also has obvious advantages in reducing the time spent administering and monitoring the antimicrobial agent and in reducing nephro- and ototoxicity. Not surprisingly, regimens that do not include aminoglycosides are preferred for elderly patients and in patients with shock. The high costs of non-aminoglycoside-containing regimens must be weighed against these concerns and the expenses of the necessary serum level monitoring.

Cephalosporins
The potential of Gram-negative agents other than aminoglycosides has been the focus of considerable interest in recent years. First-generation cephalosporins have inadequate Gram-negative bacillus activity and are unsuitable for treatment of established infection. The same constraint probably applies to second-generation cephalosporins, despite their increased antibacterial spectrum. Certainly they are unsuitable for empirical therapy in life-threatening situations. Of the newer (third- and fourth-generation) parenteral cephalosporins that are generally available, cefotaxime is the cheapest and has the best staphylococcal activity, ceftriaxone has the advantage of once daily administration, and ceftazidime has the best anti-pseudomonal activity. It appears to be superior to tobramycin in eliminating bacteria from intra-abdominal sepsis (Burrick et al., 1990). On the negative side are the emergence of plasmid-mediated EBSL resistance, the increased costs and poor anaerobe cover of cefotaxime, ceftazidime and ceftriaxone, and the potential for biliary sludging with the latter. The fourth-generation cephalosporin cefepime has excellent Gram-negative bacillus activity (including pseudomonads), and has been found to be effective in clinical trials (Berne et al., 1993). However, the practice of reserving more expensive drugs (e.g. ceftazidime and cefepime) until culture results are available may in reality only be delaying optimal therapy until the factors responsible for the initiation of multiple organ failure are beyond control. Although they are not bowel commensals, pseudomonads are underestimated early pathogens in abdominal sepsis, and are significant nosocomial pathogens. Empirical therapy should therefore include cover against them, especially if there is a history of previous antimicrobial use and/or prolonged hospitalization.

Anti-anaerobe agents
Metronidazole is clearly superior in many respects (e.g. activity, cost, ease of administration, adverse effects) to other anti-anaerobe agents. Other anti-anaerobe considerations include meropenem, imipenem, piperacillin/tazobactam and perhaps clindamycin. Although cefoxitin appeared to be useful in some earlier studies, at least following penetrating abdominal injuries (see Heseltine et al., 1986b; Hooker et al., 1991), increasing problems of resistance have now limited the use of this drug and its relatives (e.g. cefotetan) to treat serious intra-abdominal sepsis. The same is also probably true of clindamycin.

Other antimicrobial agents
Antibacterial regimens often include penicillin (with gentamicin-containing regimens) or amoxycillin to cover enterococci. These microbes have been associated with antibiotic treatment failures after surgery-treated advanced appendicitis (Berne et al., 1993). Although imipenem alone may cover enterococci, most of the other antimicrobial agents that are used to cover Gram-negative bacilli (e.g. ceftazidime,

aztreonam, gentamicin) do not have satisfactory therapeutic activity against enterococci. Piperacillin-containing regimens probably have some enterococcal activity. An interesting addition to the list of broad-spectrum antimicrobial agents has been clinafloxacin. In addition to Gram-negative enteric bacilli, this fluoroquinolone has excellent *in vitro* activity against Gram-positive cocci, including enterococci, and important intestinal obligate anaerobes. Streptococci such as *Streptococcus anginosus*, which are not susceptible to some of the Gram-negative bacillary agents, should also be considered. Empirical antifungal therapy is currently considered to be unwarranted, but should be initiated in patients who are receiving apparently adequate antibacterial agents but failing to improve, especially following upper gastrointestinal surgery (see Eggimann *et al.*, 1999).

DURATION AND METHOD OF DRUG DELIVERY

It has been suggested that specific clinical criteria (e.g. temperature and white cell count) should be applied to determine the duration of therapy in patients with established intraperitoneal infections (Nathens and Rotstein, 1994). However, this logical, widely believed and practised recommendation lacks evidence that continuing therapy aborts potentially developing infectious complications. In addition, fever may continue after the infectious microbes have been eliminated (Wittmann *et al.*, 1996). Further studies are required to define the time when antibiotics cease to serve a useful purpose. In general, if there is associated bacteraemia, therapy should last no less than 4 days, extending to 7 days in patients with major peritoneal contamination.

However, it is clear that the mortality of intra-abdominal infection is predominantly related to the magnitude of the patient's systemic response and his or her premorbid physiological reserves, which are best estimated using the Acute Physiology and Chronic Health Evaluation II (APACHE-II) scoring system (Wittmann *et al.*, 1996). With increasing APACHE-II scores, mortality from intra-abdominal infection clearly increases (e.g. an APACHE-II score of 8 has about 10% mortality, a score of 19 has a 50% mortality rate and a score of 27 has an 80% mortality rate). With APACHE-II scores of 25 or more, salvage by any method, including staged abdominal repair (STAR), is unlikely (Condon, 1996).

With intra-abdominal infections, antibiotics have invariably been administered by a parenteral route, although it seems that a switch from intravenous to oral therapy in responding patients is a feasible and cost-saving consideration. In a large multicentre trial, Solomkin *et al.* (1996) were able to demonstrate that in patients with complicated intra-abdominal infections (and a mean APACHE-II score of 8.6), the outcome was no different in those who received combined intravenous (IV) ciprofloxacin/metronidazole, IV imipenem alone, or IV ciprofloxacin/metronidazole changed to oral ciprofloxacin/metronidazole when oral feeding was resumed (3–8 days after operation). Treatment failures, which may have been dose related, were more consistently associated with the persistence of Gram-negative bacteria in the IV imipenem group. Patients who received therapy were treated for a mean period ($\pm$ SD) of 8.6 $\pm$ 3.6 days overall, with an average of 4.0 $\pm$ 3.0 days of oral treatment.

ANTIMICROBIAL RECOMMENDATIONS

Despite the numerous options available, the treatment of intra-abdominal infections is not simple. The once popular triple regimen of metronidazole, gentamicin and ampicillin (or amoxycillin), which is apparently now out of fashion, may still be utilized because it is inexpensive. Clearly, metronidazole should be included in all regimens, together with an antibiotic that is effective in eliminating *E. coli* and related coliforms and *Pseudomonas aeruginosa* from the peritoneal cavity (the classical endotoxin producers). A third-generation cephalosporin (e.g. ceftazidime) could be sufficient. Another alternative is ciprofloxacin. Increased costs of this drug in combination with metronidazole can be partially alleviated by changing both to an oral formulation as soon as is feasible (see Solomkin *et al.*, 1996). The role of new fluoroquinolones such as clinafloxacin has yet to be assessed.

We believe that empirical regimens should possess some form of enterococcal activity, especially in patients whose recovery is expected to be slow, or where the source of contamination involves the upper gastrointestinal tract. Enterococci have featured prominently in patients in whom otherwise apparently adequate treatment fails (Montravers *et al.*, 1996; Solomkin *et al.*, 1996). We are not convinced by the suggestion that enterococci are clinically insignificant other than as cofactors for *Bacteroides fragilis* in the formation of abscesses (see Wittmann *et al.*, 1996). Ciprofloxacin has marginal activity (but better than cephalosporins) against enterococci, although newer quinolones appear to be more promising in this respect. Drugs with proven activity against some enterococci include amoxycillin (with or without gentamicin), imipenem, meropenem, chloramphenicol, piperacillin and the newer fluoroquinolones (e.g. clinafloxacin).

An empirical regimen for intra-abdominal infection should include optimal cover against obligate anaerobes (metronidazole), endotoxin-producing Gram-negative bacilli (aminoglycosides, third-generation cephalosporin, fluoroquinolone), enterococci (amoxycillin or piperacillin) and *Staphylococcus aureus* if there has been recent surgery (see Table 13.6). Cover against all of these microbes may be possible with single drugs such as meropenem, imipenem or piperacillin/tazobactam. However, wherever feasible, therapy should commence with effective inexpensive drugs, and must cover any Gram-negative bacilli that are likely to be involved. This includes the environmental *Pseudomonas aeruginosa*, which is a frequent transient in the gut.

The next step depends on how rapidly the patient recovers, whether the condition deteriorates, or whether another problem is identified. For example, new blood cultures may uncover unsuspected or resistant pathogens, repeat imaging may reveal small undrainable collections (presumably containing obligate anaerobes), laboratory investigations may reveal evidence of potential central line-related sepsis, and progressively deteriorating respiratory function (or some other evidence of impending multiple organ dysfunction) may indicate that additional or alternative antimicrobial therapy may be beneficial.

Table 13.6 Therapeutic choices for intra-abdominal infections

Therapeutic regimen (IV dose)	Comment
Metronidazole (500 mg q 12 h) + tobramycin (5–7 mg/kg q 24 h) + amoxycillin (1 g q 6 h)	Adequate cover for most causal microbes; poor pharmacokinetics of tobramycin in peritoneal exudate; potential aminoglycoside toxicity requires serum monitoring; some authorities question the need for amoxycillin
Piperacillin/tazobactam (4.5 g q 8 h, Tazocin; 3.375 g q 6 h, Zosyn)	Wide-spectrum activity; concern about potentially differing diffusion of both components in seriously ill patients
Metronidazole (500 mg q 12 h) + ciprofloxacin (400–500 mg q 8 h)	Useful combination; potential to switch to oral route with both components; minimal enterococcal cover
Metronidazole (500 mg q 12 h) + ceftriaxone (1–2 g q 12 h)	Excellent cover apart from pseudomonads and enterococci
Metronidazole (500 mg q 12 h) + ceftriaxone (1–2 g q 12 h) + piperacillin (3 g q 4 h)	Adequate cover for most causal microbes
Meropenem (1–2 g q 8 h)	Adequate cover for most causal microbes; simplicity of administration; achieves adequate concentrations in abdominal tissues and fluids; well tolerated in patients with hepatic disease
Imipenem/cilastatin (1 g q 6 h)	Similar to meropenem; may result in excessive accumulation of cilastatin in patients on continuous haemofiltration; more adverse reactions than meropenem
Clinafloxacin (unknown, but possibly once daily)	A possible consideration in the treatment of intra-abdominal infections either alone or in combination with other agents

The choice of the next best antimicrobial option might be an educated guess as to the pathogenesis of the problem. Adding imipenem or meropenem to the primary regimen should provide an 'order of magnitude' better cover against anaerobes and Gram-negative bacilli, including pseudomonads. Ciprofloxacin would offer similar cover for the latter. The participation of resistant Gram-positive cocci may require the use of vancomycin or a related glycopeptide. Accumulating evidence also suggests that in patients with arrested progress, the early addition of fluconazole may be advantageous. This 'presumptive therapy' is more likely to improve outcome than waiting for positive fungal cultures and clinical evidence of yeast infection (see p. 141). Table 13.7 lists a number of clinical scenarios which represent a series of options for patients in intensive-care situations.

LIVER ABSCESS/BILIARY TRACT INFECTIONS

GENERAL POINTS

Over the past 15 years, diagnostic and interventional radiology techniques have allowed accurate localization of liver abscesses and image-guided percutaneous drainage – a far cry from the earlier clinical appraisal and operative drainage (Barakate et al., 1999). Today, percutaneous aspiration and/or drainage with ultrasound or computerized tomography (CT) guidance has become the procedure of choice for draining liver abscesses.

For some patients in whom more invasive methods are not desirable or necessary (e.g. small abscesses, or those with a background of immunodeficiency), antibiotic therapy alone can be an effective form of treatment. This should be continued until there is abscess cavity obliteration on CT, and no signs of infection.

If unilocular abscesses occur, percutaneous needle aspiration in conjunction with antibiotics seems to be effective. Multiple simple abscesses can be aspirated during the same session. Primary surgical intervention would appear to be advisable for patients with abscess rupture on presentation, for those with multiloculated abscesses or a biliary communication, and in the setting of renal impairment and hyperbilirubinaemia. Increased mortality appears to be associated with current malignancy, abscess multiloculation, failure of percutaneous drainage, haemoglobin levels of < 10 g/dL and total bilirubin of > 18 μmol/L (see Barakate et al., 1999).

MICROBIOLOGY AETIOLOGY

The most commonly reported microbes associated with pyogenic liver abscesses are *E. coli* and related coliforms (e.g. *Klebsiella*, *Proteus*), enterococci, *Streptococcus anginosus* (*Streptococcus milleri* group) and *Staphylococcus aureus*. Most studies have not investigated aspirates for obligate anaerobes, and their significance remains unknown. Empirical therapy revolves around a third-generation cephalosporin (e.g. ceftriaxone) or extended-spectrum penicillin (e.g. piperacillin) or ciprofloxacin, all together with metronidazole. Single agents such as meropenem (or imipenem) and piperacillin/tazobactam would appear to be ideally suited for initial therapy in seriously ill patients. More specific therapy should be initiated on receipt of laboratory results. Cephalosporins (and probably ciprofloxacin) are not effective against enterococci (e.g. *Enterococcus faecalis*).

The microbiology of biliary tract infections is similar to that noted above for pyogenic liver abscesses (see also Westphal and Brogard, 1999). In addition, *Enterobacter* species may form a significant component of the enteric Gram-negative isolates. In most cases, acute bacterial cholecystitis results from obstruction

Table 13.7 Potential intensive-care problems and their solutions

Scenario	Problem	Diagnosis	Current therapy	Solution
A	• Secondary peritonitis	• Fever (T_{max} 38°C); • raised white blood count (WBC); • patient looks sick	• Initial cultures grew *E. coli*, *Bacteroides fragilis* and ESBL-producing *Enterobacter*; • infiltrate on chest X-ray; • decreased pO_2	• Combined ampicillin, gentamicin and metronidazole	• Discontinue initial antibiotics; • commence meropenem plus ciprofloxacin; • abdominal CT with or without catheter or operative drainage if abscess is found; • consider bronchoscopic broncho-alveolar lavage
B	• Secondary peritonitis	• Stable • CT shows two small undrainable loculi	• No cultures obtained • Presume anaerobe (polymicrobial)	• Cefotaxime plus metronidazole	• Meropenem with or without vancomycin; • consider fluconazole
C	• Secondary peritonitis	• Some or all of the above; • oral candidosis; • ascites	• Growth of *Candida albicans* from multiple sites (possible invasive candidosis)	• Combined ampicillin, gentamicin and metronidazole	• Add in fluconazole • Possibly change antibacterial agents to meropenem with or without vancomycin

Table 13.7 Continued

Scenario	Problem	Diagnosis	Current therapy	Solution
D	• Blunt abdominal trauma; • perforated small bowel and colon; • ruptured spleen • Diffuse faecal spill; • 3 h in duration; • 1000 mL blood loss	• Mixed colonic flora contamination of peritoneum	—	• Primary resection and anastomosis × 3; • extensive peritoneal irrigation with saline; • meropenem or imipenem therapy (possibly for 7 days)
E	• Seventh day of severe acute pancreatitis • Increasing fever and WBC count; • CT shows single large lesser sac collection; • aspirate reveals Gram-negative bacilli	• Infected peri-pancreatic collection	None	• CT-guided catheter drainage or open operative debridement and drainage; • meropenem with or without vancomycin
F	• Scenario E now 6 days post-drainage; • deteriorating • Repeat CT shows incompletely drained collection	• Inadequately drained or recurrent peri-pancreatic sepsis	• Imipenem	• Operative debridement and redrainage; • upper abdominal wound left open for daily repacking; • add antifungal agent (fluconazole)

of the cystic duct by a gallstone. Acute cholangitis is caused by infection in an obstructed biliary system (e.g. gallstones in the common bile duct, prior surgery, neoplasm). Gram-negative bacteria (e.g. *E. coli*, *Klebsiella*, *Enterobacter*) are the microbes most commonly recovered from infected bile, although polymicrobial participation is common with Gram-positive cocci (e.g. enterococci) and obligate anaerobes occurring with the Enterobacteriaceae. Anaerobes tend to be associated with more severe clinical illness, the elderly, previous surgery, and malignancy, and should never be underestimated. Concurrent bacteraemia has been reported in up to 70% of patients with cholangitis.

TREATMENT STRATEGIES

Few trials have compared the efficacy of various antibiotic regimens. Therapy should cover the Enterobacteriaceae, especially *E. coli*, although it seems that even though enterococci are often present in the bile, their pathogenic role remains unclear and activity against them is unnecessary. Clearly coverage of obligate anaerobes (e.g. *Bacteroides* species, clostridia) is warranted in patients with previous bile duct–bowel anastomosis, in the elderly, in the presence of malignancy, and in seriously ill patients. In patients with acute cholecystitis or cholangitis of moderate severity, monotherapy with piperacillin or ciprofloxacin is often adequate. If patients are severely ill and/or display evidence of bacteraemia, monotherapy with meropenem or piperacillin/tazobactam would appear advisable, although the superiority of such a regimen to a combination such as ciprofloxacin plus metronidazole is unknown.

Relief of biliary obstruction is mandatory in order to minimize problems with continued obstruction. This should be undertaken after 36–48 h of antibacterial therapy, although immediate operation is indicated for gangrenous cholecystitis and perforation with peritonitis (Westphal and Brogard, 1999). Some cases of recurrent cholangitis may require long-term use of antimicrobial agents (e.g. cotrimoxazole).

ANTIBIOTIC-ASSOCIATED COLITIS

GENERAL POINTS

Although pseudomembranous colitis was recognized in the pre-antibiotic era, there is no doubt that antibiotics are the most significant precipitating cause of this disease (Fekety, 1995; Högenauer *et al.*, 1998). Initially the problem was linked to the use of oral non-absorbable broad-spectrum antibiotics (e.g. neomycin), and attributed to the bowel overgrowth of *Staphylococcus aureus*. It seems that this association may have been coincidental, and that all along the true cause of pseudomembranous colitis and antibiotic-associated diarrhoea with colitis has been the anaerobe *Clostridium difficile*. However, this may not be true for some mild cases of antibiotic-associated diarrhoea without colitis where functional disturbances of intestinal carbohydrate or bile acid metabolism, allergic and toxic effects of antibiotics on the intestinal mucosa, or antibiotic-mediated pharmacological effects on gut motility may be the most significant predisposing event (Högenauer *et al.*, 1998). The recognition of *Clostridium difficile* and its toxins as the dominant causal agent of pseudomembranous and antibiotic-associated colitis appears to date from 1977 (see Fekety, 1995). It seems that the incidence of the disease is increasing, an observation that is possibly linked to the common use of antibiotics such as broad-spectrum cephalosporins and β-lactam combinations (e.g. co-amoxyclav) (Johnson and Gerding, 1998).

The clinical spectrum of *Clostridium difficile*-associated disease ranges from asymptomatic infection, diarrhoea without colitis, non-pseudomembranous colitis with or without diarrhoea, and antibiotic-associated pseudomembranous colitis to fulminant colitis (Högenauer et al., 1998). Although it seems that *Clostridium difficile* is responsible for antibiotic-associated diarrhoea in only about 10-20% of cases, in nearly all cases of antibiotic-associated pseudomembranous colitis, tissue-culture assays for *Clostridium difficile* toxin are positive. Apart from *Clostridium difficile*, other microbes which have been incriminated in the causation of antibiotic-associated diarrhoea include *Clostridium perfringens*, *Staphylococcus aureus*, multi-drug-resistant salmonellae and *Candida* species (see Högenauer et al., 1998).

Clostridium difficile is a spore-forming, obligate anaerobe found as a normal commensal in the bowel of about 5% of healthy adults. In the neonate, colonization rates may reach 70%, while in adult hospital and nursing-home patients, colonization levels (without disease) can be as high as 60%, especially in 'contacts' of a patient with active *Clostridium difficile* colitis. Despite the presence of clostridial toxins, most infants remain asymptomatic. Cross-infection (presumably from spores) within the hospital environment obviously occurs, and the control of nosocomial outbreaks requires the implementation of appropriate hygiene measures and environmental disinfection with hypochlorite solution.

While *Clostridium difficile* colitis can occur in any age group, it is most frequent in the middle-aged or elderly, and in debilitated patients. Even short courses of antibiotics (e.g. single-dose prophylaxis) in colonized patients may precipitate active disease. Almost all antibiotics have been incriminated, with the more classical ones being clindamycin, co-amoxyclav, ampicillin/amoxycillin and a number of cephalosporins. It seems logical to assume that any antimicrobial agent which is able to pass into the intestinal lumen can induce the disease. Pseudomembranous colitis has also been reported in patients who have not received antibiotics. Whether this results from relapse/recurrence of previous antibiotic-associated disease, or is related to some other factor such as dietary changes or non-antibiotic drugs (e.g. gold salts), is unclear.

PATHOGENESIS

Colitis results from the production of toxins by *Clostridium difficile* within the intestinal lumen, an event that is probably associated with spore germination. At least two toxins have been demonstrated – toxin A and toxin B. Strains that lack genes for these toxins are non-toxigenic and cannot cause colitis. Toxin A appears to be an enterotoxin, while toxin B is a cytotoxin and is the product usually detected in laboratory tests. The toxins directly affect the colonocytes by alteration of cellular actin filaments. Release of cytokines from epithelial cells, monocytes, macrophages and neuroimmune cells of the lamina propria also contribute to the toxin-mediated inflammation of and damage to the colonic mucosa (Högenauer et al., 1998). As already stated, newborn infants colonized with toxigenic *Clostridium difficile* strains usually remain well despite the presence of large amounts of toxin A and toxin B in their stools. It has been suggested that a glycoprotein found in milk (fetulin), or a subunit of it, interferes with the action of these toxins in the caecum. In addition, the toxins do not bind well to the intestinal mucosa of neonates (Fekety, 1995). As newly born infants slowly acquire a normal intestinal flora, the rate of isolation of *Clostridium difficile* and its toxins in the stools declines towards the rates found in older children and adults.

A wide range of symptoms are associated with toxin production by *Clostridium difficile*, from profuse watery or mucoid foul-smelling diarrhoea to an acute abdomen syndrome with toxic megacolon, perforation or peritonitis (Fekety, 1995). In a significant proportion of patients, diarrhoea is not seen until weeks after the antibiotic treatment has ceased. Recurrence may be seen in 30% of cases or more. This may be related to germination of persisting spores, or to reinfection. Not surprisingly, *Clostridium difficile* colitis may be mistaken for an exacerbation of underlying chronic inflammatory bowel disease.

TREATMENT STRATEGIES

Treatment involves discontinuation of the causative antimicrobial agent where appropriate, dietary carbohydrate restriction, and the use of oral metronidazole to suppress *Clostridium difficile* growth. Doses of 400 mg three times daily for 7–10 days have been shown to be adequate for most cases. If the patient is too ill for treatment by the oral route, intravenous metronidazole can be considered (Fekety, 1995). Other potentially useful oral agents include oral vancomycin (125 mg orally 6-hourly for 7–10 days), bacitracin and fusidic acid. Apart from cost, oral vancomycin is known to select for vancomycin-resistant enterococci in a significant number of cases. However, it is the preferred agent for severe cases. Intravenous vancomycin is ineffective.

Not all patients require antimicrobial treatment. With mild to moderate disease, it is often sufficient simply to withdraw the offending antibiotic and to give supportive therapy with electrolytes and fluids (Fekety, 1995). Cholestyramine may bind toxins A and B and so be helpful. Opiates (e.g. codine) and other antiperistaltic agents should be avoided, especially in infants. Although surgery is seldom required, loop colostomy or loop ileostomy may be needed to facilitate instillation of vancomycin into the colonic lumen of patients with pseudomembranous colitis and ileus (Fekety, 1995). Some patients may require emergency colonic resection if evidence of extensive toxicity and periotonitis develops.

RENAL ABSCESS

Perinephritic abscess is a recognized complication of urinary tract infection (Sobel and Kaye, 1995), and may also be found in patients with colonic diverticular disease. The most common predisposing factors are urinary tract calculi (obstruction) and diabetes mellitus. The causal microbes are usually Gram-negative enteric bacilli, but may involve Gram-positive cocci (especially if the route of infection is haematogenous), anaerobes and fungi such as *Candida*. Multiple microbes are present in about 25% of cases (Sobel and Kaye, 1995).

Intrarenal abscess classically follows bacteraemia by staphylococci, but may occur as a consequence of acute pylonephritis.

Antimicrobial therapy in conjunction with guided percutaneous drainage (rather than open surgical drainage) should be the initial mode of therapy for perinephritic abscess, renal abscess and infected renal cysts. Antimicrobial agents that are suitable for kidney abscess include aminoglycosides (e.g. gentamicin), third-generation cephalosporins (e.g. ceftriaxone), and ciprofloxacin (which has the added benefit of oral dosing). Metronidazole provides suitable anaerobe cover. Aminoglycosides are probably more suitable for intrarenal than for perinephritic abscesses.

GYNAECOLOGICAL INFECTIONS

Necrotizing fasciitis may be an unusual complication of episiotomy. Infection involves subcutaneous tissues (i.e. the superficial fascia) and spreads in the fascial clefts overlying the deep fascia, which is usually spared (Mead, 1995). Because the skin is not primarily involved, the wound may look normal, although the patient appears seriously ill. If the deep fascia and muscle are involved (myonecrosis), *Clostridium perfringens* is commonly implicated, although *Clostridium sordelli* has also been recovered from patients with severe disease (Mead, 1995). Treatment involves surgical resection and antimicrobial agents such as metronidazole, penicillin or clindamycin.

Infection after abortion most commonly occurs in the presence of retained products of conception or operative trauma. The microbiology involves a mixture of facultative and obligate anaerobes (see Table 13.8), with the latter (e.g. *Clostridium perfringens*) often assuming a dominant role. In addition, pelvic inflammatory disease may develop in association with an intrauterine contraceptive device (IUD). In this situation, a significant causal association has been suggested for *Actinomyces* species, although some consider this to be overstated (Mead, 1995). Clearly, however, pelvic infections following abortion or in association with an IUD often involve anaerobes, although the possibility of the participation of the more common sexually transmitted pathogens (e.g. gonococci, chlamydiae) cannot be ignored.

Apart from surgery, therapy should consist of antimicrobial agents that are likely to cover the known pathogens. Suggested regimens include cefoxitin or cefotetan or clindamycin (each with doxycycline in cases where a sexually transmitted disease process is considered to be relevant), metronidazole plus ciprofloxacin (or ofloxacin), and erythromycin plus co-amoxyclav.

The composition of the normal microbial flora of the vagina is poorly documented and understood. It does appear to be dominated by obligate anaerobes and lactobacilli, although other bacteria (e.g. streptococci, *Gardnerella vaginalis*, Gram-negative bacilli and mycoplasmas) are also regularly found. Hospitalization appears to influence the vaginal flora regardless of surgery or antibiotic use, with a change towards the increased occurrence of microbes such as enterococci, the *Bacteroides fragilis* group and more resistant Gram-negative bacilli (e.g. klebsiellae, enterobacters).

Table 13.8 Microbes associated with female genital tract infections[a]

Following operative trauma and/or in the presence of retained products of conception
Obligate anaerobes
  Clostridia, especially *Clostridium perfringens*
  Peptostreptococci
  Bacteroides species (e.g. *Bacteroides fragilis*)
  Prevotellae
Others
  Streptococci (e.g. groups A and B)
  Enterococci
  Enteric Gram-negative bacilli
  Mycoplasma hominis
  (Possibly sexually transmitted diesease (STD) microbes – see below)

In association with an intrauterine contraceptive device
  Actinomyces species (e.g. *Actinomyces israelii*)
  Eubacterium species (anaerobe)

In sexually promiscuous females (cases of STD)
  Neisseria gonorrhoeae
  Chlamydia trachomatis

[a]Adapted from Finegold and Wexler (1996).

Antimicrobial prophylaxis for gynaecological surgery should take into account the patient's likely vaginal flora, so that infection is minimized in women of all ages.

Pelvic cellulitis is the most common infection following hysterectomy. Culture of the vaginal cuff may be used diagnostically in patients at risk (with vaginal discharge). However, most infections appear to respond to empirical therapy with a single agent such as co-amoxyclav, cefoxitin or cefotetan. If the patient develops well-localized collections just above the vaginal vault, drainage may facilitate cure.

An adnexal abscess may develop as a serious late post-operative complication, or after apparently successful cure of pelvic cellulitis. This rare event occurs almost exclusively in premenopausal women, often in the face of apparently adequate antibiotic prophylaxis, and may take many weeks to manifest itself. As *Bacteroides fragilis* group anaerobes have frequently been isolated from abscesses, the use of metronidazole plus an anti-Gram-negative bacillus agent (e.g. ceftriaxone, ciprofloxacin) is the usual form of chemotherapy, although piperacillin/tazobactam or imipenem alone may be sufficient. As antimicrobial therapy alone is often successful (Mead, 1995), identification of a post-operative pelvic abscess does not mandate immediate drainage. If the patient fails to respond to chemotherapy alone, drainage (percutaneously where possible) is necessary. All patients should be re-examined 2 weeks after discharge to ensure that the disease has not recurred.

chapter 14

SKIN AND SOFT-TISSUE INFECTIONS OF SURGICAL IMPORTANCE

NECROTIZING SOFT-TISSUE INFECTIONS

These infections represent diverse disease processes ranging from simple pyodermas to life-threatening spreading gangrenous infections. Since the 1930s there has been a great deal of confusion regarding the nomenclature of dermal gangrene. It was often simply referred to as Meleney's ulcer, Cullen's ulcer or Fournier's gangrene, and it was not until 1952 that the term 'necrotizing fasciitis' was used by Wilson to describe the characteristic necrotic fascia. He also emphasized the varied and often mixed aetiology.

In recent years, simplified and practical classifications of these diseases have been proposed (see Sawyer and Dunn, 1991) which utilize parameters such as anatomical location and tissue level involvement, causal agent(s) and the presence or absence of predisposing conditions (e.g. diabetes mellitus). It is also now apparent that parenteral injection of illicit drugs can produce infections that present with signs of simple cutaneous abscess, which unpredictably become extensive necrotizing soft-tissue infections (Callahan *et al.*, 1998). Undoubtedly necrotizing soft-tissue infections continue to challenge the practising surgeon, the cornerstone of therapy being immediate extensive surgical debridement and empirical antimicrobial therapy (Rode *et al.*, 1993; Swartz, 1995).

NECROTIZING FASCIITIS

This disease was first described in 1871 by Joseph Jones, a Confederate Army surgeon, who named it 'hospital gangrene' (see Hassoun *et al.*, 1995). In 1889 Fedden called it 'acute infected gangrene', while in 1918 Pfanner used the term 'necrotizing erysipelas'. A detailed description of the disease – 'haemolytic streptococcal gangrene' – was produced by Meleney in 1924 after he cultured β-haemolytic streptococci (but no anaerobes) from a series of patients. He was able to demonstrate that infection started as gangrene of subcutaneous tissue, which cut off the blood supply and resulted in rapid spread of necrosis to the overlying skin. He and other investigators in 1926 were also able to show that there was often a synergistic association between micro-aerophilic streptococci and *Staphylococcus aureus* in lesions (Meleney's synergistic gangrene). About this time, Cullen also described a patient with a progressive ulcer of the skin and fatty tissue of the abdominal wall which occurred following drainage of an abscess. A variant of the disease, known as 'Fournier's gangrene', involving necrotizing infection of the perineum had been described by Fournier in 1884.

Microbiology

The aetiology of necrotizing fasciitis usually involves at least one obligate anaerobic species (e.g. peptostreptococci, *Bacteroides* species, *Prevotella* species, clostridia, fusobacteria) in combination with one or more facultative anaerobes (e.g. Gram-negative enteric and non-enteric bacilli, staphylococci, streptococci, enterococci) or aerobic species (e.g. pseudomonads) (see Brook and Frazier, 1995; Singh *et al.*, 1996 and Table 14.1 for more complete lists). An unusual but significant pathogen found in some countries (e.g. Japan) has been the halophile *Vibrio vulnificus* (Fujisawa *et al.*, 1998). Yeasts (e.g. *Candida*) and other fungi (e.g. zygomycetes, aspergilli) have also occasionally been recovered from lesions. Necrotizing fasciitis is classically a polymicrobial infection, and not (as is often quoted) predominantly a group A streptococcal disease (Brook and Frazier, 1995).

In a recent and exhaustive survey concerning the microbiological aspects of necrotizing fasciitis (see Brook and Frazier, 1995), only aerobic or facultative anaerobes were recovered from 10% of specimens (from 81 patients), only obligate anaerobic bacteria from 22%, and mixed aerobic/anaerobic microbes from 68% of specimens. On average there were 4.6 isolates per specimen. Certain bacteria predominated in different anatomical locations which appeared to correlate with their distribution in the normal flora adjacent to the infected site (e.g. *Bacteroides fragilis* species group and clostridia from sites proximal to or inoculated by the gastrointestinal tract, oral prevotellae and fusobacteria from infection close to or inoculated by the oral cavity, *Prevotella disiens* and *Prevotella bivia* from infections proximal to the vulvovaginal area, gastrointestinal Enterobacteriaceae and group D streptococci (probably enterococci) in leg, rectal, external genitalia and trunk infections, and *Staphylococcus aureus* and group A streptococci in the extremities and head and neck region). Knowledge of this general pattern of distribution can help to determine initial empirical antimicrobial therapy (see Table 14.2).

Obligate anaerobes outnumbered more aerotolerant species at all body sites, especially the buttocks, trunk, neck, external genitalia and inguinal sites. In the survey of Brook and Frazier (1995), the predominant obligate anaerobes were peptostreptococci (101 isolates), *Bacteroides* species (47 isolates, including 35 of the *Bacteroides fragilis* group), *Prevotella* species (35 isolates), clostridia (23 isolates) and fusobacteria (15 isolates), while the more aerotolerant isolates included Enterobacteriaceae (36 isolates), streptococci (32 isolates, including 8 from group A), staphylococci (19 isolates, including 14 *Staphylococcus aureus* isolates), and pseudomonads (9 isolates). Significant clinical/microbe correlations included oedema with the *Bacteroides fragilis* group, clostridia, *Prevotella* species, *Staphylococcus aureus* and group A streptococci; gas and crepitation in tissues with clostridia and members of the Enterobacteriaceae; foul odour with *Bacteroides* species; pre-existing trauma with clostridia; underlying diabetes with *Bacteroides* species, Gram-negative enteric bacilli and *Staphylococcus aureus*; and immunosuppression and malignancy with pseudomonads and facultative anaerobic Gram-negative bacilli (Enterobacteriaceae). Apart from diabetes and peripheral vascular disease, more commonly recorded predisposing factors include trauma, alcoholism, previous infection and surgery (Kaldjian and Andriole, 1993).

Table 14.1 Microbes that are more commonly isolated from necrotizing fasciitis[a]

Microbe	Relative frequency
Obligate anaerobes[b]	
Peptostreptococci	++++
(e.g. *Peptostreptococcus magnus*)	
Bacteroides fragilis group	+++
(e.g. *Bacteroides fragilis*	
Bacteroides thetaiotaomicron)	
Prevotellae	++
(e.g. *Prevotella intermedia*,	
Prevotella bivia)	
Clostridia	++
(e.g. *Clostridium perfringens*)	
Fusobacteria	+
(e.g. *Fusobacterium nucleatum*)	
Veillonellae	+
Gram-negative aerobic and facultative anaerobic bacilli	
E. coli	++
Klebsiella pneumoniae	++
Acinetobacter anitratus[c]	++
Pseudomonas aeruginosa[c]	++
Enterobacters	+
Vibrio vulnificus[c]	+
Gram-positive cocci (excluding obligate anerobes)	
Staphylococcus aureus	+++
Streptococcus pyogenes	+
Other streptococci	+
Enterococci	+

[a]Adapted from Brook and Frazier (1995), Singh *et al.* (1996) and Fujisawa *et al.* (1998).
[b]Underestimated in earlier surveys.
[c]Non-enteric.

Cirrhosis and anti-inflammatory therapy (with both corticosteroids and non-steroidal agents) have also been implicated.

Treatment strategies

In necrotizing fasciitis, the infectious process spreads along the superficial fascial plane (not the deep underlying muscle). Its pathogenesis appears to be related to the relatively poor blood supply of the fascia and its inability to clear contaminating microbes. Once infection becomes established, microvascular thrombosis at the leading edge of the lesion may help to promote further spread and development. Excision with a margin is therefore an essential component of therapy.

Table 14.2 Some clinical features/associations with necrotizing fasciitis[a]

Anatomical site	Dominant microbes	Comment/possible clinical associations
Proximal to or inoculated from intestinal tract	*Bacteroides fragilis* species group, clostridia	Oedema, gas and crepitation (clostridia), foul odour, trauma (clostridia), diabetes (*Bacteroides*)
Close to or inoculated from oral cavity	Prevotellae and fusobacteria	Oedema
Proximal to vulvovagina	*Prevotella disiens*, *Prevotella bivia*	Oedema
Leg region	*Vibrio vulnificus*	Chronic liver dysfunction, consumption of sea foods, contact with salt water
Leg, rectal, genital and trunk regions	Gram-negative enteric bacilli, enterococci	Gas and crepitation, diabetes, immunosuppression and malignancy
Extremities and head and neck regions	*Staphylococcus aureus*, group A streptococci	Oedema, diabetes (*Staphylococcus aureus*)
Illicit intravenous or parenteral drug injection site	*Staphylococcus aureus*, streptococci	Poor nutritional state

[a]Adapted from Brook and Frazier (1995), Callahan *et al.* (1998) and Fujisawa *et al.* (1998).

Although necrotizing fasciitis can affect any part of the body, it is most common on the extremities, especially the legs. Other sites of predilection are the abdominal wall, peri-anal and groin regions, the umbilicus in the newborn, post-operative wounds, and illicit drug injection sites. Microbes enter via traumatic skin wounds or surgical incisions, or may spread haematogenously from the bowel (e.g. following perforation). In the latter case, lesions often involve the abdominal wall, lower chest and/or upper thigh.

Mortality from necrotizing fasciitis may be as high as 50%, falling to below 10% if a rapid diagnosis is made (within 4 days of the appearance of the initial symptoms). Immediate surgical debridement is mandatory. This infection often gives rise to a highly septic clinical picture and occasionally the clinical toxic shock syndrome – requiring aggressive intensive-care unit monitoring and haemodynamic resuscitation. Initial empirical antimicrobial therapy is based on the known importance of obligate anaerobes, Gram-negative bacilli (including pseudomonads), streptococci and staphylococci in this disease (see Table 14.3), and their possible association with defined clinical presentations (see above). Empirical therapy can be changed to more specific agents following the receipt of microbiological results. Possible empirical regimens include metronidazole or clindamycin for the obligate anaerobes, third-generation cephalosporins (e.g. ceftriaxone) or ciprofloxacin for the Gram-negative bacilli, and penicillin, amoxycillin, flucloxacillin, clindamycin, ceftriaxone or cefuroxime for Gram-positive cocci. A combination of clindamycin and ceftriaxone seems appropriate for monomicrobial streptococcal disease, while for the more polymicrobial form a combination of metronidazole, ciprofloxacin and cefuroxime appears reasonable. Other possibilities include monotherapy with imipenem or meropenem. On a pharmacokinetic basis, the classical benzylpenicillin and flucloxacillin would appear to be inferior in several respects to agents such as cefuroxime and clindamycin for streptococcal and/or staphylococcal disease. Selection of suitable empirical therapy should be guided by the location of the lesion and its common association with specific microbes.

In necrotizing fasciitis that involves group A streptococci, cellulitis and necrosis of the overlying skin may occur. If there is necrosis of the muscle, the disease is defined as myositis. In some cases, necrosis of the fascia and of the muscle coexist.

Table 14.3 Possible antimicrobial agents for treating necrotizing fasciitis[a]

Microbe	Antimicrobial agent	Comment/other possible considerations
Obligate anaerobes (e.g. peptostreptococci, *Bacteroides* species, prevotellae, clostridia, fusobacteria)	Metronidazole	Meropenem, imipenem, clindamycin, chloramphenicol, piperacillin/tazobactam, co-amoxyclav
Gram-negative enteric bacilli (e.g. *E. coli*, klebsiellae enterobacters)	Ciprofloxacin or ceftriaxone	Meropenem, ticarcillin/clavulanate, piperacillin/tazobactam, aminoglycosides, cefotaxime
Pseudomonas aeruginosa	Ciprofloxacin	Meropenem, piperacillin, tobramycin, ceftazidime
Other non-enteric Gram-negative bacilli (e.g. *Acinetobacter*, *Stenotrophomonas*)	Meropenem, cotrimoxazole	Highly resistant microbes, ciprofloxacin, ceftazidime
***Staphylococcus aureus*[b]**	Cefuroxime	Penicillinase-stable penicillin, cotrimoxazole, clindamycin, ciprofloxacin, co-amoxyclav, clinafloxacin
***Streptococcus pyogenes*[c]**	Cefuroxime or ceftriaxone	Penicillin, clindamycin, amoxycillin
Enterococci	Amoxycillin	Imipenem

[a]See also Hassoun *et al.* (1995).
[b]Vancomycin or other appropriate non-β-lactams if methicillin-resistant (MRSA).
[c]Ceftriaxone plus clindamycin is probably superior to penicillin for proven group A streptococcal disease.

Mortality rates increase for infections with myositis. An unusual isolate (in pure culture) from this disease has been *Serratia marcescens* (see Sawyer and Dunn, 1991). With this microbe, gentamicin, a third-generation cephalosporin or ciprofloxacin can be considered. If clostridia are involved, metronidazole, penicillin or clindamycin merit consideration.

VARIANTS OF NECROTIZING FASCIITIS

Most texts on infectious disease list a number of variants of necrotizing fasciitis that are readily recognizable by the anatomical location of the lesions.

Bacterial synergistic gangrene

Bacterial synergistic gangrene is a form of necrotizing fasciitis in which there is prominent involvement of the skin and muscle as well as the subcutaneous tissues and fascia. Most infections are located on the lower extremities or near the perineum. Initial surgery involves incision and drainage, and often radical debridement because of the deep nature of the involvement. Antibiotics should be given as for polymicrobial necrotizing fasciitis.

Fournier's gangrene

Fournier's gangrene (also known as idiopathic gangrene of the scrotum, streptococcal scrotal gangrene and perineal phlegmon) is a form of necrotizing fasciitis that occurs about the male genitalia. It may be confined to the scrotum, or it may extend to involve the perineum, penis and lower abdominal wall. Although the origin/source of infection is occasionally obscure (idiopathic), in most cases it arises from a urogenital, anorectal or dermal focus (e.g. chronic prostatitis, periurethritis, rectal perforation, post-haemorrhoidectomy, surgery for circumcision). Surgery is again the clinician's primary treatment approach, accompanied by appropriate antimicrobial therapy (as for polymicrobial necrotizing fasciitis). Microbes that are involved include obligate anaerobes, coliforms and Gram-positive cocci (e.g. staphylococci). It is often necessary to return a patient with necrotizing fasciitis to the operating-room as early as 8 h after the initial debridement. Any evidence of further extension should prompt a very early return for surgery. In addition, it is usually necessary to return on either a daily basis or every other day for a week or longer in order to achieve control of the sepsis. Occasionally, with extensive muscle involvement, amputation may be required to preserve life. The suggestion that this is primarily a pyogenic streptococcal disease (*Streptococcus pyogenes*) is false, although streptococcal gangrene may on rare occasions involve the male genital region.

Vibrio infection

As already stated, marine *Vibrio* species (e.g. *Vibrio vulnificus*) may cause soft-tissue infections, including necrotizing fasciitis. These usually occur in immunocompromised or chronically ill patients (e.g. those with cirrhosis or diabetes mellitus) who have open wounds exposed to sea water, who have consumed sea foods, or in patients who sustain skin wounds while handling shellfish. Typically a fulminant cellulitis develops with secondary deep tissue invasion (Sutherland and Meyer, 1994; Halow *et al.*, 1996; Chuang *et al.*, 1998; Fujisawa *et al.*, 1998). Lesions, often on the legs, display clinical features distinct from those of necrotizing fasciitis caused by more classical pathogens (e.g. streptococci). Treatment requires aggressive wound debridement and antibiotics, usually with a combination of a third-generation cephalosporin and minocycline or an aminoglycoside, although doxycycline alone may be adequate in minor cases. Mortality rates are high.

Opportunistic fungal infections
In recent years, invasive opportunistic fungal infections involving the cutaneous regions have been reported with increasing frequency. These often occur in debilitated and/or immunocompromised individuals (e.g. following cutaneous trauma, burns, diabetic ketoacidosis, malignancy and immunosuppression). Blood-borne spread from some systemic focus may also result in cutaneous lesions and be important in the early diagnosis of deep-seated fungal disease (Smith, 1989). Of the fungi that have been implicated, zygomycetes (mucoraceous fungi) are most common, and include species of *Absidia* and *Rhizopus* in addition to a variety of other less common zygomycete genera, and aspergilli. A fascinating aspect of some cases has been their association with contaminated adhesive dressings/bandages (elastoplast), on which the fungus was presumably present. Lesions range from vesiculopustular eruptions to ulceration and deep necrosis. Surgical excision of lesions is mandatory, and should be accompanied by aggressive antifungal therapy in the form of amphotericin B, as the cutaneous foci may serve as a primary focus for disseminated disease (or in fact arise from systemic involvement). Even with appropriate surgery and chemotherapy, the mortality rate is still very high (> 80%).

NECROTIZING MYOSITIS

As with necrotizing fasciitis, necrotizing myositis is a rapidly progressive infection with similar predisposing factors. However, in contrast to fasciitis, the primary location of infection is the skeletal muscle, and the infection is invariably monomicrobial. An important and prominent causal agent has been *Clostridium septicum*, usually in patients with bowel pathology (e.g. carcinoma, infarction), where bloodstream spread appears to have been the route of infection. The relative aerotolerance of *Clostridium septicum* (in comparison to *Clostridium perfringens*) appears to be significant in the pathogenesis. Impairment of neutrophil numbers and/or function (e.g. neutropenia, diabetes mellitus) may also be involved in the establishment of *Clostridium septicum* infections.

CLOSTRIDIAL MYONECROSIS

Clostridia are obligate anaerobic, spore-forming bacilli that are found in soil, decaying organic material and the gastrointestinal tract of animals. Of the 80 or so species in the genus, only around 20 species cause human illness, and of these *Clostridium perfringens* (often in association with other microbes, including other clostridia) is responsible for the majority of trauma-related infections. If clostridia are found in subcutaneous wounds simply as contaminants, treatment requires nothing more than debridement and local wound care. However, if conditions become favourable for their growth (e.g. anaerobic conditions, presence of foreign material, involvement of muscle), exotoxins liberated from the bacteria may result in further tissue invasion and spread.

Colon cancer or bowel infarction may predispose patients to a haematogenous variant of gas gangrene (spontaneous myonecrosis) in which the more aerotolerant *Clostridium septicum* is the usual causal agent. In one study, over 80% of patients presenting with *Clostridium septicum* infection had an underlying malignancy (Sutherland and Meyer, 1994). Predisposing factors for clostridial myonecrosis include deep wounds and devitalized tissue (often containing extraneous foreign material), and open fractures.

With myonecrosis, immediate and aggressive surgical intervention is the primary treatment, and hyperbaric oxygenation (HBO) can also be considered. Classically, penicillin has been the mainstay of antimicrobial therapy, often in combination with metronidazole or clindamycin. Of the three agents, clindamycin may be clinically the best. Where infection follows trauma, or major environmental contamination or a gut source is likely, mixed microbes may be involved (anaerobic cellulitis), and it may be prudent to include ciprofloxacin or a third-generation cephalosporin to cover any Gram-negative bacilli.

NON-CLOSTRIDIAL ANAEROBIC CELLULITIS

A clinical picture very similar to clostridial anaerobic cellulitis can follow infection by a variety of other anaerobes of gut origin (e.g. *Bacteroides* species, peptostreptococci), either alone or mixed with facultative species such as coliforms (e.g. *E. coli*), streptococci and staphylococci. Gas may be formed in the tissues. Initial antimicrobial therapy is similar to that described above for the polymicrobial form of necrotizing fasciitis. The surgical approach is the same as that used for clostridial cellulitis.

Pseudomonas aeruginosa bacteraemia (or invasion of traumatic wounds or thermal burns) may also produce a gangrenous cellulitis in immunocompromised patients. In general, we believe that pseudomonads are underestimated contributors to necrotizing cutaneous lesions, despite their inability to grow in the absence of oxygen. Empirical cover against such microbes should be initiated until it is contraindicated by laboratory results.

DIABETIC FOOT INFECTIONS

GENERAL POINTS

Diabetes is a common disease world-wide and a major cause of renal disease, cardiovascular disease, blindness, peripheral neuropathies and amputations of the lower extremities and fingers. Foot infections are a particularly significant clinical problem in diabetics, and are one of the commonest reasons for hospital admission in diabetic patients. On average, diabetic patients with such infections are in their fifth decade, have had diabetes for around 18 years, and require insulin for blood sugar control. The average hospital stay is more than 1 month, with almost 50% of these cases being hospitalized for over 3 months. Diabetic foot infections are clearly costly to both the patient and society (Bridges and Deitch, 1994). There has been a recent call to avoid amputation wherever possible in patients with diabetic foot ulcers (Anon., 1998c). Wound care, including antimicrobial agents where appropriate, is a major consideration in such an approach. Debridement, and if indicated revascularisation, results in long-term salvage of most (e.g. over 70%) threatened limbs.

In the presence of impaired blood supply or sensation, minor trauma to the foot can result in 'infections' ranging from colonization of neurotrophic ulcers to acute cellulitis (often originating in the toe and web spaces), which may rapidly progress to deep-space infection and less commonly to acute necrotizing cellulitis or fasciitis and gangrene. Toe and web-space infections which progress to deep-space infections are responsible for the highest rate of major limb amputations (because of the extensive involvement of extensor tendons and muscles of the foot). Septic arthritis involving the metatarsophalangeal joint may accompany these acute infections.

Neurotrophic ulcer, most commonly occurring over the first, second and fifth metatarsal heads, is one of the commonest presentations of diabetic foot infection. It is often independent of peripheral vascular disease, and results from localized repetitive trauma. The outwardly benign-looking ulcer can be deep and a forerunner of osteomyelitis or deep-compartment infection if a sinus tract develops between it and the deep compartment.

TREATMENT STRATEGIES

Control of localized infection involves adequate debridement of devitalized tissue, provision of a suitable environment for wound healing, modification of the weight-bearing plantar surfaces, and meticulous nail and skin care. The first three measures should be undertaken by surgeons with special interest/ expertise. With regard to antimicrobial choice, the following factors must be taken into consideration: the patient's environment (e.g. home vs. hospital); the nature of the initial injury (e.g. puncture wound, animal bite, neurotrophic ulcer); and whether or not the infection is limb-threatening. Non-limb-threatening infections (superficial, no systemic toxicity, minimal or no cellulitis or ulceration, insignificant ischaemia) involve *Staphylococcus aureus* as the major pathogen, although other bacteria (e.g. streptococci) may occasionally be isolated. Anaerobes and Gram-negative enterics are rare. In contrast, limb-threatening infections (extensive cellulitis, lymphangitis, deep ulcers, prominent ischaemia) are invariably polymicrobial, involving organisms such as *Staphylococcus aureus*, group B streptococci (*Streptococcus agalactiae*), enterococci, facultative Gram-negative bacilli, pseudomonads, and anaerobes such as *Bacteroides* species (e.g. *Bacteroides fragilis*) and peptostreptococci (see Table 14.4). Regimens that have been suggested in view of the bacteriology of non-limb-threatening infections include oral clindamycin, co-amoxyclav or cephalexin for mild infections, and metronidazole or clindamycin plus a parenteral cephalosporin (e.g. cefuroxime) for more complicated infections with cellulitis. In no instance should the potential seriousness of the infection be underestimated.

Initial therapy for limb-threatening infections must take into account the potential polymicrobial nature of the condition. While combinations of clindamycin (anaerobes, staphylococci, streptococci) together with an aminoglycoside (Gram-negative bacilli) have been used in the past, it is obvious that better alternatives (based largely on pharmacokinetics) to aminoglycosides now exist (e.g. ciprofloxacin and third-generation cephalosporins). More logical combinations include clindamycin plus ciprofloxacin, clindamycin plus ceftriaxone (although this lacks antipseudomonal cover) or, perhaps best of all, metronidazole (anaerobes) plus ciprofloxacin (Gram-negative bacilli, staphylococci) and cefuroxime (staphylococci, streptococci). Single agents (e.g. imipenem, meropenem, piperacillin/tazobactam, ticarcillin/clavulanic acid, co-amoxyclav and cefotetan) are also potential considerations.

A recent trial (Lipsky et al., 1997) compared the efficacy of the earlier fluoroquinolone ofloxacin (IV followed by oral administration) with IV ampicillin/sulbactam followed by oral co-amoxyclav in the therapy of diabetic foot infections. Despite the fact that ofloxacin is considered to have poor therapeutic activity against Gram-positive cocci and anaerobes, the overall clinical cure or improvement rate with this regimen was no different to that found with the β-lactam combination (85% vs. 83%). In this survey, Gram-positive cocci represented 63% of the isolates, mixed infections were not detected, and most of the subjects had cellulitis or an infected ulcer without osteomyelitis. This study

Table 14.4 Microbiological aspects of diabetic foot infections (often polymicrobial in limb-threatening infections)

Common microbes	Frequency[a]
Obligate anaerobes	
Bacteroides fragilis	++
Other Bacteroides species	+
Clostridium perfringens	++
Peptostreptococci	+
Aerobes/facultative anaerobes	
Staphylococcus aureus	+++
Coagulase-negative staphylococci	+
Pseudomonas aeruginosa	++
Enteric Gram-negative bacilli (e.g. Proteus)	++
Enterococci	++
Streptococci (e.g. group B)	++
Other Gram-negative bacilli (including environmental microbes)	+

[a] +++, major pathogens; ++, common; +, less common/significant.

also raises the question of the identity of the significant pathogens (compared to 'insignificant colonizers') in diabetic foot ulcers.

Antibiotics should not be stopped until the wound appears clean and all signs of cellulitis have disappeared. If osteomyelitis is present, a 4- to 6-week course of antibiotics is generally required (see p. 231). Inclusion of antibiotics such as ciprofloxacin is favoured for hypoxic osteomyelitic wounds. If the osteomyelitis involves *Staphylococcus aureus*, antimicrobial agents such as fusidic acid and rifampicin should be included.

As diabetes has been associated with impaired neutrophil chemotaxis and phagocytosis and reduced intracellular killing, supplementation of chemotherapy with granulocyte colony-stimulating factor (G-CSF, filgrastim) warrants consideration. The results of preliminary trials appear promising (Gough *et al.*, 1997).

BITE AND PUNCTURE WOUNDS

GENERAL POINTS

Nail puncture wounds of the foot are common injuries that occur in all age groups, and up to 15% of such wounds may become infected, resulting in cellulitis or localized deep-tissue abscesses. Infection may be complicated by osteochondritis with or without pyoarthrosis (Raz and Miron, 1995). The commonest (> 90%) infecting microbe is *Pseudomonas aeruginosa*, followed by *Staphylococcus aureus*. Pseudomonads appear to be common inhabitants of the inner sole of trainers and gym shoes, and are apparently inoculated into the foot tissues by nails or similar sharp objects penetrating through the sole. Treatment should initially involve extensive incision and debridement, drainage of pus, and oral ciprofloxacin for up to 14 days (Raz and Miron, 1995).

Animal bites, usually from cats or dogs, are common world-wide and often require the recipient to seek medical attention. It has been estimated that serious complications occur in up to 10% of patients with dog bite wounds and as many as 30% of patients with cat bite wounds (see Citron *et al.*, 1996). Figures from the UK suggest that around 200 000 people suffer dog bites each year; in the USA this figure has been estimated at 4.7 million (1.8% of the population). In addition, 400 000 people annually are bitten by cats in the USA (Goldstein, 1998).

Dog bites most frequently occur in males under 20 years of age, whereas cat bites are most common in 30- to 40-year-old women. More bites occur in the warmer months of the year, and are provoked in an animal either owned by or known to the victim. Bite wounds range from lacerations and evulsions to punctures, and may be accompanied by extensive trauma and crush injury. Infectious complications occur in 15–20% of injuries and include cellulitis, abscess formation, septic arthritis, osteomyelitis and bacteraemia. Between 1979 and 1994, attacks by dogs (predominantly pit bull terriers and Rottweilers) resulted in 279 deaths in the USA (Goldstein, 1998).

MICROBIAL AETIOLOGY

The bacteriology of animal bite wounds is diverse, and reflects the oral flora of the animal involved (see Table 14.5). Infections involve both aerobes and obligate anaerobes, usually in combination. Although *Staphylococcus aureus*, *Pasteurella multocida* and anaerobic cocci (peptostreptococci) have been widely accepted as the pathogens most commonly involved in animal bite wounds, other unusual Gram-negatives (e.g. *Weeksella zoohelcum*, *Capnocytophaga canimorsus*, *Neisseria weaveri*) and obligate anaerobes (e.g. *Porphyromonas salivosa*, *Prevotella heparinolytica*, *Bacteroides tectum*) clearly play important roles. *Eikenella corrodens* is a Gram-negative bacillus which has been recovered from 'bites' of human origin (Goldstein, 1995, 1997; Goldstein *et al.*, 1999).

Pasteurella species recovered from cat and dog bite wounds have been differentiated into new species and *Pasteurella multocida* subspecies (see Goldstein, 1998). *Pasteurella multocida* subspecies *multocida* and subspecies *septica* are most often isolated from serious infections, with the latter revealing tropism for the central nervous system. *Pasteurella canis* (previously *Pasteurella multocida* biotype 6 or 'dog type') has only been isolated from bites of canine origin, with a number of other species (e.g. *Pasteurella dagmatis, Pasteurella stomatis*) regularly recovered from cat scratch and bite wounds.

With the recent advances in anaerobe technology, our understanding of the important role of obligate anaerobes in many infectious processes involving the oral cavity and gastrointestinal tract has increased dramatically. Anaerobes are, after all, by far the dominant microbes in the normal flora. Although *Porphyromonas* species represent a significant component of the anaerobic flora of the oral cavity of cats, and have been recovered from almost 100% of diseased cat and dog gingival pockets, and from abscesses sustained in cat fights, they have not been regularly associated with human infections following cat bites. However, in 1996 Citron *et al.* were able to isolate 40 strains of *Porphyromonas* (formerly pigmented *Bacteroides* species) from 29 of 102 cat and dog bite wounds in humans. *Porphyromonas salivosa*, *Porphyromonas gingivalis* (the most frequent *Porphyromonas* species in the human oral cavity) and *Porphyromonas canoris* were the most frequent isolates. Some of the isolates appeared to represent previously undescribed species. It was noteworthy that the isolates required at least 5 days of incubation before significant growth was visible.

Several saccharolytic *Bacteroides* and *Prevotella* species have been regularly isolated from dog and cat bite wounds. These include *Bacteroides tectum* (often mistaken for *Prevotella bivia*), *Prevotella heparinolytica*, *Prevotella zoogleoformans*, *Prevotella buccae* and *Prevotella oris*. Because of the slow growth of many of these species, microbiologists are advised to keep plates for 7 days before they are discarded (Goldstein, 1998).

TREATMENT STRATEGIES

The principles of therapy consist of irrigation, cautious debridement, elevation, recognition and monitoring of common complications, and appropriate antimicrobial therapy (Goldstein, 1998). The choice of antimicrobial agent favours those

Table 14.5 Microbiological aspects of bite-wound infections (microbes may originate from the environment, the victim's skin or the biter's oral cavity)[a]

More significant and/or common microbes	Comment
Aerobes/facultative anaerobes	
Actinobacillus species	*Actinobacillus lignieresii* (horse bites), *Actinobacillus actinomycetemcomitans*
Capnocytophaga canimorsus[c]	Fatal disease in asplenia and those with liver disease; very slow growing; dog bite
Eikenella corrodens[c]	Human and dog bite
Flavobacterium species	Pig bite
Haemophilus species	*Haemophilus felis*, *Haemophilus aphrophilus*
Moraxella species	
Neisseria species	*Neisseria weaveri*, *Neisseria canis*
Pasteurella multocida[c]	20–50% of cases
Other *Pasteurella* species	Often listed as *Pasteurella multocida* (e.g. *P. canis*, *P. dagmatis*, *P. stomatis*, *P. haemolytica*)
Pseudomonas aeruginosa	Environmental microbe
Staphylococcus aureus[c]	20–40% of cases
Other staphylococci	Including coagulase-positive *Staphylococcus intermedius*
Streptococci[c]	Around 50% of cases; mainly β- and α-haemolytic
Weeksella zoohelcum	Unusual Gram-negative
Obligate anaerobes	Around 10% of cases; always polymicrobial
Actinomyces species	
Bacteroides fragilis	
Bacteroides tectum[c]	
Bacteroides forsythus[c]	
Clostridium perfringens	
Eubacterium species	*Eubacterium lentum*, *Eubacterium moniliforme*
Fusobacterium nucleatum[c]	
Other fusobacteria	
Leptotrichia buccalis	
Porphyromonas species[b,c]	*Porphyromonas asaccharolytica*, *Porphyromonas salivosa*
Prevotella species[b,c]	*Prevotella bivia*, *P. intermedia*, *P. heparinolytica*, *P. melaninogenica*, *P. oris-buccae*, *P. zoogleoformans*
Peptostreptococcus species[c]	*Peptostreptococcus anaerobius*, *Peptostreptococcus magnus*
Veillonella parvula	

[a]Adapted from Goldstein (1997) and Goldstein *et al.* (1999).
[b]Many of these were originally listed as 'oral *Bacteroides*' species.
[c]More common and/or significant pathogens for empirical antibiotic cover.

that are active against *Staphylococcus aureus*, pasteurellae, streptococci and obligate anaerobes such as *Bacteroides tectum*, *Prevotella heparinolytica*, *Porphyromonas salivosa*, fusobacteria and peptostreptococci. The oral Gram-negative anaerobic bacilli (*Porphyromonas*, *Fusobacterium*, *Prevotella* and *Bacteroides* species) have usually been regarded as susceptible to penicillin, although in the last decade increasing rates of resistance to this and other antimicrobial agents, such as clindamycin and cefoxitin, have become evident. However, oral co-amoxyclav with or without a more proven anti-anaerobe agent (e.g. metronidazole, clindamycin) or cefoxitin alone still remain obvious therapeutic choices.

Other agents which can theoretically be considered include fluoroquinolones like clinafloxacin, gatifloxacin and levofloxacin, which have excellent activity against almost all of the aerobic oral bacteria mentioned above, and in the case of the first two many of the significant oral anaerobes. Fusobacteria, including *Fusobacterium nucleatum*, are exceptions, with all current quinolones showing poor activity against these anaerobes (Goldstein, 1998). The potential role of newer macrolides/azalides (e.g. azithromycin, clarithromycin) and ketolides is unknown, although compared to erythromycin they show increased activity against many of the causative microbes.

chapter 15

LOCOMOTOR/ORTHOPAEDIC INFECTIONS

OSTEOMYELITIS

The term 'osteomyelitis' refers to medullary infections involving trabecular bone and marrow. Although some authorities prefer the term 'osteitis', which is more logical, convention has seen the retention of osteomyelitis, a term that first appeared in the French literature in the mid-nineteenth century (Norden *et al.*, 1994). In acute disease, the vascular supply to the bone is compromised and proteolytic enzymes released from phagocytes aid microbial spread. The pus formed also spreads into vascular channels, raising the intra-osseous pressure and impairing blood flow. When both the medullary and periosteal blood supplies are compromised, large areas of dead devascularized bone (sequestra) may be formed. Within this necrotic and ischaemic tissue, bacteria may be difficult to eradicate even with intensive antibiotic therapy and surgery. Chronic disease may develop from the acute form and result in a focus of infected dead bone or scar tissue, an ischaemic soft tissue envelope, and a refractory clinical course. Cure is not guaranteed in osteomyelitis, as the infection may recur years after apparent successful therapy.

There are various classifications of bone infections. One of these, devised by Waldvogel and colleagues, is based on whether the osteomyelitis is haematogenous in origin or arises secondary to a contiguous focus of infection (see Mader and Calhoun, 1995). This classification will be followed for the purpose of this chapter.

HAEMATOGENOUS OSTEOMYELITIS

Haematogenous osteomyelitis is characteristically a disease of children, although it may occur at any stage of life. The typical infection is juxtaphyseal in the metaphysis of a growing long bone. In adults the majority of haematogenous infections occur in the spine (Norden *et al.*, 1994). The pathogenesis appears to involve minor trauma (which produces a small haematoma, vascular obstruction and subsequent bone necrosis) that results in an area of bone susceptible to inoculation from a transient bacteraemia. In recent years it seems that the incidence of childhood infections of the long bones associated with *Staphylococcus aureus* bacteraemia has decreased, while there has been an increasing frequency of haematogenous *Staphylococcus aureus* osteomyelitis of the vertebral column, especially in older patients (i.e. those over 50 years of age) (Jensen *et al.*, 1997). While this may reflect an increased rate of bacteraemia among older patients, other factors (e.g. degenerative changes) may be important. Most cases do not have an obvious portal of entry, while the male-to-female ratio is even higher than that seen in osteomyelitis of other bones where males dominate.

Microbial aetiology

A single pathogenic microbe is invariably recovered from the bone lesion. These microbes tend to vary with the patient's age, no doubt reflecting the varying causation of bacteraemias in different age groups. However, the importance of *Staphylococcus aureus* in all age groups is a striking feature. It accounts for at least 80% of cases of acute haematogenous osteomyelitis in children under 15 years of age, and 50–60% of all cases of haematogenous osteomyelitis in all age groups (Jensen *et al.*, 1997). Thus *Staphylococcus aureus*, group B streptococci (*Streptococcus agalactiae*) and coliforms (e.g. *E. coli*) are common in infants, *Staphylococcus aureus*, group A streptococci (*Streptococcus pyogenes*) and, until recently, *Haemophilus influenzae* in children over 1 year of age, *Staphylococcus aureus* in adults, and Gram-negative bacilli and increasingly *Staphylococcus aureus* in the elderly (Jensen *et al.*, 1997; Lew and Waldvogel, 1997). A major factor in the establishment of bone infections is the ability of the causal microbes to adhere to osseous or associated structures (e.g. fibronectin, laminin, collagen). This is enhanced where pre-existing damage occurs; normal bone is highly resistant to infection. A list of the microbes more commonly associated with osteomyelitis is shown in Table 15.1.

Vertebral osteomyelitis (usually involving two adjacent vertebrae and the intervertebral disc) involves *Staphylococcus aureus* in most cases, although *Pseudomonas aeruginosa* is a significant causal agent in intravenous drug users (often cervical vertebrae) or in those with long-term urinary catheters (often lumbar vertebrae). Where disease involves blood-borne spread from infected sites such as the respiratory tract, genito-urinary tract, skin and soft tissues, or an infected intravenous site, the microbes responsible are invariably those which commonly infect that site (e.g. coagulase-negative staphylococci from an infected cannula).

Treatment strategies

Antibiotics and surgery must be considered in the treatment of all cases of musculoskeletal infection (Norden *et al.*, 1994). The attending clinician must answer the following questions regarding surgery.
1. Is surgery necessary for this patient?
2. When should surgery be performed?
3. What operative procedure is indicated?

With regard to antimicrobial agents, many ponderables exist.
1. Which antibiotics should be used for the initial treatment?
2. Which antibiotics should be used for specific treatment?
3. What dose of antibiotic should be given and by what route?
4. For how long should the antibiotic be given?

If the results of investigations suggest osteomyelitis, antimicrobial therapy should be commenced either empirically on the basis of the likely microbe, depending on the patient's age and any predisposing conditions, or on microscopy-visible and/or

Table 15.1 Microbes that are more commonly recovered from osteomyelitis[a]

Microbe	Relative frequency	Comment
Staphylococcus aureus	++++	Most common microbe
Coagulase-negative staphylococci	++	Foreign-body-associated infections
Gram-negative enteric bacilli	++	Nosocomially acquired; elderly
Streptococci and oral anaerobes	++	Associated bites/tooth-inflicted injuries
Pasteurella multocida, Eikenella corrodens	+	Human or animal bites
Pseudomonas aeruginosa	+	Nosocomial; IV drug abuse; cervical or lumbar vertebrae
Salmonellae, *Streptococcus pneumoniae*	+	Sickle-cell disease
Gut anaerobes	+	Diabetic foot lesions
Mycobacterium tuberculosis	+	
Fungi (e.g. aspergilli, *Candida albicans, Scedosporium prolificans*)	+	Immunocompromised; catheter related; IV drug abuse
Bartonella henselae	+	HIV infection

[a]Adapted from Lew and Waldvogel (1997).

isolated microbes. If a significant abscess (either sub-periosteal or in the overlying soft tissues) is identified, it should be drained by surgery or, if readily accessible, by aspiration through a wide-bore needle. If an abscess has not been identified, antibiotics and supportive therapy should be followed by regular and critical review. Persisting systemic signs indicate a high probability of an abscess, for which a rigorous search should therefore be made (Norden *et al.*, 1994). An algorithm for the investigation and management of this disease is shown in Figure 15.1.

Whether and when surgical intervention is indicated may be a difficult decision, and the place of surgery has yet to be precisely defined by controlled trials. The balance of available evidence supports the policy of selective surgery, with positive indications including the presence of a significant soft-tissue or sub-periosteal abscess at the time of presentation, or the development of such an abscess following initial conservative therapy, and suspected joint involvement. At surgery, necrotic soft tissues should be excised and copious irrigation used to wash out or dilute any remaining bacteria. Complications of haematogenous osteomyelitis include recurrent or chronic infection, involvement of an adjacent joint, pathological fracture, and disturbance of growth (Norden *et al.*, 1994).

In the case of chronic osteomyelitis, surgical management includes adequate drainage, extensive debridement of all necrotic tissue, obliteration of dead spaces, adequate soft-tissue coverage and restoration of an effective blood supply

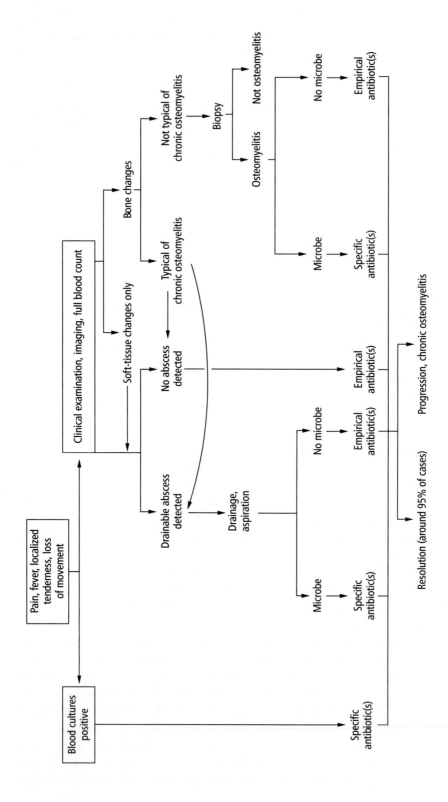

Figure 15.1 Algorithm for investigation and management of acute haematogenous osteomyelitis. Adapted from Norden et al. (1994).

(Mader *et al.*, 1997). Even when all of the necrotic tissue has been removed, the remaining bed of tissue must be regarded as contaminated with the infecting microbe(s). Unfortunately, adequate bone debridement may leave a large bony defect or dead space. This must be replaced with durable vascularized tissue and the wound closed. Although local tissue flaps may be used to fill dead spaces, an alternative is to use cancellous bone grafts beneath local or transferred tissues. Obviously careful pre-operative planning is critical. Adequate soft-tissue cover of bone is necessary to arrest osteomyelitis.

Antibiotic-impregnated acrylic beads may be used in an attempt to sterilize and temporarily maintain dead space. The beads are usually removed within 2 to 4 weeks and replaced with a cancellous bone graft. The most commonly used antibiotics in beads are vancomycin, ofloxacin, tobramycin and gentamicin. Local delivery of antibiotics such as amikacin and clindamycin into dead space has been performed with an implantable pump (Mader *et al.*, 1997).

If movement is present at the site of infection, permanent stabilization of the skeletal structures using plates, screws and/or external fixation is required. Because of the risk of secondary infection, external fixation is preferred to internal devices, which readily become colonized. New methods of external fixation have recently been developed (see Mader *et al.*, 1997).

After cultures have been obtained, a parenteral antimicrobial regimen is started in order to cover the clinically suspect pathogens (see Table 15.2). This initial empirical regimen is continued or changed following the results of laboratory tests. Possibilities for *Staphylococcus aureus* include flucloxacillin (12 g/day IV divided 4-hourly), cefazolin (up to 4 g/day IV 6-hourly), fusidic acid (up to 1.5 g/day IV divided 8-hourly) or clindamycin (up to 1.2 g/day IV divided 6-hourly). The latter two drugs are suitable for patients who are allergic to β-lactams (see Turnidge and Grayson, 1993). Fusidic acid, rifampicin, ciprofloxacin and to some extent cefazolin all appear to penetrate bone better than most other agents, and some authorities suggest a combination of flucloxacillin and fusidic acid (or rifampicin), at least in the first couple of weeks. Another widely advocated combination for *Staphylococcus aureus* is a combination of ciprofloxacin and rifampicin, although its superiority over more conventional β-lactam-containing regimens is largely unproven. Newer fluoroquinolones may be superior to ciprofloxacin in this combination. An unappreciated option may be cotrimoxazole, especially for methicillin-resistant *Staphylococcus aureus* (Stein *et al.*, 1998).

Treatment is usually for 4–6 weeks after the last major debridement surgery in adults, although shorter courses may be adequate in children. Where feasible, a change to oral therapy after 2–4 weeks, or a combined parenteral/oral combination, can be considered. Additional information on the duration of antibiotic regimens can be found in the excellent review by Mader *et al.* (1997). Because of the limited bioavailability of high-dose oral β-lactams and poor patient tolerance, early intravenous/oral switch therapy with these drugs is not recommended in adults with acute osteomyelitis. Antimicrobial agents that are suitable for infections involving Gram-negative bacilli include ceftriaxone (2 g once daily) and ciprofloxacin (up to 750 mg orally 12-hourly). Cephalosporins such as ceftriaxone and cefodizime appear to

Table 15.2 Possible antimicrobial considerations for infections of bones and joints[b]

Causative microbe	Antimicrobial agent[a]	Alternatives[a]
Staphylococcus aureus		
Penicillin susceptible	Penicillin G	Cephazolin, fusidic acid, clindamycin
Penicillin resistant	Penicillinase-stable penicillin, (e.g. flucloxacillin, nafcillin)	Cephazolin, fusidic acid, clindamycin, ciprofloxacin or clinafloxacin both plus rifampicin
Methicillin-resistant (MRSA)	Vancomycin	Depends on local sensitivity patterns – one or combinations from cotrimoxazole, gentamicin, fusidic acid, rifampicin, ciprofloxacin, clinafloxacin
Staphylococcus epidermidis	Penicillinase-stable penicillin, netilmicin, vancomycin	Cephazolin, clindamycin, ciprofloxacin
Streptococci	Ceftriaxone, penicillin G	Cefuroxime, cephazolin, clindamycin, erythromycin
Enteric Gram-negative bacilli	Ciprofloxacin, ceftriaxone	Piperacillin/tazobactam, meropenem
Pseudomonas aeruginosa	Ciprofloxacin plus tobramycin	Ceftazidime or piperacillin or cefepime, all with tobramycin
Obligate anaerobes	Metronidazole	Clindamycin, co-amoxyclav, imipenem, meropenem
Mixed aerobes/anaerobes	Co-amoxyclav,[c] piperacillin/tazobactam	Ticarcillin/clavulanate, imipenem, meropenem
Candida	Fluconazole	Amphotericin B
Filamentous fungi/moulds	Amphotericin B	

[a]Check local sensitivity patterns.
[b]Adapted from Lew and Waldvogel (1997), Norden *et al.* (1994) and Mader *et al.* (1997).
[c]May have inadequate Gram-negative cover.

penetrate well into the bones of patients undergoing hip arthroplasty (Scaglione *et al.*, 1997). Possible choices of antimicrobial agent are summarized in Table 15.2.

POST-TRAUMATIC AND CONTIGUOUS OSTEOMYELITIS

Microbes may be inoculated directly into bone by trauma, animal bites or during surgery, or they may extend from an adjacent soft-tissue infection. In contrast to the haematogenous form, multiple microbes may be recovered from infected bone. However, coagulase-positive and coagulase-negative staphylococci remain the dominant pathogens (75% of cases), followed by Gram-negative bacilli and obligate anaerobes. Infections usually appear insidiously 1 month or more after 'inoculation'. In patients with diabetes or vascular insufficiency, osteomyelitis is frequently found in the feet (see p. 219).

Chronic infection may develop from both haematogenous and contiguous osteomyelitis, and the nidus of the persistent contamination must be removed before the infection regresses.

Treatment of contiguous and chronic osteomyelitis is complicated by the fact that there is poor diffusion around the infected necrotic bone. Adequate surgical intervention (e.g drainage, debridement, removal of dead space) and culture is mandatory. Therapy involving polymicrobial cover can then be initiated. This is subsequently assessed and/or altered according to laboratory results. Treatment alternatives include single agents such as piperacillin/tazobactam or similar β-lactam combinations, imipenem or meropenem, or combinations of ciprofloxacin or ceftriaxone with clindamycin or metronidazole. The role of broad-spectrum fluoroquinolones, such as clinafloxacin, has yet to be assessed. Antimicrobial agents are continued for 4–6 weeks after the last major debridement surgery. The availability of oral agents (e.g. ciprofloxacin) is a major advance, although patients should receive at least 2 weeks of parenteral therapy before being switched to an oral regimen. An algorithm for the investigation and management of post-traumatic osteomyelitis is shown in Figure 15.2.

UNUSUAL FORMS OF OSTEOMYELITIS

Osteomyelitis is a complication of intravenous drug addiction. Lesions most commonly involve the vertebrae (cervical), pubis and clavicles. Microbes which have been recovered from blood cultures and/or infected bone include staphylococci (both *Staphylococcus aureus* and *Staphylococcus epidermidis*), Gram-negative bacilli (including pseudomonads) and *Candida*. All of these are 'regular' contaminants of dirty syringes.

Bone infection is a significant complication of sickle-cell disease and vascular diseases (e.g. atherosclerosis). Unusual but common pathogens in this setting include *Salmonella* species, other Gram-negative bacilli and *Streptococcus pneumoniae*.

Salmonellae

As early as 1876, salmonellae were recognized as a cause of osteomyelitis, when periostitis involving mainly the long bones was noted during the convalescent stage of typhoid fever (see Santos and Sapico, 1998). About 13 years later the entity called typhoid spine was recognized. Of the reported cases of salmonella osteomyelitis, about 20% appear to have vertebral involvement. It is not known what influence the ability of bacteria such as *Salmonella typhi* to survive within phagocytic cells and the reticulo-endothelial system (including bone marrow) has on the pathogenesis of this entity (Norden *et al.*, 1994).

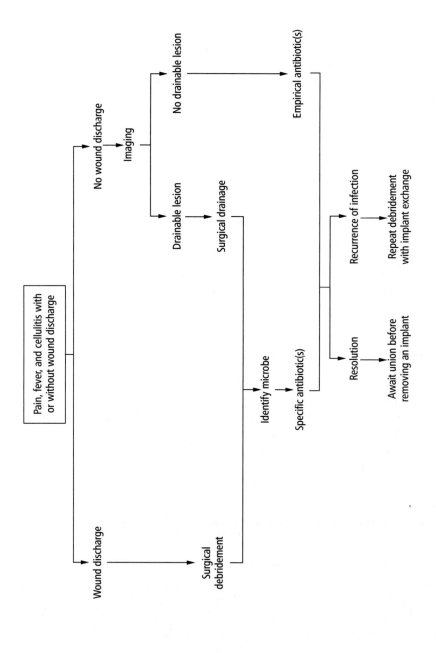

Figure 15.2 Algorithm for the investigation and management of post-traumatic or post-surgical osteomyelitis. Adapted from Norden et al. (1994).

Vertebral osteomyelitis involving salmonellae primarily involves the lumbar region (about 70% of cases), is invariably monomicrobial and is predominantly found in the male sex. Blood cultures are positive in about 50% of cases, with stool and urine being other potential diagnostic sources of the offending microbe. A wide variety of salmonellae have been implicated, of which *Salmonella typimurium*, *Salmonella typhi*, *Salmonella enteritidis* and *Salmonella cholerasuis* seem to be most common, and the possibility of lower animal products (e.g. meat) being the source of the microbe clearly exists (Santos and Sapico, 1998). Infected abdominal aortic aneurysms appear to be a feature of the disease in older patients. Although primarily treated medically with prolonged courses of antibiotics (mean duration of about 60 days), patients with infected aneurysms may require resection with thorough debridement, extra-anatomical bypass grafting and prolonged antibiotic therapy (Santos and Sapico, 1998). Even with such an approach, mortality levels are still high in older patients. Potentially useful antimicrobial agents include fluoroquinolones such as ciprofloxacin (probably the agent of choice), ceftriaxone and, where sensitivities suggest, amoxycillin or cotrimoxazole.

Mycobacterial

Haematogenous spread of *Mycobacterium tuberculosis* early in the course of primary infection may give rise to skeletal infection, but rarely does such disease arise from spread from an adjacent caseating lesion (e.g. lymph node). In children and adolescents, metaphysis of long bones is most frequently involved, while in adults, the axial skeleton followed by the proximal femur, knee and small bones of the hands and feet are the main bones involved. Infection is usually localized to one site. Therapy involves a prolonged course of combined antimicrobial agents (e.g. isoniazid, rifampicin, pyrazinamide), and in some cases surgery.

Fungal

A variety of fungi (e.g. *Candida*, cryptococci, zygomycetes, *Scedosporium prolificans*, *Coccidioides*) have been associated with bone infections. These usually occur as a complication of catheter-related or contaminated drug injection fungaemias, or following prolonged periods of neutropenia. Therapy involves surgical debridement and appropriate antifungal chemotherapy (see p. 126). Prolonged courses of oral fluconazole have shown promise with disease involving *Candida albicans*. Some fungi (e.g. *Scedosporium prolificans*) appear to be refractory to many of the currently available antifungal agents.

PROSTHETIC INFECTIONS

Joint replacement surgery (e.g. hip, knee, shoulder) has been one of the major medical advances during the last 25 years, with millions of individuals world-wide benefiting from indwelling prosthetic articulations. Unfortunately, up to 5% of indwelling prostheses become infected. The cost of treating such infections has been estimated to be up to 80 million dollars per year in the USA alone (Brause, 1995).

The pathogenesis of prosthetic joint infections is similar to that of osteomyelitis (i.e. haematogenous or contiguous spread). The latter follows wound sepsis adjacent to the prosthesis or contamination during the operative procedure. Any factors that delay wound healing (e.g. necrosis, wound infection, suture abscess) increase the risk of infection. Certain patient populations also seem to be more prone to developing post-operative prosthetic device infections (e.g. those who are overweight, very old, have underlying diseases such as diabetes mellitus or rheumatoid arthritis, are debilitated and have a poor nutritional status, or are receiving corticosteroids). In a recent 22-year (1969–1991) study at the Mayo Clinic on risk factors for prosthetic joint infection, the major risk factors identified were the development of a surgical site infection not involving the prosthesis, a National Nosocomial Infections Surveillance System surgical patient risk index score of 1 or 2, the presence of malignancy, and a history of joint arthroplasty (Berbari et al., 1998). Infections usually occur in osseous tissues adjacent to the foreign device (i.e. at the bone–cement interface).

Not surprisingly, coagulase-negative staphylococci (e.g. the skin commensal *Staphylococcus epidermidis*), with their added ability to adhere to foreign materials, are the most common causal agent (representing around 30% of infections), followed by coagulase-positive staphylococci (25%), Gram-negative bacilli, including pseudomonads (20%), β-haemolytic streptococci, including non-group-A types (12%), oral streptococci (8%) and obligate anaerobes (see Table 15.3 and Brause, 1995). However, the potential spectrum of causative microbes is unlimited, and may include a number of organisms that are often wrongly discarded as contaminants (e.g. *Bacillus* species, fungi). A slow-growing or 'small colony variant' of *E. coli* has been recovered from a chronic hip infection (Roggenkamp et al., 1998). Generally infection is monomicrobial, although mixed participation may occur. Over the last 20 years there seems to have been little change in the group of pathogens responsible for early infection of orthopaedic prostheses, despite major changes in prophylactic regimens.

Microbes that adhere to surfaces and/or grow in thick biofilms show increased phenotypic resistance to antimicrobial agents. Presumably this is in some way associated with failure of the antibiotic to penetrate the mucoid biofilm. This phenomenon occurs with prosthetic joint infections, and has complicated their chemotherapy. In addition, some microbes (e.g. *Staphylococcus aureus*) may survive for long periods of time within osteoblasts, protected to some degree from many antibiotics. These strains may grow only slowly ('small colony variants') in culture. Simple surgical drainage without removal of the involved prosthesis, followed by antimicrobial therapy has a high failure rate. Complete removal of all foreign materials that are likely to be colonized by the microbes (e.g. prosthesis, metallics, cement) is therefore an important component of any treatment protocol. This must be accompanied by appropriate antimicrobial therapy, selected after consultation between the orthopaedic surgeon and the clinical microbiologist.

One suggested protocol involves removal of the prosthesis and cement followed by a 6-week course of an appropriate bactericidal agent chosen on the basis of laboratory sensitivity tests. Reimplantation is performed at the conclusion of the antibiotic course. Such a two-stage procedure has been reported to have a success rate of 90% or higher (Brause, 1995). If treatment involves removal of the

Table 15.3 Bacteriology of prosthetic joint infections[a]

Bacteria	Relative frequency	Comment
Staphylococci	+ + + +	
Staphylococcus aureus	+ +	Nasal cavity, skin source
Coagulase-negative species	+ +	Skin source
Streptococci	+ +	
β-Haemolytic	+	Oral cavity, skin-lesion source
Viridans group	+	Dento-gingival source
Enterococci	+	Urinary/intestinal source
Gram-negative bacilli	+ +	Genito-urinary or intestinal source
Obligate anaerobes (e.g. peptostreptococci)	+	Dento-gingival or intestinal source

[a]Adapted from Brause (1995).

prosthesis and associated materials, followed immediately by reimplantation of a new prosthesis using antibiotic-impregnated cement and then a course of antibiotic, success rates are less than with the two-stage procedure, especially if *Staphylococcus aureus* is involved (Lew and Waldvogel, 1997).

If a combination of situations occurs (e.g. the infected prosthesis cannot be safely removed, the pathogen involved is considered to be relatively avirulent and highly susceptible to an appropriate tolerable oral antibiotic, and the prosthesis is not loose), continuous lifelong oral suppressive therapy can be considered as an alternative to the usual medico-surgical treatment (long-term parenteral antimicrobial agents in combination with surgical debridement, and removal of the foreign material or its subsequent one- or two-stage replacement). In a recent study from France involving patients infected with multi-resistant staphylococci (susceptible only to glycopeptides and cotrimoxazole), long-term oral ambulatory treatment with cotrimoxazole appeared to be an effective alternative to conventional medico-surgical treatment (Stein *et al.*, 1998). High doses (trimethoprim, 20 mg/kg body weight/day and sulphamethoxazole, 100 mg/kg/day) were used for 6 months for prosthetic hip infections (with removal of any unstable prosthesis after 5 months of treatment), for 9 months for prosthetic knee infections (with removal of unstable devices after 6 months) and for 6 months for infected osteosynthetic devices (with removal of the device after 3 months if necessary). Their overall treatment success rate was 67% (26 of 39 patients). Eight patients stopped the treatment because of side-effects, while one patient was non-compliant. Resistant bacteria developed in three patients in whom treatment failed.

Another recent study has suggested that debridement and prolonged suppressive antibiotic therapy (up to 8 weeks of parenteral treatment followed by prolonged oral therapy) may be an effective alternative to removal of infected orthopaedic prostheses in some patients (Segreti *et al.*, 1998). Although this study indicates that attempts to salvage infected devices in selected patients are entirely appropriate, well-designed studies are clearly needed to define both the patient population that is likely to benefit and the optimal antimicrobial regimens to be employed (Karchmer, 1998). The combination of a fluoroquinolone and rifampicin for the treatment of staphylococcal orthopaedic prosthetic infections warrants serious consideration.

Because of the catastrophic repercussions of an infected prosthesis, prevention is far better than cure. Pre-operative conditions which may result in transient bacteraemias and prosthesis colonization (e.g. pyogenic oral lesions, cystitis) should be identified and eliminated and peri-operative antibiotic prophylaxis should be utilized (see p. 119). Some hospitals insist that patients colonized with troublesome hospital-acquired microbes must be excluded from (or near to) wards caring for patients who are having elective prosthetic surgery. An extensive study mounted by the Medical Research Council and the Public Health Laboratory Service in the UK has shown that a combination of special operative techniques (e.g. 'clean-air' theatres, meticulous antisepsis) plus antimicrobial prophylaxis can reduce the infection rate to about 0.1% (see Slack, 1995). The incorporation of antibiotics (e.g. gentamicin, vancomycin, ciprofloxacin) into bone cements warrants serious consideration (Tunney *et al.*, 1998). In addition, once the prosthesis is in place, any subsequent procedures which may result in transient bacteraemias (e.g. dental surgery, genito-urinary tract manipulations, cystoscopy), must be covered by appropriate antibiotic prophylaxis.

MICROBIOLOGICAL ASPECTS OF LARGE-JOINT SEPSIS

The microbes that are reported to cause bacterial arthritis vary according to the population being studied, and particularly with the age of the patient and whether the joint is native or prosthetic. Some of these features are summarized in Table 15.4, while a more extended list can be found in the excellent review by Atkins and Bowler (1998). Important considerations in the microbiological diagnosis include blood cultures, a peripheral leucocyte count, and aspiration of the joint in adults (ultrasound in children) followed by Gram stain and culture. Multiple specimens, especially with prosthetic joints, may be required to recover the causal microbe and/or to assess the significance of any isolates (Atkins and Bowler, 1998).

Table 15.4 Common microbes associated with septic arthritis[a]

Microbe	Group (relative frequency)			
	Adults	Children	Neonate	Prosthetic
Staphylococcus aureus	++++	+++	+++	++
Coagulase-negative staphylococci				+++
Streptococcus pneumoniae	++	++		
β-Haemolytic streptococci	++[c]	++[c]	++[d]	
(e.g. *Streptococcus pyogenes*)				
Neisseria gonorrhoeae	+++		+	
Enterobacteriaceae	++[e]	+	++	+
(e.g. *E. coli*, salmonellae)				
Pseudomonas aeruginosa	+		+	+
Haemophilus influenzae	+	+[b]	+	
Mycobacterium tuberculosis	+	+		+
Kingella kingae		+		
Obligate anaerobes	+			+
Fungi (e.g. *Candida*)	+			+

[a]Adapted from Atkins and Bowler (1998).
[b]Should be rare after introduction of Hib vaccine in children.
[c]Mainly group A (*Streptococcus pyogenes*).
[d]Mainly group B (*Streptococcus agalactiae*).
[e]When aged over 60 years, or other predisposing factors.

chapter 16

INFECTION IN THE IMMUNOCOMPROMISED AND TRANSPLANT PATIENT

HOST BARRIERS TO INFECTION

Well-recognized and virulent pathogens such as *Staphylococcus aureus*, *Streptococcus pneumoniae* and *Bacteroides fragilis* cause infections in individuals with normal as well as impaired host defence mechanisms. In contrast, microbes with low virulence – the so-called 'opportunistic pathogens' – are only capable of causing disease in severely debilitated and immunocompromised patients. These diseases are known as opportunistic infections. A classical example of an opportunistic pathogen is the coagulase-negative staphylococcus which for many years was simply regarded as a normal commensal incapable of initiating pathological processes. However, it is now obvious that this heterogenous group includes a wide range of species, many of which are undoubtedly pathogens, albeit of low virulence (Howard, 1994; Patel and Paya, 1997; Singh, 1998). To cause disease they must either be present in enormous numbers or find a suitable compromised host or artificial attachment material.

Normal host defences can conveniently be subdivided into those related to the normal structure and function of the skin and mucous membranes (mucocutaneous defences), those related to blood and tissue fluids (humoral defences), and those related to the white blood cell populations (cellular defences).

Mucocutaneous defences include the physical barriers of the skin and mucous membranes, the dryness, acidity, normal flora and secretions associated with the skin, and a variety of factors associated with mucous membranes which may be anatomically restricted (e.g. cilia, cough reflex, enzyme production, gastric acidity, peristalsis and normal flora). Defence mechanisms associated with blood and tissue fluids are relatively poorly understood and characterized, but include various antimicrobial peptides and enzymes, complement and antibody. Defences associated with the white cell population have received most attention, and the important role of phagocytic cells (e.g. neutrophils, monocytes/macrophages) and lymphocyte populations in defence against many infectious processes is well established. Deficiencies in one or more of these normal host defences can be loosely referred to as immune deficiencies.

While some immune deficiencies are inherited and/or congenital (e.g. chronic granulomatous disease), others appear spontaneously or are iatrogenic in origin.

Local (mucocutaneous) defences may be compromised by thermal burns, wounds (traumatic or surgical), excessive hydration, intubation, intestinal or airways obstruction, achlorhydria and antibiotic use. Cellular mechanisms may be compromised simply by insufficient numbers of functional cells (e.g. neutropenia, following chemotherapy, HIV infection), or by defects in function attributable to genetic factors or the use of chemotherapeutic agents for some other disease.

It is widely accepted that specific pathogens may be associated with particular host defects. Thus impaired neutrophil function or lack of numbers favours infection by many bacteria and fungi (Engels et al., 1999). Problems in mounting a good antibody response (e.g. following splenectomy) may predispose to infection by several capsulated bacteria, including *Streptococcus pneumoniae*, while defects in the cell-mediated limb of the immune response favour survival of many intracellular pathogens, including bacteria, viruses, a variety of fungi and protozoa (see Table 16.1). In many of these examples, a functional cell-mediated immune (CMI) response is also essential for satisfactory resolution of the associated disease process. Without an adequate CMI response, the disease becomes chronic and basically incurable (e.g. cryptococcal meningitis in AIDS patients).

Heart, liver, kidney, pancreas, lung and other solid-organ transplant recipients require non-specific immunosuppression to prevent rejection of the graft. This non-specific process results in a deficit of host defences, especially cell-mediated immunity (Howard, 1994; Singh, 1998). Infections that occur early (within the first 4 weeks) after organ transplantation are usually bacterial and related to the surgical procedure itself, rather than to any immunocompromise (e.g. urinary tract

Table 16.1 Association of infection by certain microbes with specific host abnormalities

Abnormality	Infectious disease/microbe favoured
Neurophil dysfunction (e.g. impaired intracellular killing, lack of numbers – chemotherapy)	Hepatosplenic candidosis (disseminated/invasive candidosis); septicaemia in neutropenic patients (e.g. by *Pseudomonas aeruginosa*)
Immunosuppression following organ transplantation (impaired cell-mediated immunity)	*Candida albicans*, herpes simplex virus
Impaired CD_4 lymphocyte activity (e.g. HIV-infected patients)	*Mycobacterium tuberculosis*, *Cryptococcus neoformans*, *Pneumocystis carinii*, cytomegalovirus (CMV), oesophageal candidosis
Deficient IgG production (e.g. following splenectomy)	*Streptococcus pneumoniae*
Colonized IV catheter	Invasive *Staphylococcus epidermidis* disease, disseminated candidosis

infections, wound infections, chest infections and vascular catheter-related infections) (Engels et al., 1999). The only fungal pathogens that are commonly encountered are *Candida*, and the only virus is herpes simplex virus (HSV). The possibility exists that the allograft harbours microbes and is itself a source of infection (La Rocco and Burgert, 1997). This is particularly serious if cytomegalovirus (CMV) is transmitted to a recipient who has not previously been exposed to the virus.

Infections with opportunistic viral and fungal pathogens dominate the initial 6-month post-transplant period (see Figure 16.1). Risk factors for fungal infections have been outlined elsewhere (Patel et al., 1996b). Cytomegalovirus, a member of the herpes virus family, is the most common virus infection encountered, while infection with hepatitis B and hepatitis C may be prominent in liver transplant patients who harbour the virus prior to surgery. If problems of persistent rejection or chronic viral infection persist for more than 6 months, the risk from opportunistic infections remains high. Not surprisingly, classical clinical presentations may be modified, and a rapid diagnosis and early aggressive and specific therapy are required.

After transplantation, non-specific immune suppression alters the acute (and chronic) inflammatory response, the mechanical and immune barriers to microbial invasion, chemotaxis, phagocytosis and T- and B-lymphocyte function. Azathioprine and other antimetabolites (e.g. mycophenolic acid) produce granulocytopenia and leucopenia. Corticosteroids impair transcription of the interleukin (IL)-1β gene, cause rapid degradation of IL-1 mRNA, and reduce IL-6 transcription. Steroids inhibit gene expression of IL-2, tumour necrosis factor (TNF) and interferon (IFN). Monoclonal antibodies react with T-cell surface receptors. Cyclosporin (CSA) and tacrolimus (FK506) inhibit T-cell synthesis of cytokines IL-2, IL-3, IFN, IL-6 and IL-7. Helper and cytotoxic T-cells are the primary targets while suppressors are induced (e.g. tacrolimus inhibits suppressor T-cells) (Suthanthiran and Strom, 1994).

CSA has complex pharmacokinetics which are compounded by multiple drug interactions. Inhibition or stimulation of its metabolism occurs through cytochrome P_{450} enzyme systems. Frequent determinations of CSA blood levels are required to protect both renal function and the allograft. The co-administration of some antibiotics (e.g. aminoglycosides) and azole antifungal agents (e.g. ketoconazole) results in elevation of blood CSA concentrations, while lowering of CSA levels occurs with rifampicin.

BACTERIAL INFECTIONS

Bacterial infections are classically peri-operative, related to nosocomial exposure (catheters and lines), and occur predominantly in the urinary tract, abdomen or lung, with the site partly dependent on the organ(s) transplanted. Rotation and removal of intravenous lines are important for preventing infection. All transplant recipients should be given peri-operative broad-spectrum prophylactic antibiotics.

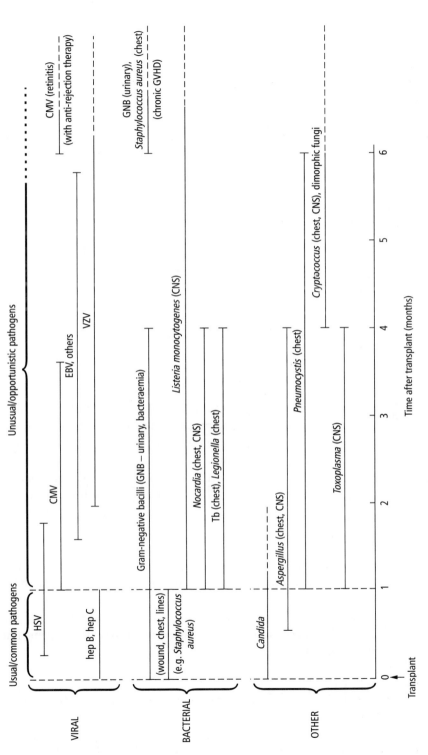

Figure 16.1 Time-course of infections in transplant recipients. HSV, herpes simplex virus; CMV, cytomegalovirus; EBV, Epstein-Barr virus; VZV, varicella zoster virus; hep B, hepatitis B virus; hep C, hepatitis C virus; CNS, central nervous system. Adapted from Howard (1994).

LUNG

Any transplant patient with fever, radiological infiltrate and hypoxia should have bronchoscopy with lavage, brushing and specific stains. Co-trimoxazole (trimethoprim/sulphamethoxazole) for *Pneumocystis carinii* pneumonia (PCP), a β-lactam (e.g. cefuroxime) for the commoner respiratory pathogens plus erythromycin for legionellae may be started empirically, and ganciclovir may be added if cytomegalovirus (CMV) is suspected. Treatment should also include physiotherapy and bronchodilators.

ABDOMEN

Perforations of the gastrointestinal tract have a current mortality of around 30% (Bardaxoglou *et al.*, 1993; Benoit *et al.*, 1993). About 50% of these problems occur during induction, immunosuppression or rejection episodes. Delay in diagnosis of colonic perforation (due to diverticulitis) is a major factor contributing to mortality – plain X-rays and soluble contrast enema and computed tomography scanning are the best diagnostic adjuncts to high levels of clinical suspicion. Early surgery which involves resection without anastomosis is important. The use of soft drains and post-operative percutaneous drainage of abscesses complement antibiotic therapy with metronidazole and a suitable Gram-negative bacillus agent (e.g. ciprofloxacin, a third-generation cephalosporin or daily gentamicin, see p. 201).

URINARY TRACT

Reflux is a very common side-effect of the ureteroneocystostomy used in renal transplantation. Reflux has not been a significant cause of infection or graft loss in adult recipients. Although urinary infection is common and often catheter-related, the usual high urine output (due to poor concentrating ability) promotes flushing and minimizes pyelonephritis. Patients with significant urinary tract infections should therefore be investigated for urinary tract obstruction. Antibiotics which are concentrated in the urine are commonly prescribed. Urinary tract infection is very common after pancreas transplant operations, particularly with 'duct–bladder' anastomosis.

CENTRAL NERVOUS SYSTEM

Listeria monocytogenes is the microbe most commonly involved, while nocardiae and *Mycobacterium tuberculosis* infections share clinical characteristics with fungal infections. Listeriae may cause focal or diffuse infections with apparently normal CSF findings. Relapse is partly related to the intracellular nature of the infection or truncated therapy. Antibiotic therapy with amoxycillin or penicillin plus gentamicin may be employed.

VIRAL INFECTIONS

CYTOMEGALOVIRUS

The herpes virus, cytomegalovirus (CMV), is ubiquitous in the human population, and reactivation of latent infection may be demonstrated in the majority of transplant recipients (Fiala et al., 1975; Patel and Paya, 1997; Singh, 1998). CMV reactivation has been associated with encephalitis, pneumonitis, leucopenia, fever, arthralgia, retinitis and hepatitis (see Fiala et al., 1975). Serious bacterial infections appear to enhance the replication of CMV. As primary infections are associated with significant mortality, screening for CMV should be universal. Diagnosis by virus culture is most reliable, but may take a considerable time. More rapid diagnoses can be performed using some of the recently developed molecular techniques (e.g. polymerase chain reaction (PCR) assays). Patients at high risk (e.g. positive donors, negative recipients) should be treated with prophylactic parenteral ganciclovir followed by oral acyclovir (Patel et al., 1996a). CMV resistance to antiviral agents has been demonstrated.

HERPES SIMPLEX, HERPES ZOSTER AND EPSTEIN–BARR VIRUSES

Simplex and zoster infections become a problem when they are primary, disseminated or are located within the central nervous system. Acyclovir therapy has significantly improved their prognosis. Epstein–Barr virus infection has been incriminated in post-transplant lymphoproliferative disorders. This virus may respond to ganciclovir. It has been suggested that infection by the novel herpes virus, human herpes virus-6 (HHV-6), is a significant cause of fever of unknown origin in solid-organ transplant patients, and a predictor of subsequent CMV infection (see Singh, 1998).

HEPATITIS B, C AND G VIRUSES

Hepatitis virus antigenaemia is becoming more prevalent in dialysis and transplant units. Chronic infection with either hepatitis B virus (HBV) or hepatitis C virus (HCV), or both, may be well tolerated after renal transplantation (Chatterjee et al., 1974). Interferon treatment has not been shown to be beneficial. The nucleoside analogue, lamivudine, may be of value in liver transplant patients who have recurrent HBV infection (Ben-Ari et al., 1997), while famciclovir (an oral form of the purine nucleoside penciclovir) has also been employed in the prophylaxis and therapy of HBV infection after transplantation (Singh, 1998). However, a proportion of patients with HBV and HCV infections develop significant progressive liver disease. Few clinicians would support the transplant of an organ from an HBV- or HCV-positive donor to a negative recipient. The existence of multiple genotypes of HCV with differing biological behaviours, as well as the generation of antigenic diversity, limit the capacity of the immune system to generate protective immunity (Fishman et al., 1996). This can be compared with hepatitis B vaccination, which is successful except for those with impaired responsiveness – dialysis patients have reduced immunity. Ribavirin, a guanosine analogue that is used primarily to treat respiratory syncytial virus (RSV) pneumonitis, normalizes liver enzymes in non-transplant patients with HCV infections, and may be of benefit for recurrent infection after transplantation (Cattral et al., 1996). Interferon is used to treat chronic hepatitis B and C infections.

The novel flavivirus, hepatitis G virus (HGV), although frequently detected in transplant recipients, appears to have minimal clinical impact, and influences neither the graft nor the patient outcome (Singh, 1998).

The use of hepatitis B immunoglobulin (passive immunization) for 6 months following transplantation in order to reduce viral surface-antigen blood levels in liver transplant patients who have received transplants because of hepatitis B virus disease has had some success in reducing post-operative complications (Sawyer et al., 1998).

FUNGAL INFECTIONS

Fungal infections have a significant mortality, particularly after bone-marrow, renal and liver transplantation (Nampoory et al., 1996; Patel et al., 1996b; Patel and Paya, 1997; Singh et al., 1997b). A high transplant transfusion requirement and post-transplant bacterial infection (possibly therapy) are significant risk factors. In one survey of over 2000 liver transplant recipients, 26 patients developed invasive aspergillosis, with around one-third of these occurring after re-transplantation. A remarkable finding was the early onset of the disease, with the median time of onset being 17 days post-transplantation (see Singh, 1998). Aspergillus infections are associated with the highest mortality. Chemotherapy with amphotericin B or itraconazole can be considered, but appears to have little effect unless it is accompanied by surgical resection of pulmonary lesions, or some form of immunomodulation. The prevalence of invasive candidosis may be reduced by selective decontamination of the digestive tract using oral (non-absorbed) nystatin or amphotericin B, or by the use of oral fluconazole. This approach is most often used after hepatic transplantation. Systemic amphotericin B has also been suggested as prophylaxis in high risk situations. However, complications associated with its renal toxicity and intravenous administration have severely restricted its use.

Cryptococcus neoformans may cause a subacute meningitis accompanied by headache and fever, and this may be associated with or preceded by pulmonary lesions. The cerebral spinal fluid (CSF) findings resemble those seen in tuberculosis. However, detection of cryptococcal antigen and/or isolation of the fungus are pathognomonic. Numerous characteristic encapsulated yeasts are often readily visible in the CSF. Therapy with amphotericin B is required, and is often combined with flucytosine or fluconazole. The most popular current regimen appears to be a short course of amphotericin B with or without concurrent flucytosine, followed by a course of fluconazole to complete a total of 12 weeks of therapy (see p. 137). Although intrathecal amphotericin B has been advocated, such a regimen has not received universal support.

Surgical treatment may be useful in localized fungal infections involving the lungs or central nervous system. It is often difficult to achieve adequate therapeutic drug levels at the site of fungal growth, as fungi tend to invade and occlude blood vessels, resulting in suboptimal drug delivery. Resolution of infection also requires appropriate white cell function – something which is often absent in transplant recipients.

TOXOPLASMOSIS

The diagnosis of infection by the protozoan *Toxoplasma gondii*, requires a high index of suspicion because clinical manifestations of fever, pneumonitis and neurological symptoms are common to a number of opportunistic infections. The donor may be the source of the microbe. More useful diagnostic considerations include PCR assays of body fluids or tissue, special stains and serology. Early empirical initiation of therapy with pyrimethamine and sulphadiazine (or cotrimoxazole) plus folinic acid may improve survival (Renoult *et al.*, 1997). Clindamycin has been substituted for sulphadiazine in some studies.

REFERENCES

Abbott, S.L. and Janda, J.M. (1997) *Enterobacter cancerogenus* ('*Enterobacter taylorae*') infections associated with severe trauma or crush injuries. *American Journal of Clinical Pathology* 107, 359–61.

Acar, J.F. and Goldstein, F.W. (1997) Trends in bacterial resistance to fluoroquinolones. *Clinical Infectious Diseases* 24 (Suppl. 1), S67–S73.

Adair, C.G., Gorman, S.P., O'Neill, F.B. et al. (1993) Selective decontamination of the digestive tract (SDD) does not prevent the formation of microbial biofilms on endotracheal tubes. *Journal of Antimicrobial Chemotherapy* 31, 689–97.

Ahmed, A.O., van Belkum, A., Fahal, A.H. et al. (1998) Nasal carriage of *Staphylococcus aureus* and epidemiology of surgical-site infections in a Sudanese university hospital. *Journal of Clinical Microbiology* 36, 3614–18.

Anaissie, E.J., Mattiuzzi, G.N., Miller, C.B. et al. (1998) Treatment of invasive fungal infections in renally impaired patients with amphotericin B colloidal dispersion. *Antimicrobial Agents and Chemotherapy* 42, 606–11.

Anon. (1996a) Penicillin allergy. *Drug and Therapeutics Bulletin* 34, 87–8.

Anon. (1996b) Meropenem – an advantageous antibiotic? *Drug and Therapeutics Bulletin* 34, 53–5.

Anon. (1997a) Major advances in the treatment of HIV-1 infection. *Drug and Therapeutics Bulletin* 35, 25–9.

Anon. (1997b) Combination antibacterial therapy: appropriate in certain situations. *Drugs Therapy Perspectives* 9(5), 10–13.

Anon. (1998a) The broad-spectrum quinolones are here. *Drugs Therapy Perspectives* 12(6), 1–7.

Anon. (1998b) Cefirome: another broadsword in the war against severe hospital-acquired infections? *Drugs Therapy Perspectives* 11(2), 1–5.

Anon. (1998c) Avoid amputation if possible in patients with diabetic foot ulcers. *Drugs Therapy Perspectives* 11(3), 13–16.

Anon. (1999a) Selective digestive decontamination: still useful in patients with liver disease? *Drugs Therapy Perspectives* 13(7), 10–12.

Anon. (1999b) Tackling antimicrobial resistance. *Drug and Therapeutics Bulletin* 37, 9–16.

Anon. (1999c) Rifapentine offers greater convenience in the treatment of pulmonary tuberculosis. *Drugs Therapy Perspectives* 13(7), 1–4.

Anon. (1999d) Lipid-based formulations of amphotericin B: new clothes for the emperor? *Drugs Therapy Perspectives* 13(11), 1–5.

Appelbaum, P.C. (1993) Comparative susceptibility profile of piperacillin/tazobactam against anaerobic bacteria. *Journal of Antimicrobial Chemotherapy* 31 (Suppl. A), 29–38.

Archer, G.L. and Climo, M.W. (1994) Antimicrobial susceptibility of coagulase-negative staphylococci. *Antimicrobial Agents and Chemotherapy* 38, 2231–7.

Archibald, L., Phillips, L., Monnet, D. *et al.* (1997) Antimicrobial resistance in isolates from inpatients and outpatients in the United States: increasing importance of the intensive-care unit. *Clinical Infectious Diseases* **24**, 211–15.

Arnow, P.M., Carandang, G.C., Zabner, R. and Irwin, M.E. (1996) Randomized controlled trial of selective bowel decontamination for prevention of infections following liver transplantation. *Clinical Infectious Diseases* **22**, 997–1003.

Arthur, M., Reynolds, P. and Courvalin, P. (1996) Glycopeptide resistance in enterococci. *Trends in Microbiology* **4**, 401–7.

Atkins, B.L. and Bowler I.C.J.W. (1998) The diagnosis of large joint sepsis. *Journal of Hospital Infection* **40**, 263–74.

Aubier, M., Verster, R., Regamey, C., Geslin, P., Vercken, J.-B. and the Sparfloxacin European Study Group (1998) Once-daily sparfloxacin versus high-dosage amoxycillin in the treatment of community-acquired, suspected pneumococcal pneumonia in adults. *Clinical Infectious Diseases* **26**, 1312–20.

Bailey, T.C., Little, J.R., Littenberg, B. *et al.* (1997) A meta-analysis of extended-interval dosing versus multiple daily dosing of aminoglycosides. *Clinical Infectious Dieseases* **24**, 786–95.

Balfour, H.H. (1999) Antiviral drugs. *New England Journal of Medicine* **340**, 1255–68.

Bamberger, D.M. (1996) Outcome of medical treatment of bacterial abscesses without therapeutic drainage: review of cases reported in the literature. *Clinical Infectious Diseases* **23**, 592–603.

Bamberger, D.M., Herndon, B.L., Dew, M. *et al.* (1997) Efficacies of ofloxacin, rifampin, and clindamycin in treatment of *Staphylococcus aureus* abscesses and correlation with results of an *in-vitro* assay of intracellular bacterial killing. *Antimicrobial Agents and Chemotherapy* **41**, 1178–81.

Bandyk, D.F. and Esses, G.E. (1994) Prosthetic graft infection. *Surgical Clinics of North America* **74**, 571–90.

Barakate, M.S., Stephen, M.S., Waugh, R.C. *et al.* (1999) Pyogenic liver abscess: a review of 10 years' experience in management. *Australian and New Zealand Journal of Surgery* **69**, 205–9.

Bardaxoglou, E., Maddern, G., Ruso, L. *et al.* (1993) Gastrointestinal surgical emergencies following kidney transplantation. *Transplant International* **6**, 148–52.

Barie, P.S., Christou, N.V., Dellinger, E.P. *et al.* (1990) Pathogenicity of the enterococcus in surgical infections. *Annals of Surgery* **212**, 155–9.

Barnish, M.J. (1994) Peritonitis in patients undergoing continuous ambulatory peritoneal dialysis. *Complications in Surgery* **11**, 25–32.

Barone, J. A., Moskovitz, B.L., Guarnieri, J. *et al.* (1998) Enhanced bioavailability of itraconazole in hydroxypropyl-β-cyclodextrin solution versus capsules in healthy volunteers. *Antimicrobial Agents and Chemotherapy* **42**, 1862–5.

Barry, A.L., Fuchs, P.C. and Brown, S.D. (1998) Antipneumococcal activities of a ketolide (HMR 3647), a streptogramin (quinupristin-dalfopristin), a macrolide (erythromycin), and a lincosamide (clindamycin) *Antimicrobial Agents and Chemotherapy* **42**, 945–6.

Barton, M.D. (1998) Does the use of antibiotics in animals affect human health? *Australian Veterinary Journal* 76, 177–80.

Baum, M.L., Anish, D.S., Chalmers, T.C. et al. (1981) A survey of clinical trials of antibiotic prophylaxis in colon surgery: evidence against further use of no-treatment controls. *New England Journal of Medicine* 305, 795–9.

Bayer, A.S., Li, C. and Ing, M. (1998) Efficacy of trovafloxican, a new quinolone antibiotic, in experimental staphylococcal endocarditis due to oxacillin-resistant strains. *Antimicrobial Agents and Chemotherapy* 42, 1837–41.

Beck-Sagué, C.M., Jarvis, W.R. and the National Nosocomial Infections Surveillance System (1993) Secular trends in the epidemiology of nosocomial fungal infections in the United States, 1980–1990. *Journal of Infectious Diseases* 167, 1247–51.

Ben-Ari, Z., Shmueli, D., Mor, E. et al. (1997) Beneficial effect of lamivudine in recurrent hepatitis B after liver transplantation. *Transplantation* 63, 393–6.

Benoit, G., Moukarzel, M., Verdelli, G. et al. (1993) Gastrointestinal complications in renal transplantation. *Transplant International* 6, 45–9.

Berbari, E.F., Hanssen, A.D., Duffy, M.C. et al. (1998) Risk factors for prosthetic infection: case–control study. *Clinical Infectious Diseases* 27, 1247–54.

Bernard, H.R. and Cole, W.R. (1964) The prophylaxis of surgical infection: the effect of prophylactic antimicrobial drugs on the incidence of infection following potentially contaminated operations. *Surgery* 56, 151–7.

Berne, T.V., Yellin, A.E., Appleman, M.D. et al. (1993) A clinical comparison of cefepime and metronidazole versus gentamicin and clindamycin in the antibiotic management of surgically treated advanced appendicitis. *Surgery, Gynecology and Obstetrics* 177 (Suppl.), 18–22.

Berne, T.V., Yellin, A.E., Appleman, M.D. et al. (1996) Meropenem versus tobramycin with clindamycin in the antibiotic management of patients with advanced appendicitis. *Journal of the American College of Surgeons* 182, 403–7.

Bert, F., Maubec, E., Bruneau, B. et al. (1998) Multi-resistant *Pseudomonas aeruginosa* outbreak associated with contaminated tap water in a neurosurgery intensive-care unit. *Journal of Hospital Infection* 39, 53–62.

Blanton, R.E., Wachira, T.M., Zeyhle, E.E. et al. (1998) Oxfendazole treatment for cystic hydatid disease in naturally infected animals. *Antimicrobial Agents and Chemotherapy* 42, 601–5.

Blondeau, J.M., Yaschuk, Y. and the Canadian Multicentre Study Group (1997) Canadian multicentre Susceptibility Study, with a focus on cephalosporins, from 15 Canadian medical centres. *Antimicrobial Agents and Chemotherapy* 41, 2773–5.

Blythe, D., Keenlyside, D., Dawson, S.J. and Galloway, A. (1998) Environmental contamination due to methicillin-resistant *Staphylococcus aureus* (MRSA). *Journal of Hospital Infection* 38, 67–70.

Boswell, G.W., Buell, D. and Bekersky, I. (1998) AmBisome (liposomal amphotericin B): a comparative review. *Journal of Clinical Pharmacology* 38, 583–92.

Bottone, E.J., Weitzman, I. and Hanna, B.A. (1979) *Rhizopus rhizopodiformis*: emerging etiological agent of mucormycosis. *Journal of Clinical Microbiology* 9, 530–7.

Bradley, J.S. and Scheld, W.M. (1997) The challenge of penicillin-resistant *Streptococcus pneumoniae* meningitis: current antibiotic therapy in the 1990s. *Clinical Infectious Diseases* **24 (Suppl. 2)**, S213–S221.

Brause, B.D. (1995) Infections with prostheses in bones and joints. In Mandell, G.L., Bennett, J.E. and Dolin, R. (eds), *Mandell, Douglas and Bennett's principles and practice of infectious diseases*, 4th edn. Vol. 1. New York: Churchill Livingstone, 1051–5.

Breasted, J.H. (1930) *The Edwin Smith Surgical Papyrus*. Vol. 1. Chicago, IL: University of Chicago Press.

Brennan, S.S., Smith, G.M.R., Evans, M. and Pollock, A.V. (1982) The management of the perforated appendix: a controlled clinical trial. *British Journal of Surgery* **69**, 510–12.

Bretagne, S., Bart-Delabesse, E., Wechsler, J. et al. (1997) Fatal primary cutaneous aspergillosis in a bone-marrow transplant recipient: nosocomial acquisition in a laminar-air-flow room. *Journal of Hospital Infection* **36**, 235–9.

Bridges, R. McI. and Deitch, E.A. (1994) Diabetic foot infections – pathophysiology and treatment. *Surgical Clinics of North America* **74**, 537–55.

Brogden, R.N. and Peters, D.H. (1994) Teicoplanin – a reappraisal of its antimicrobial activity, pharmacokinetic properties and therapeutic efficacy. *Drugs* **47**, 823–54.

Brook, I. (1989) A 12-year study of aerobic and anaerobic bacteria in intra-abdominal and postsurgical abdominal wound infections. *Surgery, Gynecology and Obstetrics* **169**, 387–92.

Brook, I. (1995) *Prevotella* and *Porphyromonas* infections in children. *Journal of Medical Microbiology* **42**, 340–7.

Brook, I. and Gillmore, J.D. (1993) *In-vitro* susceptibility and *in-vivo* efficacy of antimicrobials in the treatment of intra-abdominal sepsis in mice. *Journal of Antimicrobial Chemotherapy* **31**, 393–401.

Brook, I. and Ledney, G.D. (1994) The treatment of irradiated mice with polymicrobial infection caused by *Bacteroides fragilis* and *Escherichia coli*. *Journal of Antimicrobial Chemotherapy* **33**, 243–52.

Brook, I. and Frazier, E.H. (1995) Clinical and microbiological features of necrotizing fasciitis. *Journal of Clinical Microbiology* **33**, 2382–7.

Brook, I. and Frazier, E.H. (1999) Aerobic and anaerobic microbiology of surgical-site infection following spinal fusion. *Journal of Clinical Microbiology* **37**, 841–3.

Brook, I., Frazier, E.H. and Yeager, J.K. (1998) Microbiology of infected eczema herpeticum. *Journal of the American Academy of Dermatology* **38**, 627–9.

Bryson, H.M. and Brogden, R.N. (1994) Piperacillin/tazobactam – a review of its antibacterial activity, pharmacokinetic properties and therapeutic potential. *Drugs* **47**, 506–35.

Buchholz, H.W. and Engelbrecht, E. (1970) Über die Depotwirkung einiger Antibiotika bei Vermischung mit dem Kunstharz Palaco. *Chirurg* **41**, 511–15.

Burke, J.F. (1961) The effective period of preventive antibiotic action in experimental incisions and dermal lesions. *Surgery* **50**, 161–8.

Burrick, M.P., Heim-Duthoy, K.L., Yellin, A.E. et al. (1990) Ceftazidime/clindamycin versus tobramycin/clindamycin in the treatment of intra-abdominal infections. *The American Surgeon* 56, 613–7.

Callahan, T.E., Schecter, W.P. and Horn, J.K. (1998) Necrotizing soft tissue infection masquerading as cutaneous abscess following illicit drug injection. *Archives of Surgery* 133, 812–18.

Cannon, R.D. (1997) *Candida albicans*. *Japanese Journal of Medical Mycology* 38, 297–302.

Cars, O. (1997) Colonisation and infection with resistant Gram-positive cocci: epidemiology and risk factors. *Drugs* 54 (Suppl. 6), 4–10.

Casadewall, B. and Courvalin, P. (1999) Characterization of the *van*D glycopeptide resistance gene cluster from *Enterococcus faecium* BM4339. *Journal of Bacteriology* 181, 3644–8.

Casewell, M.W. (1997) Control of infection in liver transplant patients. *Antibiotics and Chemotherapy* 1, 13–15.

Catchpole, C., Wise, R. and Fraise, A. (1997) MRSA bacteraemia. *Journal of Hospital Infection* 35, 159–61.

Catchpole, C. and Hastings, J.G.M. (1995) Measuring pre- and post-dose vancomycin levels: time for a change? *Journal of Medical Microbiology* 45, 309–11.

Cattral, M.S., Krajden, M., Wanless, I.R. et al. (1996) A pilot study of ribavirin therapy for recurrent hepatitis C virus infection after liver transplantation. *Transplantation* 61, 1483–8.

Chambers, H.F. and Neu, H.C. (1995a) Other β-lactam antibiotics. In Mandell, G.L., Bennett, J.E. and Dolin, R. (eds), *Mandell, Douglas and Bennett's principles and practice of infectious diseases*, 4th edn. Vol. 1. New York: Churchill Livingstone, 264–72.

Chambers, H.F. and Neu, H.C. (1995b) Penicillins. In Mandell, G.L., Bennett, J.E. and Dolin, R. (eds), *Mandell, Douglas and Bennett's principles and practice of infectious diseases*, 4th edn. Vol. 1. New York: Churchill Livingstone, 233–45.

Chan, C.T. and Gold, W.L. (1998) Intramedullary abscess of the spinal cord in the antibiotic era: clinical features, microbial etiologies, trends in pathogenesis, and outcomes. *Clinical Infectious Diseases* 27, 619–26.

Chatterjee, S.M., Payne, J.E., Bischel, M.D. et al. (1974) Successful renal transplantation in patients positive for hepatitis B antigen. *New England Journal of Medicine* 291, 62–5.

Chow, A.W. (1995) Infections of the oral cavity, neck, and head. In Mandell, G.L., Bennett, J.E. and Dolin, R. (eds), *Mandell, Douglas and Bennett's principles and practice of infectious diseases*, 4th edn. Vol. 1. New York: Churchill Livingstone, 593–606.

Chuang, Y.-C., Ko, W.-C., Wang, S.-T. et al. (1998) Minocycline and cefotaxime in the treatment of experimental murine *Vibrio vulnificus* infection. *Antimicrobial Agents and Chemotherapy* 42, 1319–22.

Citron, D.M. and Appleman, M.D. (1997) Comparative *in-vitro* activities of trovafloxacin (CP-99, 219) against 221 aerobic and 217 anaerobic bacteria isolated from patients with intra-abdominal infections. *Antimicrobial Agents and Chemotherapy* **41**, 2312–16.

Citron, D.M., Gerardo, S.H., Claros, M.C. et al. (1996) Frequency of isolation of *Porphyromonas* species from infected dog and cat bite wounds in humans and their characterization by biochemical tests and arbitrarily primed polymerase chain reaction fingerprinting. *Clinical Infectious Diseases* **23 (Suppl. 1)**, S78–S82.

Classen, D.C., Evans, R.S., Pestotnik, S.L. et al. (1992) The timing of prophylactic administration of antibiotics and the risk of surgical wound infection. *New England Journal of Medicine* **326**, 281–6.

Clemons, K.V. and Stevens, D.A. (1998) Comparison of fungizone, amphotec, ambisome and abelcet for treatment of systemic murine cryptococcosis. *Antimicrobial Agents and Chemotherapy* **42**, 899–902.

Cloeren, M. and Perl, T.M. (1998) Occupationally acquired infections and the healthcare worker. *Current Opinion in Infectious Diseases* **11**, 475–82.

Clunie, G.J.A. (1997) Surgical infection and its prevention. In Clunie, G.J.A., Tjandra, J.J. and Francis, D.M.A. (eds), *Textbook of surgery*. Carlton, Australia: Blackwell Science, 30–5.

Cockerill, F.R. (1999) Genetic methods for assessing antimicrobial resistance. *Antimicrobial Agents and Chemotherapy* **43**, 199–212.

Cockerill, F.R., Hughes, J.G., Vetter, E.A. et al. (1997) Analysis of 281 797 consecutive blood cultures performed over an eight-year period: trends in microorganisms isolated and the value of anaerobic culture of blood. *Clinical Infectious Diseases* **24**, 403–18.

Cohn, S.M. and Fisher, B.T. (1996) Do surgeons have a role as infectious disease consultants? *Archives of Surgery* **131**, 990–3.

Collignon, P.J. (1995) Hospital-acquired infections: a skeleton in the closet of medical progress. *Medical Journal of Australia* **163**, 228.

Condon, R.E. (1996a) The STAR approach to the complicated abdomen. *Infectious Diseases in Clinical Practice* **5 (Suppl. 2)**, S74–S76.

Condon, R.E. (1996b) Current antibiotic-related issues in treatment of appendicitis. *Infectious Diseases in Clinical Practice* **5 (Suppl. 1)**, S2–S8.

Cooper, G.S., Shlaes, D.M., Jacobs, M.R. and Salata, R.A. (1993) The role of *Enterococcus* in intra-abdominal infections: case–control analysis. *Infectious Diseases in Clinical Practice* **2**, 332–9.

Cornwell, E.E., Belzberg, H., Berne, T.V. et al. (1997) Pharmacokinetics of aztreonam in critically ill surgical patients. *American Journal of Health-System Pharmacy* **54**, 537–40.

Coukell, A.J. and Brogden, R.N. (1998) Liposomal amphotericin B: therapeutic use in the management of fungal infections and visceral leishmaniasis. *Drugs* **55**, 585–612.

Couper, M.R. (1997) Strategies for the rational use of antimicrobials. *Clinical Infectious Diseases* **24 (Suppl. 1)**, S154–S156.

Cox, R.A. and Conquest, C. (1997) Strategies for the management of healthcare staff colonized with epidemic methicillin-resistant *Staphylococcus aureus*. *Journal of Hospital Infection* **35**, 117–27.

Cruse, P.J.E. and Foord, R. (1980) The epidemiology of wound infection: a 10-year prospective study of 62 939 wounds. *Surgical Clinics of North America* **60**, 27–40.

Culver, D.H., Horan, T.C., Gaynes, R.P. et al. (1991) Surgical wound infection rates by wound class, operative procedure, and patient risk index. *American Journal of Medicine* **91 (Suppl. 3B)**, 152S–157S.

D'Amelio, L.F., Wagner, B, Azimuddin, S. et al. (1995) Antibiotic patterns associated with fungal colonization in critically ill surgical patients. *American Surgeon* **61**, 1049–53.

Dahlberg, P.S., Sielaff, T.D. and Dunn, D.L. (1996) Emerging resistance in staphylococci and enterococci. *Infectious Diseases in Clinical Practice* **5 (Suppl. 2)**, S49–S56.

Damjanovic, V., Connolly, C.M., van Saene, H.K.F. et al. (1993) Selective decontamination with nystatin for control of a *Candida* outbreak in a neonatal intensive-care unit. *Journal of Hospital Infection* **24**, 245–59.

Davies, J. and Wright, G.D. (1997) Bacterial resistance to aminoglycoside antibiotics. *Trends in Microbiology* **5**, 234–40.

de Groot, H.G.W., Hustinx, P.A., Lampe, A.S. and Oosterwijk, W.M. (1993) Comparison of imipenem/cilastatin with the combination of aztreonam and clindamycin in the treatment of intra-abdominal infections. *Journal of Antimicrobial Chemotherapy* **32**, 491–500.

de Pauw, B.E., Donnelly, J.P., Verweij, P.E. and Meis, J.F.G.M. (1998) Current management of fungal infections in immunocompromised patients and polyene lipid complexes for treatment of invasive fungal infection. *Journal of Infectious Diseases and Antimicrobial Agents* **15**, 85–96.

de Vries, P.J., Verkooyen, R.P., Leguit, P. and Verbrugh, H.A. (1990) Prospective randomised study of once-daily versus thrice-daily netilmicin in patients with intra-abdominal infections. *European Journal of Clinical Microbiology and Infectious Diseases* **9**, 161–8.

Dellinger, R.P. (1993) Airway management and nosocomial infection. *Critical Care Medicine* **21**, 1109–10.

den Hoed, P.T., Boelhouwer, R.U., Veen, H.F. et al. (1998) Infections and bacteriological data after laparoscopic and open gallbladder surgery. *Journal of Hospital Infection* **39**, 27–37.

Denning, D.W. (1998) Invasive aspergillosis. *Clinical Infectious Diseases* **26**, 781–805.

Dever, L.L. and Johanson, W.G. (1993) An update on selective decontamination of the digestive tract. *Current Opinion in Infectious Diseases* **6**, 744–50.

Dew, R.B. and Susla, G.M. (1996) Once-daily aminoglycoside treatment. *Infectious Diseases in Clinical Practice* **5**, 12–24.

Dezfulian, M., Bitar, R.A. and Bartlett, J.G. (1993) Comparative efficacy of ceftriaxone in experimental infections involving *Bacteroides fragilis* and *Escherichia coli*. *Chemotherapy* **39**, 355–60.

Donskey, C.J. and Rice, L.B. (1999) The influence of antibiotics on spread of vancomycin-resistant enterococci: the potential role of selective use of antibiotics as a control measure. *Clinical Microbiology Newsletter* **21**, 57–65.

Duckworth, G.J., Cookson, B.D., Humphreys, H. and Heathcock, R. on behalf of working party (1998) Revised guidelines for the control of methicillin-resistant

Staphylococcus aureus infection in hospitals. *Journal of Hospital Infection* 39, 253–90.

Edmiston, C.E. and Walker, A.P. (1996) Microbiology of intraabdominal infections. *Infectious Diseases in Clinical Practice* 5 (**Suppl. 1**), S15–S19.

Ednie, L.M., Spangler, S.K., Jacobs, M.R. and Appelbaum, P.C. (1997a) Susceptibilities of 228 penicillin- and erythromycin-susceptible and -resistant pneumococci to RU 64004, a new ketolide, compared with susceptibilities to 16 other agents. *Antimicrobial Agents and Chemotherapy* 41, 1033–6.

Ednie, L.M., Spangler, S.K., Jacobs, M.R. and Appelbaum, P.C. (1997b) Antianaerobic activity of the ketolide RU 64004 compared to activities of four macrolides, five β-lactams, clindamycin and metronidazole. *Antimicrobial Agents and Chemotherapy* 41, 1037–41.

Ednie, L.M., Jacobs, M.R. and Appelbaum, P.C. (1997c) Comparative antianaerobic activities of the ketolides HMR 3647 (RU 66647) and HMR 3004 (RU 64004). *Antimicrobial Agents and Chemotherapy* 41, 2019–22.

Ednie, L.M., Jacobs, M.R. and Appelbaum, P.C. (1998) Comparative activities of clinafloxacin against Gram-positive and -negative bacteria. *Antimicrobial Agents and Chemotherapy* 42, 1269–73.

Edwards, J.E. (1991) Invasive *Candida* infections – evolution of a fungal pathogen. *New England Journal of Medicine* 324, 1060–2.

Eggiman, P., Francioli, P., Bille, J. et al. (1999) Fluconazole prophylaxis prevents intraabdominal candidiasis in high-risk surgical patients. *Critical Care Medicine* 27, 1066–72.

Eklund, A.-E., Nord, C.E. and the Swedish Study Group (1993) A randomized multicentre trial of piperacillin/tazobactam versus imipenem/cilastatin in the treatment of severe intra-abdominal infections. *Journal of Antimicrobial Chemotherapy* 31 (**Suppl. A**), 79–85.

Eltringham, I. (1997) Mupirocin resistance and methicillin-resistant *Staphylococcus aureus* (MRSA). *Journal of Hospital Infection* 35, 1–8.

Emery, C.L. and Weymouth, L.A. (1997) Detection and clinical significance of extended-spectrum β-lactamases in a tertiary-care medical centre. *Journal of Clinical Microbiology* 35, 2061–7.

Engels, E.A., Ellis, C.A., Supran, S.E. et al. (1999) Early infection in bone marrow transplantation: quantitative study of clinical factors that affect risk. *Clinical Infectious Diseases* 28, 256–66.

Espinèl-Ingroff, A. (1998) In vitro activity of the new triazole voriconazole (UK-109,496) against opportunistic filamentous and dimorphic fungi and common and emerging yeast pathogens. *Journal of Clinical Microbiolgy* 36, 198–202.

Fabian, T.C. (1996) Principles of antibiotic therapy for penetrating abdominal trauma. In *Abdominal infections: new approaches and management.* Report of a ZENECA Pharmaceuticals Symposium, 6 October 1996, San Francisco.

Fabian, T.C. and Boldreghini, S.J. (1985) Antibiotics in penetrating abdominal trauma: comparison of ticarcillin plus clavulanic acid with gentamicin plus clindamycin. *American Journal of Medicine* 79, 157–60.

Farr, B.M. (1995) Rifamycins. In Mandell, G.L., Bennett, J.E. and Dolin, R. (eds), *Mandell, Douglas and Bennett's principles and practice of infectious diseases*, 4th edn. Vol. 1. New York: Churchill Livingstone, 317–29.

Feinberg, M.B., Carpenter, C., Fauci, A.S. *et al.* (1998) Report of the NIH panel to define principles of therapy of HIV infection and guidelines for the use of antiretroviral agents in HIV-infected adults and adolescents. *Annals of Internal Medicine* **128**, 1057–100.

Fekety, R. (1995) Antibiotic-associated colitis. In Mandell, G.L., Bennett, J.E. and Dolin, R. (eds), *Mandell, Douglas and Bennett's principles and practice of infectious diseases*, 4th edn. Vol. 1. New York: Churchill Livingstone, 978–87.

Fernandez, C., Gaspar, C., Torrellas, A. *et al.* (1995) A double-blind, randomized, placebo-controlled clinical trial to evaluate the safety and efficacy of mupirocin calcium ointment for eliminating nasal carriage of *Staphylococcus aureus* among hospital personnel. *Journal of Antimicrobial Chemotherapy* **35**, 399–408.

Ferroni, A., Nguyen, L., Pron, B. *et al.* (1998) Outbreak of nosocomial urinary tract infections due to *Pseudomonas aeruginosa* in a paediatric surgical unit associated with tap-water contamination. *Journal of Hospital Infection* **39**, 301–7.

Fiala, M., Payne, J.E., Berne, T.V. *et al.* (1975) Epidemiology of cytomegalovirus infection after transplantation and immunosuppression. *Journal of Infectious Diseases* **132**, 421–33.

Fichtenbaum, C.J. and Powderly, W.C. (1998) Refractory mucosal candidiasis in patients with human immunodeficiency virus infection. *Clinical Infectious Diseases* **26**, 556–65.

Finch, R.G. (1995) Adverse reactions to antibiotics. In Greenwood, D. (ed.), *Antimicrobial chemotherapy*, 3rd edn. Oxford: Oxford University Press, 206–18.

Finegold, S.M. (1995a) Overview of clinically important anaerobes. *Clinical Infectious Diseases* **20 (Suppl. 2)**, S205–S207.

Finegold, S.M. (1995b) Anaerobic infections in humans: an overview. *Anaerobe* **1**, 3–9.

Finegold, S.M. (1995c) Lung abscess. In Mandell, G.L., Bennett, J.E. and Dolin, R. (eds), *Mandell, Douglas and Bennett's principles and practice of infectious diseases*, 4th edn. Vol. 1. New York: Churchill Livingstone, 641–6.

Finegold, S.M. and Johnson, C.C. (1995) Peritonitis and intra-abdominal infections. In Blaser, M.J., Smith, P.D., Ravdin, J.I. *et al.* (eds), *Infections of the gastrointestinal tract*. New York: Raven Press, 369–403.

Finegold, S.M. and Mathisen, G.E. (1995) Metronidazole. In Mandell, G.L., Bennett, J.E. and Dolin, R. (eds), *Mandell, Douglas and Bennett's principles and practice of infectious diseases*, 4th edn. Vol. 1. New York: Churchill Livingstone, 329–34.

Finegold, S.M. and Wexler, H.M. (1996) Present status of therapy for anaerobic infections. *Clinical Infectious Diseases* **23 (Suppl. 1)**, S9–S14.

Fishman, J.A., Rubin, R.H., Koziel, M.J. and Periera, B.J.G. (1996) Hepatitis C virus and organ transplantation. *Transplantation* **62**, 147–54.

Fitoussi, F., Doit, C., Benali, K. *et al.* (1998) Comparative *in-vitro* killing activities of meropenem, imipenem, ceftriaxone plus vancomycin at clinically achievable cerebrospinal fluid concentrations against penicillin-resistant *Streptococcus pneumoniae* isolates from children with meningitis. *Antimicrobial Agents and Chemotherapy* **42**, 942–4.

Flanagan, P.G. and Barnes, R.A. (1998) Fungal infection in the intensive-care unit. *Journal of Hospital Infection* **38**, 163–77.

Flexner, C. (1998) HIV-protease inhibitors. *New England Journal of Medicine* **338**, 1281–92.

Ford, C.W., Hamel, J.C., Stapert, D. *et al.* (1997) Oxazolidinones: new antibacterial agents. *Trends in Microbiology* **5**, 196–200.

Ford, C.D., Reilly, W., Wood, J. *et al.* (1998) Oral antimicrobial prophylaxis in bone-marrow transplant recipients: randomized trial of ciprofloxacin versus ciprofloxacin-vancomycin. *Antimicrobial Agents and Chemotherapy* **42**, 1402–5.

Francis, D.M.A. (1997) Surgical aspects of HIV infection. In Clunie, G.J.A., Tjandra, J.J. and Francis, D.M.A. (eds), *Textbook of surgery*. Carlton, Australia: Blackwell Science, 66–71.

Fraser, V.J., Jones, M., Dunkel, J. *et al.* (1992) Candidemia in a tertiary care hospital: epidemiology, risk factors, and predictors of mortality. *Clinical Infectious Diseases* **15**, 414–21.

French, G.L. (1998) Enterococci and vancomycin resistance. *Clinical Infectious Diseases* **27 (Suppl. 1)**, S75–S83.

Fry, D.E. (1996) Bloodborne pathogens: implications for the surgical environment in the year 2000. *Infectious Diseases in Clinical Practice* **5 (Suppl. 2)**, S63–S67.

Fujisawa, N., Yamada, H., Kohda, H. *et al.* (1998) Necrotizing fasciitis caused by *Vibrio vulnificus* differs from that caused by streptococcal infection. *Journal of Infection* **36**, 313–16.

Fung-Tomc, J.C. (1997) Fourth-generation cephalosporins. *Clinical Microbiology Newsletter* **19**, 129–36.

Garibaldi, R.A., Cushing, D. and Lerer, T. (1991) Risk factors for postoperative infection. *American Journal of Medicine* **91 (Suppl. 3B)**, 158S–163S.

Garrison, R.N. and Wilson, M.A. (1994) Intravenous and central catheter infections. *Surgical Clinics of North America* **74**, 557–69.

Gastinne, H., Wolff, M, Delatour, F. *et al.* (1992) A controlled trial in intensive-care units of selective decontamination of the digestive tract with nonabsorbable antibiotics. *New England Journal of Medicine* **326**, 594–9.

Gavalda, J., Torres, C., Tenorio, C. *et al.* (1999) Efficacy of ampicillin plus ceftriaxone in treatment of experimental endocarditis due to *Enterococcus faecalis* strains highly resistant to aminoglycosides. *Antimicrobial Agents and Chemotherapy* **43**, 639–646.

Gilbert, D.N. (1995) Aminoglycosides. In Mandell, G.L., Bennett, J.E. and Dolin, R. (eds), *Mandell, Douglas and Bennett's principles and practice of infectious diseases*, 4th edn. Vol. 1. New York: Churchill Livingstone, 279–306.

Gil-Grande, L.A., Rodriguez-Caabeiro, F., Prieto, J.G. *et al.* (1993) Randomised controlled trial of efficacy of albendazole in intra-abdominal hydatid disease. *Lancet* **342**, 1269–72.

Gill, M.A., Chenella, F.C., Heseltine, P.N.R. et al. (1986a) Economic considerations in perforated and gangrenous appendicitis: imipenem–cilastatin versus clindamycin and gentamicin. *Current Therapeutic Research* 40, 393–402.

Gill, M.A., Cheetham, T.C., Chenella, F.C. et al. (1986b) Matched case–control study of adjusted versus nonadjusted gentamicin dosing in perforated and gangrenous appendicitis. *Therapeutic Drug Monitoring* 8, 451–6.

Gold, H.S. and Moellering, R.C. (1996) Antimicrobial-drug resistance. *New England Journal of Medicine* 335, 1445–53.

Goldmann, D.A. and Huskins, W.C. (1997) Control of nosocomial antimicrobial-resistant bacteria: a strategic priority for hospitals worldwide. *Clinical Infectious Diseases* 24 (Suppl. 1), S139–S145.

Goldstein, E.J.C. (1995) Bites. In Mandell, G.L., Bennett, J.E. and Dolin, R. (eds), *Mandell, Douglas and Bennett's principles and practice of infectious diseases*, 4th edn. Vol. 2. New York: Churchill Livingstone, 2765–7.

Goldstein, E.J.C. (1996) Anaerobic bacteremia. *Clinical Infectious Diseases*, 23 (Suppl. 1), S97–S101.

Goldstein, E.J.C. (1997) Animal bite infections. In Mandell, G.L. (ed.), *Essential atlas of infectious diseases for primary care. Part 2*. Philadelphia, PA: Churchill Livingstone, 2.7–2.9.

Goldstein, E.J.C. (1998) New horizons in the bacteriology, antimicrobial susceptibility and therapy of animal bite wounds. *Journal of Medical Microbiology* 47, 95–7.

Goldstein, E.J.C. and Ueno, K. (1996) Introduction. *Clinical Infectious Diseases* 23 (Suppl. 1), S1.

Goldstein, E.J.C., Citron, D.M., Cherubin, C.E. and Hillier, S.L. (1993) Comparative susceptibility of the *Bacteroides fragilis* group species and other anaerobic bacteria to meropenem, imipenem, piperacillin, cefoxitin, ampicillin/sulbactam, clindamycin and metronidazole. *Journal of Antimicrobial Chemotherapy* 31, 363–72.

Goldstein, E.J.C., Citron, D.M., Merriam, C.V. et al. (1999) Activity of gatifloxacin compared to those of five other quinolones versus aerobic and anaerobic isolates from skin and soft tissue samples of human and animal bite wound infections. *Antimicrobial Agents and Chemotherapy* 43, 1475–9.

Goodman, J.L., Winston, D.J., Greenfield, R.A. et al. (1992) A controlled trial of fluconazole to prevent fungal infection in patients undergoing bone marrow transplantation. *New England Journal of Medicine* 326, 845–51.

Gorbach, S.L. (1993a) Treatment of intra-abdominal infections. *Journal of Antimicrobial Chemotherapy* 31 (Suppl. A), 67–78.

Gorbach, S.L. (1993b) Intra-abdominal infections. *Clinical Infectious Diseases* 17, 961–7.

Gorbach, S.L. (1994) Antibiotic treatment of anaerobic infections. *Clinical Infectious Diseases* 18 (Suppl. 4), S305–S310.

Gorbach, S.L. (1996) Antimicrobial resistance in the 1990s. *Infectious Diseases in Clinical Practice* 5 (Suppl. 1), S32–S36.

Gough, A., Clapperton, M., Rolando, N. et al. (1997) Randomised placebo-controlled trial of granulocyte-colony-stimulating factor in diabetic foot infection. *Lancet* 350, 855–9.

Graninger, W., Presteril, E., Schneeweiss, B. et al. (1993) Treatment of *Candida albicans* fungaemia with fluconazole. *Journal of Infection* **26**, 133–46.

Greenberg, R.N., Cayavec, P., Danko, L.S. et al. (1994) Comparison of cefoperazone plus sulbactam with clindamycin plus gentamicin as treatment for intra-abdominal infections. *Journal of Antimicrobial Chemotherapy* **34**, 391–401.

Greenwood, D. (1992) Antimicrobial agents. In Greenwood, D., Slack, R. and Pentherer, J. (eds), *Medical microbiology*, 14th edn. Edinburgh: Churchill Livingstone, 67–77.

Greenwood, D. (1995a) Inhibitors of bacterial cell wall synthesis. In Greenwood, D. (ed.), *Antimicrobial chemotherapy*, 3rd edn. Oxford: Oxford University Press, 13–31.

Greenwood, D. (1995b) Inhibitors of bacterial protein synthesis. In Greenwood, D. (ed.), *Antimicrobial chemotherapy*, 3rd edn. Oxford: Oxford University Press, 32–8.

Greenwood, D. (1995c) Synthetic antibacterial agents and miscellaneous antibiotics. In Greenwood, D. (ed.), *Antimicrobial chemotherapy*, 3rd edn. Oxford: Oxford University Press, 49–61.

Greenwood, D. (1995d) Pharmacokinetics. In Greenwood, D. (ed.), *Antimicrobial chemotherapy*, 3rd edn. Oxford: Oxford University Press, 188–97.

Greenwood, D. (1998) Resistance to antimicrobial agents: a personal view. *Journal of Medical Microbiology* **47**, 751–5.

Gregory, J.E., Golden, A. and Haymaker, W. (1943) Mucormycosis of the central nervous system: report of three cases. *Bulletin of the Johns Hopkins Hospital* **73**, 405–19.

Griego, R.D. and Zitelli, J.A. (1998) Intra-incisional prophylactic antibiotics for dermatologic surgery. *Archives of Dermatology* **134**, 688–92.

Grubbauer, H.M. (1999) Antimicrobial-bonded catheters: important aspects. *Critical Care Medicine* **27**, 1050–1.

Gwaltney, J.M. (1995) Sinusitis. In Mandell, G.L., Bennet, J.E. and Dolin, R. (eds), *Mandell, Douglas and Bennett's principles and practice of infectious diseases*, 4th edn. Vol. 1. New York: Churchill Livingstone, 585–90.

Gyssens, I.C. (1999) Preventing postoperative infections: current treatment recommendations. *Drugs* **57**, 175–85.

Gyssens, I.C., Geerligs, I.E.J., Nannini-Bergman, M.G. et al. (1996) Optimizing the timing of antimicrobial prophylaxis in surgery: an intervention study. *Journal of Antimicrobial Chemotherapy* **38**, 301–8.

Haley, R.W., Culver, D.H., Morgan, W.M. et al. (1985) Identifying patients at high risk of surgical wound infection. *American Journal of Epidemiology* **121**, 206–15.

Halls, G.A. (1993) The management of infections and antibiotic therapy: a European survey. *Journal of Antimicrobial Chemotherapy* **31**, 985–1000.

Halow, K.D., Harner, R.C. and Fontenelle, L.J. (1996) Primary skin infections secondary to *Vibrio vulnificus*: the role of operative intervention. *Journal of the American College of Surgeons* **183**, 329–34.

Hammond, J.M.J., Potgeiter, P.D., Saunders, G.L. and Forder, A.A. (1992) Double-blind study of selective decontamination of the digestive tract in intensive care. *Lancet* **340**, 5–9.

Hancock, R.E.W. and Chapple, D.S. (1999) Peptide antibiotics. *Antimicrobial Agents Chemotherapy* **43**, 1317–23.

Harbarth, S., Dharan, S., Liassine, N. *et al.* (1999) Randomized, placebo-controlled, double-blind trial to evaluate the efficacy of mupirocin for eradicating carriage of methicillin-resistant *Staphylococcus aureus*. *Antimicrobial Agents and Chemotherapy* **43**, 1412–16.

Haria, M. and Lamb, H.M. (1997) Trovafloxacin. *Drugs* **54**, 435–45.

Hassoun, B.S., Ferris, E.B., Pien, F.D. and Youngblood, D.A. (1995) Necrotizing fasciitis. *Complications in Surgery* **120**, 15–21.

Hau, T. (1998) Biology and treatment of peritonitis: the historic development of current concepts. *Journal of the American College of Surgeons* **186**, 475–84.

Hazen, K.C. (1995) New and emerging yeast pathogens. *Clinical Microbiology Reviews* **8**, 462–78.

Hebart, H., Kanz, L., Jahn, G. and Einsele, H. (1998) Management of cytomegalovirus infection after solid-organ or stem-cell transplantation – current guidelines and future prospects. *Drugs* **55**, 59–72.

Heimbach, D. (1993) Editorial response: use of hyperbaric oxygen. *Clinical Infectious Diseases* **17**, 239–40.

Hentges, D.J. (1993) The anaerobic microflora of the human body. *Clinical Infectious Diseases* **16 (Suppl. 4)**, S175–S180.

Heseltine, P.N.R., Yellin, A.E., Appleman, M.D. *et al.* (1986a) Imipenem therapy for perforated and gangrenous appendicitis. *Surgery, Gynecology and Obstetrics* **162**, 43–8.

Heseltine, P.N.R., Berne, T.V., Yellin, A.E. *et al.* (1986b) The efficacy of cefoxitin vs. clindamycin/gentamicin in surgically treated stab wounds of the bowel. *Journal of Trauma* **26**, 241–5.

Hiramatsu, K., Hanaki, H., Ino, T. *et al.* (1997) Methicillin-resistant *Staphylococcus aureus* clinical strain with reduced vancomycin susceptibility. *Journal of Antimicrobial Chemotherapy* **40**, 135–46.

Hoellman, D.B., Visalli, M.A., Jacobs, M.R. and Appelbaum, P.C. (1998) Activities and time-kill studies of selected penicillins, β-lactamase inhibitor combinations, and glycopeptides against *Enterococcus faecalis*. *Antimicrobial Agents and Chemotherapy* **42**, 857–61.

Hoffmann, J., Rolff, M., Lomborg, V. and Franzmann, M. (1991) Ultraconservative management of appendiceal abscess. *Journal of the Royal College of Surgeons of Edinburgh* **36**, 18–20.

Högenauer, C., Hammer, H.F., Krejs, G.J. and Reisinger, E.C. (1998) Mechanisms and management of antibiotic-associated diarrhea. *Clinical Infectious Diseases* **27**, 702–10.

Holzapfel, L, Chevret, S., Madinier, G. *et al.* (1993) Influence of long-term oro- or nasotracheal intubation on nosocomial maxillary sinusitis and pneumonia: results of a prospective, randomized, clinical trial. *Critical Care Medicine* **21**, 1132–8.

Hooker, K.D., DiPiro, J.T. and Wynn J.J. (1991) Aminoglycoside combinations versus beta-lactams alone for penetrating abdominal trauma: a meta-analysis. *Journal of Trauma* 31, 1155–60.

Hooper, D.C. (1995) Quinolones. In Mandell, G.L., Bennett, J.E. and Dolin, R. (eds), *Mandell, Douglas and Bennett's principles and practice of infectious diseases*, 4th edn. Vol. 1. New York: Churchill Livingstone, 364–76.

Houlihan, H.H., Mercier, R.-C. and Rybak, M.J. (1997) Pharmacodynamics of vancomycin alone and in combination with gentamicin at various dosing intervals against methicillin-resistant *Staphylococcus aureus*-infected fibrin–platelet clots in an *in vitro* infection model. *Antimicrobial Agents and Chemotherapy* 41, 2497–501.

Howard, R.J. (1994) Infections in the immunocompromised patient. *Surgical Clinics of North America* 74, 609–20.

Howe, R.A., Brown, N.M. and Spencer, R.C. (1996) The new threats of Gram-positive pathogens: re-emergence of things past. *Journal of Clinical Pathology* 49, 444–9.

Huang, Y.-C., Lin, T.-Y., Leu, H.-S. *et al.* (1998) Yeast carriage on the hands of hospital personnel working in intensive-care units. *Journal of Hospital Infection* 39, 47–51.

Hudson, I.R.B. (1994) The efficacy of intranasal mupirocin in the prevention of staphylococcal infections: a review of recent experience. *Journal of Hospital Infection* 27, 81–98.

Humphreys, H. and Duckworth, G. (1997) Methicillin-resistant *Staphylococcus aureus* (MRSA) – a re-appraisal of control measures in the light of changing circumstances. *Journal of Hospital Infection* 36, 167–70.

Irish, D., Eltringham, I., Teall, A. *et al.* (1998) Control of an outbreak of an epidemic methicillin-resistant *Staphylococcus aureus* also resistant to mupirocin. *Journal of Hospital Infection* 39, 19–26.

Ishida, H., Ishida, Y., Kurosaka, Y. *et al.* (1998) *In vitro* and *in vivo* activities of levofloxacin against biofilm-producing *Pseudomonas aeruginosa*. *Antimicrobial Agents and Chemotherapy* 42, 1641–5.

Jacoby, G.A. (1998) Editorial response: epidemiology of extended-spectrum β-lactamases. *Clinical Infectious Diseases* 27, 81–3.

Jenkins, D.J. (1998) Hydatidosis – a zoonosis of unrecognised increasing importance? *Journal of Medical Microbiology* 47, 281–2.

Jensen, A.G., Espersen, F., Skinhøj, P. *et al.* (1997) Increasing frequency of vertebral osteomyelitis following *Staphylococcus aureus* bacteraemia in Denmark 1980–1990. *Journal of Infection* 34, 113–18.

Johnson, A.P (1998) Antibiotic resistance among clinically important Gram-positive bacteria in the UK. *Journal of Hospital Infection* 40, 17–26.

Johnson, S. and Gerding, D.N. (1998) *Clostridium difficile*-associated diarrhea. *Clinical Infectious Diseases* 26, 1027–36.

Jones, S.G. and Fraise, A.P. (1997) Coping with nosocomial infection: a non-antibiotic approach. *British Journal of Hospital Medicine* 58, 217–20.

Kaldjian, L.C. and Andriole, V.T. (1993) Necrotizing fasciitis: use of computed tomography for noninvasive diagnosis. *Infectious Diseases in Clinical Practice* 2, 325–9.

Karchmer, A.W. (1995) Cephalosporins. In Mandell, G.L., Bennett, J.E. and Dolin, R. (eds), *Mandell, Douglas and Bennett's principles and practice of infectious diseases*, 4th edn. Vol. 1. New York: Churchill Livingstone, 247–64.

Karchmer, A.W. (1998) Editorial response: salvage of infected orthopedic devices. *Clinical Infectious Diseases* 27, 714–16.

Kashuba, A.D.M., Bertino, J.S. and Nafziger, A.N. (1998) Dosing of aminoglycosides to rapidly attain pharmacodynamic goals and hasten therapeutic response by using individualized pharmacokinetic monitoring of patients with pneumonia caused by Gram-negative organisms. *Antimicrobial Agents and Chemotherapy* 42, 1842–4.

Kauffman, C.A. and Carver, P.L. (1997) Use of azoles for systemic antifungal therapy. *Advances in Pharmacology* 39, 143–89.

Kennedy, J.M. and van Riji, A.M. (1998) Effects of surgery on the pharmacokinetic parameters of drugs. *Clinical Pharmacokinetics* 35, 293–312.

Kernodle, D.S., Classen, D.C., Stratton, C.W. and Kaiser, A.B. (1998a) Association of borderline oxacillin-susceptible strains of *Staphylococcus aureus* with surgical wound infections. *Journal of Clinical Microbiology* 36, 219–22.

Kernodle, D.S., Voladri, R.K.R. and Kaiser, A.B. (1998b) β-Lactamase production diminishes the prophylactic efficacy of ampicillin and cefazolin in a guinea pig model of *Staphylococcus aureus* wound infection. *Journal of Infectious Diseases* 177, 701–6.

Kibbler, C.C., Quick, A. and O'Neill, A.-M. (1998) The effect of increased bed numbers on MRSA transmission in acute medical wards. *Journal of Hospital Infection* 39, 213–19.

King, C.T., Rogers, P.D., Cleary, J.D. and Chapman, S.W. (1998) Antifungal therapy during pregnancy. *Clinical Infectious Diseases* 27, 1151–60.

Klemm, K. (1979) Gentamycin-PMMA-Kugeln in der behanlung abszedierender knochen- und weichteilinfektionen. *Zentralblatt für Chirurgie* 104, 934–42.

Kluytmans, J. (1998) Reduction of surgical site infections in major surgery by elimination of nasal carriage of *Staphylococcus aureus*. *Journal of Hospital Infection* 40, S25–S29.

Kluytmans, J., van Belkum, A. and Verbrugh, H. (1997) Nasal carriage of *Staphylcocccus aureus*: epidemiology, underlying mechanisms, and associated risks. *Clinical Microbiology Reviews* 10, 505–20.

Knowles, D.J.C. (1997) New strategies for antibacterial drug design. *Trends in Microbiology* 5, 379–83.

Kollef, M.H. (1999) The prevention of ventilator-associated pneumonia. *New England Journal of Medicine* 340, 627–34.

Krcmery, V., Jesenska, Z., Spanik, S. *et al.* (1997) Fungaemia due to *Fusarium* spp. in cancer patients. *Journal of Hospital Infection* 36, 223–8.

Lacy, M.K., Nicolau, D.P., Nightingale, C.H. and Quintiliani, R. (1998) The pharmacodynamics of aminoglycosides. *Clinical Infectious Diseases* 27, 23–7.

La Rocco, M.T. and Burgert, S.J. (1997) Infection in the bone marrow transplant recipient, and role of the microbiology laboratory in clinical transplantation. *Clinical Microbiology Reviews* 10, 277–97.

Latgé, J.P. (1999) *Aspergillus fumigatus* and aspergillosis. *Clinical Microbiology Reviews* 12, 310–50.

Leader, W.G., Chandler, M.H.H. and Castiglia, M. (1995) Pharmacokinetic optimisation of vancomycin therapy. *Clinical Pharmacokinetics* 28, 327–42.

Leclercq, R. (1997) Enterococci acquire new kinds of resistance. *Clinical Infectious Diseases* 24 (Suppl. 1), S80–S84.

Lecuona, M., Torress-Lana, À., Delgado-Rodriguez, M. *et al.* (1998) Risk factors for surgical site infections diagnosed after hospital discharge. *Journal of Hospital Infection* 39, 71–4.

Lee, K., Chong, Y., Jeong, S.H. *et al.* (1996) Emerging resistance of anaerobic bacteria to antimicrobial agents in South Korea. *Clinical Infectious Diseases* 23 (Suppl. 1), S73–S77.

Lerner, P.I. (1996) Editorial response: medical resolution of proven and putative bacterial abscesses without surgical drainage. *Clinical Infectious Diseases* 23, 604–7.

Levison, M.A. and Zeigler, D. (1991) Correlation of APACHE II score, drainage technique and outcome in postoperative intra-abdominal abscess. *Surgery, Gynecology and Obstetrics* 172, 89–94.

Levy, I., Rubin, L.G., Vasishtha, S. *et al.* (1998) Emergence of *Candida parapsilosis* as the predominant species causing candidemia in children. *Clinical Infectious Diseases* 26, 1086–8.

Lew, D.P. and Waldvogel, F.A. (1997) Osteomyelitis. *New England Journal of Medicine* 336, 999–1007.

Lewis, K. (1998) Audit of timing of antibiotic prophylaxis in hip and knee arthroplasty. *Journal of the Royal College of Surgeons of Edinburgh* 43, 339–40.

Lewis, R.T., Goodall, R.G., Marien, B. *et al.* (1987) Biliary bacteria, antibiotic use, and wound infection in surgery of the gallbladder and common bile duct. *Archives of Surgery* 122, 44–7.

Lingnau, W., Berger, J., Javorsky, F. *et al.* (1998) Changing bacterial ecology during a five-year period of selective intestinal decontamination. *Journal of Hospital Infection* 39, 195–206.

Lipsky, B.A., Baker, P.D., Landon, G.C. and Fernau, R. (1997) Antibiotic therapy for diabetic foot infection: comparison of two parenteral-to-oral regimens. *Clinical Infectious Diseases* 24, 643–8.

Lister, P.D., Sanders, W.E. and Sanders, C.C. (1998) Cefepime–aztreonam: a unique double β-lactam combination for *Pseudomonas aeruginosa*. *Antimicrobial Agents and Chemotherapy* 42, 1610–19.

Lode, H., Borner, K. and Koeppe, P. (1998) Pharmacodynamics of fluoroquinolones. *Clinical Infectious Diseases* 27, 33–9.

Logan, J.M.J., Orange, G.V. and Maggs, A.F. (1999) Identification of the cause of a brain abscess by direct 16S ribosomal DNA sequencing. *Journal of Infection* 38, 45–7.

Lowy, F.D. (1998) *Staphylococcus aureus* infections. *New England Journal of Medicine* 339, 520–32.

Lucas, G.M., Lechtzin, N., Puryear, D.W. et al. (1998) Vancomycin-resistant and vancomycin-susceptible enterococcal bacteremia: comparison of clinical features and outcomes. *Clinical Infectious Diseases* **26**, 1127–33.

Lynch, A.C. and Stubbs, R.S. (1999) Hydatid disease in New Zealand – what remains and how should we treat it? *New Zealand Medical Journal* **112**, 131–4.

McArdle, C.S. (1994) Oral prophylaxis in biliary tract surgery. *Journal of Antimicrobial Chemotherapy* **33**, 200–2.

McArdle, C.S., Morran, C.G., Anderson, J.R. et al. (1995) Oral ciprofloxacin as prophylaxis in gastroduodenal surgery. *Journal of Hospital Infection* **30**, 211–16.

McClean, K.L., Sheehan, G.J. and Harding, G.K.M. (1994) Intra-abdominal infection: a review. *Clinical Infectious Diseases* **19**, 100–16.

McDonald, C.K. and Kuritzkes, D.R. (1997) Human immunodeficiency virus type 1 protease inhibitors. *Archives of Internal Medicine* **157**, 951–9.

McDonald, M. (1997) The epidemiology of methicillin-resistant *Staphylococcus aureus*: surgical relevance 20 years on. *Australian and New Zealand Journal of Surgery* **67**, 682–5.

McDonald, M., Grabsch, E., Marshall, C. and Forbes, A. (1998) Single-versus multiple-dose antimicrobial prophylaxis for major surgery: a systematic review. *Australian and New Zealand Journal of Surgery* **68**, 388–95.

MacGowan, A. (1998) Novel strategies for the use of antibiotics. *Current Opinion in Infectious Diseases* **11**, 471–3.

McLaws, M.-L., Murphy, C. and Keogh, G. (1997) The validity of surgical wound infection as a clinical indicator in Australia. *Australian and New Zealand Journal of Surgery* **67**, 675–8.

McNeil, M.M., Lasker, B.A., Lott, T.J. and Jarvis, W.R. (1999) Post-surgical *Candida albicans* infections associated with an extrinsically contaminated intravenous anesthetic agent. *Journal of Clinical Microbiology* **37**, 1398–403.

Mader, J.T. and Calhoun, J. (1995) Osteomyelitis. In Mandell, G.L., Bennett, J.E. and Dolin, R. (eds), *Mandell, Douglas and Bennett's principles and practice of infectious diseases*, 4th edn. Vol. 1. New York: Churchill Livingstone, 1039–51.

Mader, J.T., Mohan, D. and Calhoun, J. (1997) A practical guide to the diagnosis and management of bone and joint infections. *Drugs* **54**, 253–64.

Malangoni, M.A. (1998) Surgical site infections: the cutting edge. *Current Opinion in Infectious Diseases* **11**, 465–9.

Maller, R., Ahrne, H., Holmen, C. et al. (1993) Once- versus twice-daily amikacin regimen: efficacy and safety in systemic Gram-negative infections. *Journal of Antimicrobial Chemotherapy* **31**, 939–48.

Malloch, D. and Salkin, I.F. (1984) A new species of *Scedosporium* associated with osteomyelitis in humans. *Mycotaxon* **21**, 247–55.

Marasco, S. and Woods, S. (1998) The risk of eye-splash injuries in surgery. *Australian and New Zealand Journal of Surgery* **68**, 785–7.

Mark, P.E., Abraham, G., Careau, P. et al. (1999) The *ex vivo* antimicrobial activity and colonization rate of two antimicrobial-bonded central venous catheters. *Critical Care Medicine* **27**, 1128–31.

Markus, S. and Buday, M.D. (1989) Culturing indwelling central venous catheters *in situ*. *Infections in Surgery* **May**, 157–62.

Marra, F., Partovi, N. and Jewesson, P. (1996) Aminoglycoside administration as a single daily dose: an improvement to current practice or a repeat of previous errors? *Drugs* **52**, 344–70.

Marshall, C.G., Lessard, I.A.D., Park, I.-S. and Wright, G.D. (1998) Glycopeptide antibiotic resistance genes in glycopeptide-producing organisms. *Antimicrobial Agents and Chemotherapy* **42**, 2215–20.

Martin, C., Cotin, A., Giraud, A. *et al*. (1998) Comparison of concentrations of sulbactam-ampicillin administered by bolus injections or bolus plus continuous infusion in tissues of patients undergoing colorectal surgery. *Antimicrobial Agents and Chemotherapy* **42**, 1093–7.

Martineau, L. and Shek, P.N. (1999) Efficacy of liposomal antibiotic therapy in a rat infusion model of *Escherichia coli* peritonitis. *Critical Care Medicine* **27**, 1153–8.

Martínez-Vázquez, C., Fernández-Ulloa, J., Bordón, J. *et al*. (1998) *Candida albicans* endophthalmitis in brown heroin addicts: response to early vitrectomy preceded and followed by antifungal therapy. *Clinical Infectious Diseases* **27**, 1130–33.

Masterton, R.G. (1999) Worthwhile infection control information? A report of the 38th Interscience Conference on Antimicrobial Agents and Chemotherapy (ICAAC), San Diego, California: 24–27 September 1998. *Journal of Hospital Infection* **42**, 269–74.

Matsuo, H., Hayashi, J., Ono, K. *et al*. (1997) Administration of aminoglycosides to hemodialysis patients immediately before dialysis: a new dosing modality. *Antimicrobial Agents and Chemotherapy* **41**, 2597–601.

Mauerhan, D.R., Nelson, C.L., Smith, D.L. *et al*. (1994) Prophylaxis against infection in total joint arthroplasty. *Journal of Bone and Joint Surgery* **76A**, 39–45.

Mead, P.B. (1995) Infections of the female pelvis. In Mandell, G.L., Bennett, J.E. and Dolin, R. (eds), *Mandell, Douglas and Bennett's principles and practice of infectious diseases*, 4th edn. Vol. 1. New York: Churchill Livingstone, 1090–8.

Medeiros, A.A. (1997) Evolution and dissemination of β-lactamases accelerated by generations of β-lactam antibiotics. *Clinical Infectious Diseases* **24 (Suppl. 1)**, S19–S45.

Miller, G.H., Sabatelli, F.J., Hare, R.S. *et al*. (1997) The most frequent aminoglycoside resistance mechanisms – changes with time and geographic area: a reflection of aminoglycoside usage patterns? *Clinical Infectious Diseases* **24 (Suppl. 1)**, S46–S62.

Mini, E., Nobili, S. and Periti, P. (1997) Methicillin-resistant staphylococci in clean surgery – is there a role for prophylaxis? *Drugs* **54 (Suppl. 6)**, 39–52.

Mochida, C., Hirakata, Y., Matsuda, J. *et al*. (1998) Antimicrobial susceptibility testing of *Bilophila wadsworthia* isolates submitted for routine laboratory examination. *Journal of Clinical Microbiology* **36**, 1790–2.

Moellering, R.C. (1998) Antibiotic resistance: lessons for the future. *Clinical Infectious Diseases*, **27 (Suppl. 1)**, S135–S140.

Montgomery, R.S. and Wilson, S.E. (1996) Intra-abdominal abscesses: image-guided diagnosis and therapy. *Clinical Infectious Diseases* **23**, 28–36.

Montravers, P., Gauzit, R., Muller, C. et al. (1996) Emergence of antibiotic-resistant bacteria in cases of peritonitis after intra-abdominal surgery affects the efficacy of empirical antimicrobial therapy. *Clinical Infectious Diseases* **23**, 486–94.

Morgan, M.S., Lytle, J. and Bryson, P.J.V. (1995) The place of hyperbaric oxygen in the treatment of gas gangrene. *British Journal of Hospital Medicine* **53**, 424–6.

Morris, D.L. (1997) Infections of the liver. In Clunie, G.J.A., Tjandra, J.J. and Francis, D.M.A. (eds), *Textbook of surgery*. Carlton, Australia: Blackwell Science, 320–3.

Morris, W.T. (1994) Effectiveness of ceftriaxone versus cefoxitin in reducing chest and wound infections after upper abdominal operations. *American Journal of Surgery* **167**, 391–5.

Morris, W.T., Innes, D.B., Richardson, R.A. et al. (1980) The prevention of post-appendicectomy sepsis by metronidazole and cefazolin: a controlled double-blind trial. *Australian and New Zealand Journal of Surgery* **50**, 429–33.

Moyle, G.J., Gazzard, B.G., Cooper, D.A. and Gatell, J. (1998) Antiretroviral therapy for HIV infection: a knowledge-based approach to drug selection and use. *Drugs* **55**, 383–404.

Murdoch, D.A. (1998) Gram-positive anaerobic cocci. *Clinical Microbiology Reviews* **11**, 81–120.

Murray, B.E. (1997) Antibiotic resistance. *Advances in Internal Medicine* **42**, 339–67.

Murray, B.E. (1998) Diversity among multidrug-resistant enterococci. *Emerging Infectious Diseases* **4**, 37–47.

Nampoory, M.R.N., Khan, Z.U., Johny, K.V. et al. (1996) Invasive fungal infections in renal transplant recipients. *Journal of Infection* **33**, 95–101.

Nash, G., Foley, F.D., Goodwin, M.N. et al. (1971) Fungal burn wound infection. *Journal of the American Medical Association* **215**, 1664–6.

Nathens, A.B. and Rotstein, O.D. (1994) Therapeutic options in peritonitis. *Surgical Clinics of North America* **74**, 677–91.

Nathwani, D. and Wood, M.J. (1993) Penicillins: a current review of their clinical pharmacology and therapeutic use. *Drugs* **45**, 866–94.

Nicas, T.I., Zeckel, M.L. and Braun, D.K. (1997) Beyond vancomycin: new therapies to meet the challenge of glycopeptide resistance. *Trends in Microbiology* **5**, 240–9.

Nichols, R.L. (1994) Antibiotic prophylaxis in surgery. *Current Opinion in Infectious Diseases* **7**, 647–52.

Nichols, R.L. (1996) Update: antibiotic prophylaxis in surgery. *Infectious Diseases in Clinical Practice* **5** (Suppl. 2), S77–S84.

Nichols, R.L. and Smith, J.W. (1993) Risk of infection, infecting flora and treatment considerations in penetrating abdominal trauma. *Surgery, Gynecology and Obstetrics* **177** (Suppl. 1), 50–4.

Nichols, R.L. and Smith, J.W. (1994) Anaerobes from a surgical perspective. *Clinical Infectious Diseases* **18** (Suppl. 4), S280–S286.

Nicolau, D.P., Nie, L., Tessier, P.R. et al. (1998) Prophylaxis of acute osteomyelitis with absorbable ofloxacin-impregnated beads. *Antimicrobial Agents and Chemotherapy* **42**, 840–2.

Norden, C., Gillespie, W.J. and Nade, S. (1994) *Infections in bones and joints*. Boston, MA: Blackwell Scientific Publications.

Offner, F., Cordonnier, C., Ljungman, P. et al. (1998) Impact of previous aspergillosis on the outcome of bone marrow transplantation. *Clinical Infectious Diseases* 26, 1098–103.

Oie, S. and Kamiya, A. (1996) Survival of methicillin-resistant *Staphylococcus aureus* (MRSA) on naturally contaminated dry mops. *Journal of Hospital Infection* 34, 145–9.

Palzkill, T. (1998) β-Lactamases are changing their activity spectrums. *ASM News* 64, 90–5.

Patel, R. and Paya, C.V. (1997) Infections in solid-organ transplant recipients. *Clinical Microbiology Reviews* 10, 86–124.

Patel, R., Snydman, D.R., Rubin, R.H. and Patterson, G.A. (1996a) Cytomegalovirus prophylaxis in solid-organ transplant recipients. *Transplantation* 61, 1279–89.

Patel, R., Portela, D., Badley, A.D. et al. (1996b) Risk factors of invasive *Candida* and non-*Candida* fungal infections after liver transplantation. *Transplantation* 62, 926–34.

Patey, O., Varon, E., Prazuck, T. et al. (1994) Multicentre survey in France of the antimicrobial susceptibilities of 116 blood culture isolates of the *Bacteroides fragilis* group. *Journal of Antimicrobial Chemotherapy* 33, 1029–34.

Patterson, J.M.M., Novak, C.B., Mackinnon, S.E. and Patterson, G.A. (1998) Surgeons' concern and practices of protection against bloodborne pathogens. *Annals of Surgery* 228, 266–72.

Peacock, J.E., Herington, D.A. and Cruz, J.M. (1993) Amphotericin B therapy: past, present, and future. *Infectious Diseases in Clinical Practice* 2, 81–93.

Pechère, J.-C. (1996) Streptogramins: a unique class of antibiotics. *Drugs* 51 (Suppl. 1), 13–19.

Pederzoli, P., Bassi, C. and Vesentini, S. (1993) A randomized multicentre clinical trial of antibiotic prophylaxis of septic complications in acute necrotizing pancreatitis with imipenem. *Surgery, Gynecology and Obstetrics* 176, 480–3.

Pefanis, A., Thauvin-Eliopoulos, C., Holden, J. et al. (1994) Activity of fleroxacin alone and in combination with clindamycin or metronidazole in experimental intra-abdominal abscesses. *Antimicrobial Agents and Chemotherapy* 38, 252–5.

Perry, C.M. and Markham, A. (1999) Piperacillin/tazobactam: an updated review of its use in the treatment of bacterial infections. *Drugs* 57, 805–43.

Pfaller, M.A., Jones, R.N., Doern, G.V. et al. (1998a) International surveillance of bloodstream infections due to *Candida* species: frequency of occurrence and antifungal susceptibilities of isolates collected in 1997 in the United States, Canada, and South America for the SENTRY program. *Journal of Clinical Microbiology* 36, 1886–9.

Pfaller, M.A., Jones, R.N., Doern, G.V. et al. (1998b) Bacterial pathogens isolated from patients with bloodstream infection: frequency of occurrence and antimicrobial susceptibility patterns from the SENTRY program (United States and Canada, 1997). *Antimicrobial Agents and Chemotherapy* 42, 1762–70.

Playforth, M.J., Smith, G.M.R., Evans, M. and Pollock, A.V. (1988) Antimicrobial bowel preparation: oral, parenteral, or both? *Diseases of the Colon and Rectum* 31, 90–3.

Poirier, J.-M., Hardy, S., Isnard, F. et al. (1997) Plasma itraconazole concentrations in patients with neutropenia: advantages of a divided daily dosage regimen. *Therapeutic Drug Monitoring* 19, 525–9.

Poulsen, K.B. and Meyer, M. (1996) Infection registration underestimates the risk of surgical wound infections. *Journal of Hospital Infection* 33, 207–16.

Poynard, T., Bedossa, P., Chevallier, M. et al. (1995) A comparison of three interferon alfa-2b regimens for the long-term treatment of chronic non-A, non-B hepatitis. *New England Journal of Medicine* 332, 1457–62.

Quebbeman, E.J. (1996) Rituals in the operating room: are they necessary? *Infectious Diseases in Clinical Practice* 5 (Suppl. 2), S68–S70.

Quintiliani, R., Nicolau, D.P. and Nightingale, C.H. (1996) Clinical relevance of penicillin-resistant *Streptococcus pneumoniae*, with particular attention to therapy with ceftizoxime, cefotaxime, and ceftriaxone. *Infectious Diseases in Clinical Practice* 5 (Suppl. 1), S37–S41.

Rasmussen, B.A., Bush, K. and Tally, F.P. (1997) Antimicrobial resistance in anaerobes. *Clinical Infectious Diseases* 24 (Suppl. 1), S110–S120.

Raz, P. and Miron, D. (1995) Oral ciprofloxacin for treatment of infection following nail puncture wounds of the foot. *Clinical Infectious Diseases* 21, 194–5.

Reiss, R., Eliashiv, A. and Deutsch, A.A. (1982) Septic complications and bile cultures in 800 consecutive cholecystectomies. *World Journal of Surgery* 6, 195–9.

Renoult, E., Georges, E., Biava, M.F. et al. (1997) Toxoplasmosis in kidney transplant recipients: a life-threatening but treatable disease. *Transplantation Proceedings* 29, 821–2.

Rex, J.H., Pfaller, M.A. and Galgiani, J.N. (1997) Development of interpretive breakpoints for antifungal susceptibility testing: conceptual framework and analysis of *in vitro* – *in vivo* correlation data for fluconazole, itraconazole and *Candida* infections. *Clinical Infectious Diseases* 24, 235–47.

Ribaud, P., Chastang, C., Latgé, J.-P. et al. (1999) Survival and prognostic factors of invasive aspergillosis after allogeneic bone marrow transplantation. *Clinical Infectious Diseases* 28, 322–30.

Richardson, M.D. (1997) Lipid complexes of amphotericin B: the competitive picture. *Journal of Medical Microbiology* 46, 185–7.

Robinson, J.D. (1997) Gentamicin monitoring in pediatric patients. *Annals of Pharmacotherapy* 31, 1539–40.

Robinson, M.T. (1998) Avoiding interactions with antibiotics. *New Ethics Journal* June, 61–6.

Robson, R.A. (1997) New antiviral agents. *New Ethics Journal* June, 9–12.

Rode, H., Brown, R.A. and Millar, A.J.W. (1993) Surgical skin and soft tissue infections. *Current Opinion in Infectious Diseases* 6, 683–90.

Roggenkamp, A., Sing, A., Horneff, M. et al. (1998) Chronic prosthetic hip infection caused by a small-colony variant of *Escherichia coli*. *Journal of Clinical Microbiology* 36, 2530–4.

Romanelli, V.A., Howie, M.B., Myerowitz, P.D. et al. (1993) Intraoperative and postoperative effects of vancomycin administration in cardiac surgery patients: a prospective, double-blind, randomized trial. *Critical Care Medicine* 21, 1124–31.

Rotstein, C., Bow, E.J., Laverdiere, M. *et al.* (1999) Randomized placebo-controlled trial of fluconazole prophylaxis for neutropenic cancer patients: benefit based on purpose and intensity of cytotoxic therapy. *Clinical Infectious Diseases* **28**, 331–40.

Rotter, M.L. (1998) Semmelweis' sesquicentennial: a little-noted anniversary of handwashing. *Current Opinion in Infectious Diseases*, **11**, 457–60.

Rybak, M.J. and McGrath, B.J. (1996) Combination antimicrobial therapy for bacterial infections: guidelines for the clinician. *Drugs* **52**, 390–405.

Ryono, R.A., Jones, K.S., Coleman, R.W. and Holodniy, M. (1996) Prescribing practices and cost of antibacterial prophylaxis for surgery at a US Veterans Affairs Hospital. *PharmacoEconomics* **10**, 630–43.

Salama, H., Abdel-Wahab, M.F. and Strickland, G.T. (1995) Diagnosis and treatment of hepatic hydatid cysts with the aid of echo-guided percutaneous cyst puncture. *Clinical Infectious Diseases* **21**, 1372–6.

Salonen, J.H., Eerola, E. and Meurman, O. (1998) Clinical significance and outcome of anaerobic bacteremia. *Clinical Infectious Diseases* **26**, 1413–17.

Salyers, A.A. and Amábile-Cuevas, C.F. (1997) Why are antibiotic resistance genes so resistant to elimination? *Antimicrobial Agents and Chemotherapy* **41**, 2321–5.

Sanders, W.E. and Sanders, C.C. (1997) *Enterobacter* spp.: pathogens poised to flourish at the turn of the century. *Clinical Microbiology Reviews* **10**, 220–41.

Sanderson, P.J. (1993) Antimicrobial prophylaxis in surgery: microbiological factors. *Journal of Antimicrobial Chemotherapy* **31 (Suppl. B)**, 1–9.

Sands, K., Vineyard, G. and Platt, R. (1996) Surgical site infections occurring after hospital discharge. *Journal of Infectious Diseases* **173**, 963–70.

Santos, E.M. and Sapico, F.L. (1998) Vertebral osteomyelitis due to salmonellae: report of two cases and review. *Clinical Infectious Diseases* **27**, 287–95.

Santos, K.R.N., Bravo Neto, G.P., Fonseca, L.S. and Gontijo Filho, P.P. (1997) Incidence surveillance of wound infection in hernia surgery during hospitalization and after discharge in a university hospital. *Journal of Hospital Infection* **36**, 229–33.

Sauaia, A., Moore, F.A., Moore, E.E. and Lezotte, D.C. (1996) Early risk factors for postinjury multiple organ failure. *World Journal of Surgery* **20**, 392–400.

Saunders, G.L., Hammond, J.M.J., Potgieter, P.D. *et al.* (1994) Microbiological surveillance during selective decontamination of the digestive tract (SDD). *Journal of Antimicrobial Chemotherapy* **34**, 529–44.

Saunders, N.J. (1995) Vancomycin administration and monitoring reappraisal. *Journal of Antimicrobial Chemotherapy* **36**, 279–82.

Sawyer, M.D. and Dunn, D.L. (1991) Deep soft tissue infections. *Current Opinion in Infectious Diseases* **4**, 649–54.

Sawyer, R.G. and Pruett, T.L. (1994) Wound infections. *Surgical Clinics of North America* **74**, 519–36.

Sawyer, R.G., McGory, R.W., Gaffey, M.J. *et al.* (1998) Improved clinical outcomes with liver transplantation for hepatitis B-induced chronic liver failure using passive immunization. *Annals of Surgery* **227**, 841–50.

Scaglione, F., de Martini, G., Peretto, L. et al. (1997) Pharmacokinetic study of cefodizime and ceftriaxone in sera and bones of patients undergoing hip arthroplasty. *Antimicrobial Agents and Chemotherapy* **41**, 2292–4.

Schardey, H.M., Joosten, U., Finke, U. et al. (1997) The prevention of anastomotic leakage after total gastrectomy with local decontamination: a prospective, randomized, double-blind, placebo-controlled multicentre trial. *Annals of Surgery* **225**, 172–80.

Schentag, J.J., Hyatt, J.M. and Carr, J.R. (1998) Genesis of methicillin-resistant *Staphylococcus aureus* (MRSA), how treatment of MRSA infections has selected for vancomycin-resistant *Enterococcus faecium*, and the importance of antibiotic management and infection control. *Clinical Infectious Diseases* **26**, 1204–14.

Schoonover, L.L., Occhipinti, D.J., Rodvold, K.A. and Danziger, L.H. (1995) Piperacillin/tazobactam: a new beta-lactam/beta-lactamase inhibitor combination. *Annals of Pharmacotherapy* **29**, 501–14.

Schülin, T., Wennersten, C.B., Moellering, R.C. and Eliopoulos, G.M. (1997) *In vitro* activity of RU 64004, a new ketolide antibiotic, against Gram-positive bacteria. *Antimicrobial Agents and Chemotherapy* **41**, 1196–202.

Seeberger, M.D., Staender, S., Oertli, D. et al. (1998) Efficacy of specific aseptic precautions for preventing propofol-related infections: analysis by a quality-assurance programme using the explicit outcome method. *Journal of Hospital Infection* **39**, 67–70.

Segreti, J., Nelson, J.A. and Trenholme, G.M. (1998) Prolonged suppressive antibiotic therapy for infected orthopedic prostheses. *Clinical Infectious Diseases* **27**, 711–13.

Seguin, J.C., Walker, R.D., Caron, J.P. et al. (1999) Methicillin-resistant *Staphylococcus aureus* outbreak in a veterinary teaching hospital: potential human-to-animal transmission. *Journal of Clinical Microbiology* **37**, 1459–63.

Sheehan, D.J., Hitchcock, C.A. and Sibley, C.M. (1999) Current and emerging azole antifungal agents. *Clinical Microbiology Reviews* **12**, 40–79.

Sieradzki, K., Villari, P. and Tomasz, A. (1998) Decreased susceptibilities to teicoplanin and vancomycin among coagulase-negative methicillin-resistant clinical isolates of staphylococci. *Antimicrobial Agents and Chemotherapy* **42**, 100–7.

Sieradzki, K., Roberts, R.B., Haber, S.W. and Tomasz, A. (1999) The development of vancomycin resistance in a patient with methicillin-resistant *Staphylococcus aureus* infection. *New England Journal of Medicine* **340**, 517–23.

Singh, G., Ray, P., Sinha, S.K. et al. (1996) Bacteriology of necrotizing infections of soft tissues. *Australian and New Zealand Journal of Surgery* **66**, 747–50.

Singh, N. (1998) Infections in solid-organ transplant recipients. *Current Opinion in Infectious Diseases* **11**, 411–17.

Singh, N., Chang, F.Y., Gayowski, T. and Marino, I.R. (1997a) Infections due to dematiaceous fungi in organ transplant recipients: case report and review. *Clinical Infectious Diseases* **24**, 369–74.

Singh, N., Gayowski, T., Wagener, M.M. et al. (1997b) Invasive fungal infections in liver transplant recipients receiving tacrolimus as the primary immunosuppressive agent. *Clinical Infectious Diseases* **24**, 179–84.

Slack, R.C.B. (1995) Chemoprophylaxis. In Greenwood, D. (ed.), *Antimicrobial chemotherapy*, 3rd edn. Oxford: Oxford University Press, 219–30.
Slama, T.G. (1992) Current thoughts and controversies of antibiotic prophylaxis. *Current Opinion in Infectious Diseases* 5, 787–93.
Smith, A.M. and Klugman, K.P. (1998) Alterations in PBP 1A essential for high-level penicillin resistance in *Streptococcus pneumoniae*. *Antimicrobial Agents and Chemotherapy* 42, 1329–33.
Smith, J.M.B. (1989) *Opportunitistic mycoses of man and other animals*. Wallingford: CAB International Mycological Institute.
Smith, J.M.B. and Payne, J.E. (1994) Antimicrobial therapy in selected surgical patients. *Australian and New Zealand Journal of Surgery* 64, 658–66.
Smith, T.L., Pearson, M.L., Wilcox, K.R. et al. (1999) Emergence of vancomycin resistance in *Staphylococcus aureus*. *New England Journal of Medicine* 340, 493–501.
Snydman, D.R., McDermott, L., Cuchural, G.J. et al. (1996) Analysis of trends in antimicrobial resistance patterns among clinical isolates of *Bacteroides fragilis* group species from 1990 to 1994. *Clinical Infectious Diseases* 23 (Suppl. 1), S54–S65.
Sobel, J.D. and Kaye, D. (1995) Urinary tract infections. In Mandell, G.L., Bennett, J.E. and Dolin, R. (eds) *Mandell, Douglas and Bennett's principles and practice of infectious diseases*, 4th edn. Vol. 1. New York: Churchill Livingstone, 662–90.
Solomkin, J.S. (1988) Use of new beta-lactam antibiotics for surgical infections. *Surgical Clinics of North America* 68, 1–24.
Solomkin, J.S., Flohr, A. and Simmons, R.L. (1982a) *Candida* infections in surgical patients: dose requirements and toxicity of amphotericin B. *Annals of Surgery* 195, 177–85.
Solomkin, J.S., Flohr, A.M. and Simmons, R.L. (1982b) Indications for therapy for fungemia in postoperative patients. *Archives of Surgery* 117, 1272–5.
Solomkin, J.S., Dellinger, E.P., Christou, N.V. and Busuttil, R.W. (1990) Results of a multicentre trial comparing imipenem/cilastatin to tobramycin/clindamycin for intra-abdominal infections. *Annals of Surgery* 212, 581–91.
Solomkin, J.S., Reinhart, H.H., Dellinger, E.P. et al. (1996) Results of a randomized trial comparing sequential intravenous/oral treatment with ciprofloxacin plus metronidazole to imipenem/cilastatin for intra-abdominal infections. *Annals of Surgery* 223, 303–15.
Souli, M. and Giamarellou, H. (1998) Effects of slime produced by clinical isolates of coagulase-negative staphylococci on activities of various antimicrobial agents. *Antimicrobial Agents and Chemotherapy* 42, 939–41.
Stacey, A., Burden, P., Croton, C. and Jones, E. (1998) Contamination of television sets by methicillin-resistant *Staphylococcus aureus* (MRSA). *Journal of Hospital Infection* 39, 243–4.
Stark, C.A., Edlund, C., Sjöstedt, S. et al. (1993) Antimicrobial resistance in human oral and intestinal anaerobic microfloras. *Antimicrobial Agents and Chemotherapy* 37, 1665–9.

Stein, A., Bataille, J.F., Drancourt, M. et al. (1998) Ambulatory treatment of multidrug-resistant *Staphylococcus*-infected orthopedic implants with high-dose oral co-trimoxazole (trimethoprim-sulfamethoxazole). *Antimicrobial Agents and Chemotherapy* **42**, 3086–91.

Stevens, D.A. (1997) New directions in antifungal therapy. *Japanese Journal of Medical Mycology* **38**, 141–4.

Stevens, D.A., Diaz, M., Negroni, R. et al. (1997) Safety evaluation of chronic fluconazole therapy. *Chemotherapy* **43**, 371–7.

Stevens, D.L., Madaras-Kelly, K.J. and Richards, D.M. (1998) *In-vitro* antimicrobial effects of various combinations of penicillin and clindamycin against four strains of *Streptococcus pyogenes*. *Antimicrobial Agents and Chemotherapy* **42**, 1266–8.

Stratov, I., Gottlieb, T., Bradbury, R. and O'Kane, G.M. (1998) Candidaemia in an Australian teaching hospital: relationship to central line and TPN use. *Journal of Infection* **36**, 203–7.

Such, J. and Runyon, B.A. (1998) Spontaneous bacterial peritonitis. *Clinical Infectious Diseases* **27**, 669–76.

Sugar, A.M. and Liu, X.P. (1998) Interactions of itraconazole with amphotericin B in the treatment of murine invasive candidiasis. *Journal of Infectious Diseases* **177**, 1660–63.

Sullivan, D. and Coleman, D. (1998) *Candida dubliniensis*: characteristics and identification. *Journal of Clinical Microbiology* **36**, 329–34.

Suthanthiran, M. and Strom, T.B. (1994) Renal transplantation. *New England Journal of Medicine* **331**, 365–76.

Sutherland, M.E. and Meyer, A.A. (1994) Necrotizing soft-tissue infections. *Surgical Clinics of North America* **74**, 591–606.

Swartz, M.N. (1995) Cellulitis and subcutaneous tissue infections. In Mandell, G.L., Bennett, J.E. and Dolin, R. (eds), *Mandell, Douglas and Bennett's principles and practice of infectious diseases*, 4th edn. Vol. 1. New York: Churchill Livingstone, 909–29.

Swenson, C.E., Perkins, W.R., Roberts, P. et al. (1998) *In vitro* and *in vivo* antifungal activity of amphotericin B lipid complex: are phospholipases important? *Antimicrobial Agents and Chemotherapy* **42**, 767–71.

Symmers, W. St. C. (1968) Silicone mastitis in 'topless' waitresses and some other varieties of foreign-body mastitis. *British Medical Journal* **3**, 19–22.

Tarkowski, A. and Wagner, H. (1998) Arthritis and sepsis caused by *Staphylococcus aureus*: can the tissue injury be reduced by modulating the host's immune system? *Molecular Medicine Today* **4**, 15–18.

Tegeder, I., Bremer, F., Oelkers, R. et al. (1997) Pharmacokinetics of imipenem–cilastatin in critically ill patients undergoing continuous venovenous hemofiltration. *Antimicrobial Agents and Chemotherapy* **41**, 2640–45.

Telford, G.L. (1996) Recommendations for reducing the risk of viral infections for surgeons. *Infectious Diseases in Clinical Practice* **5** (**Suppl. 2**), S71–S73.

Tetteroo, G.W.M., Wagenvoort, J.H.T. and Bruining, H.A. (1994) Bacteriology of selective decontamination: efficacy and rebound colonization. *Journal of Antimicrobial Chemotherapy* **34**, 139–48.

Thauvin-Eliopoulos, C., Tripodi, M.-F., Moellering, R.C. and Eliopoulos, G.M. (1997) Efficacies of piperacillin–tazobactam and cefepime in rats with experimental intra-abdomoninal abscesses due to an extended-spectrum β-lactamase-producing strain of *Klebsiella pneumoniae*. *Antimicrobial Agents and Chemotherapy* **41**, 1053–7.

Thom, S. (1993) Editorial response: a role for hyperbaric oxygen in clostridial myonecrosis. *Clinical Infectious Diseases* **17**, 238.

Thomson, A.H., Duncan, N., Silverstein, B. *et al.* (1996) Antimicrobial practice: development of guidelines for gentamicin dosing. *Journal of Antimicrobial Chemotherapy* **38**, 885–93.

Thys, J.P., Vanderhoeft, P., Herchuelz, A. *et al.* (1988) Penetration of aminoglycosides in uninfected pleural exudates and in pleural empyemas. *Chest* **93**, 530–2.

Tighe, M.J., Kite, P., Fawley, W.N. *et al.* (1996a) An endoluminal brush to detect the infected central venous catheter *in situ*: a pilot study. *British Medical Journal* **313**, 1528–9.

Tighe, M.J., Kite, P., Thomas, D. *et al.* (1996b) Rapid diagnosis of catheter-related sepsis using the acridine orange leukocyte cytosporin test and an endoluminal brush. *Journal of Parenteral and Enteral Nutrition* **20**, 215–18.

Tunkel, A.R. (1995) Topical antibacterials. In Mandell, G.L., Bennett, J.E. and Dolin, R. (eds), *Mandell, Douglas and Bennett's principles and practice of infectious diseases*, 4th edn. Vol. 1. New York: Churchill Livingstone, 381–9.

Tunney, M.M., Ramage, G., Patrick, S. *et al.* (1998) Antimicrobial susceptibility of bacteria isolated from orthopedic implants following revision hip surgery. *Antimicrobial Agents and Chemotherapy* **42**, 3002–5.

Turnidge, J. (1995) What to use instead of flucloxacillin. *Australian Prescriber* **18**, 54–6.

Turnidge, J.D. (1998) The pharmacodynamics of β-lactams. *Clinical Infectious Diseases* **27**, 10–22.

Turnidge, J. and Grayson, M.L. (1993) Optimum treatment of staphylococcal infections. *Drugs* **45**, 353–66.

Vallés, J., León, C. and Alvarez-Lerma, F. (1997) Nosocomial bacteremia in critically ill patients: a multicentre study evaluating epidemiology and prognosis. *Clinical Infectious Diseases* **24**, 387–95.

van Saene, H.K.F., Damjanovic, V., Williets, T. *et al.* (1998) Pathogenesis of ventilator-associated pneumonia: is the contribution of biofilm clinically significant? *Journal of Hospital Infection* **38**, 231–40.

Verweij, P.E., Erjavec, Z., Sluiters, W. *et al.* (1998) Detection of antigen in sera of patients with invasive aspergillosis: intra- and inter-laboratory reproducibility. *Journal of Clinical Microbiology* **36**, 1612–16.

Vial, T., Biour, M., Descotes, J. and Trepo, C. (1997) Antibiotic-associated hepatitis: update from 1990. *Annals of Pharmacotherapy* **31**, 204–20.

Visalli, M.A., Jacobs, M.R. and Appelbaum, P.C. (1998) Activities of three quinolones, alone and in combination with extended-spectrum cephalosporins or gentamicin, against *Stenotrophomonas maltophilia*. *Antimicrobial Agents and Chemotherapy* **42**, 2002–5.

Vora, S., Purimetla, N., Brummer, E. and Stevens, D.A. (1998) Activity of voriconazole, a new triazole, combined with neutrophils or monocytes against *Candida albicans*: effect of granulocyte colony-stimulating factor and granulocyte-macrophage colony-stimulating factor. *Antimicrobial Agents and Chemotherapy* **42**, 907–10.

Vuorisalo, S., Pokela, R. and Syrjälä, H. (1997) Is single-dose antibiotic prophylaxis sufficient for coronary artery bypass surgery? An analysis of peri- and postoperative serum cefuroxime and vancomycin levels. *Journal of Hospital Infection* **37**, 237–47.

Wagstaff, A.J. and Balfour, J.A. (1997) Grepafloxacin. *Drugs* **53**, 817–24.

Wald, A., Leisenring, W., van Burik, J.-A. and Bowden, R.A. (1997) Epidemiology of *Aspergillus* infections in a large cohort of patients undergoing bone marrow transplantation. *Journal of Infectious Diseases* **175**, 1459–66.

Walker, A.P., Krepel, C.J., Gohr, C.M. and Edmiston, C.E. (1994) Microflora of abdominal sepsis by locus of infection. *Journal of Clinical Microbiology* **32**, 557–8.

Walsh, T.J., Hiemenz, J.W., Seibel, N.L. *et al.* (1998) Amphotericin B lipid complex for invasive fungal infections: analysis of safety and efficacy in 556 cases. *Clinical Infectious Diseases* **26**, 1383–96.

Wells, G.R., Taylor, E.W., Lindsay, G. *et al.* (1989) Relationship between bile colonization, high-risk factors and postoperative sepsis in patients undergoing biliary tract operations while receiving a prophylactic antibiotic. *British Journal of Surgery* **76**, 374–7.

Welton, L.A., Thal, L.A., Perri, M.B. *et al.* (1998) Antimicrobial resistance in enterococci isolated from turkey flocks fed virginiamycin. *Antimicrobial Agents and Chemotherapy* **42**, 705–8.

Wen, H., New, R.R.C. and Craig, P.S. (1993) Diagnosis and treatment of human hydatidosis. *British Journal of Clinical Pharmacology* **35**, 565–74.

Weng, D.E., Wilson, W.H., Little, R. and Walsh, T.J. (1998) Successful medical management of isolated renal zygomycosis: case report and review. *Clinical Infectious Diseases* **26**, 601–5.

Wenisch, C., Bartunek, A., Zedtwitz-Liebenstein, K. *et al.* (1997) Prospective randomized comparison of cefodizime versus cefuroxime for perioperative prophylaxis in patients undergoing coronary artery bypass grafting. *Antimicrobial Agents and Chemotherapy* **41**, 1584–8.

Wenzel, R.P. (1993) Preoperative prophylactic antibiotics: brief historical note. *Infection Control and Hospital Epidemiology* **14**, 121.

Wenzel, R.P. and Edmond, M.B. (1998) Vancomycin-resistant *Staphylococcus aureus*: infection control considerations. *Clinical Infectious Diseases* **27**, 245–51.

Westergren, V., Lundblad, L., Hellquist, H.B. and Forsum, U. (1998) Ventilator-associated sinusitis: a review. *Clinical Infectious Diseases* **27**, 851–64.

Westphal, J.-F. and Brogard, J.-M. (1999) Biliary tract infections: a guide to drug treatment. *Drugs* **57**, 81–91.

Wexler, H.M., Molitoris, E., Molitoris, D. and Finegold, S.M. (1998) *In vitro* activity of levofloxacin against a selected group of anaerobic bacteria isolated from skin and soft tissue infections. *Antimicrobial Agents and Chemotherapy* **42**, 984–6.

White, T.C. (1998) Antifungal drug resistance in *Candida albicans*. *ASM News* **63**, 427–33.

Wilson, S.E. (1993) A critical analysis of recent innovations in the treatment of intra-abdominal infection. *Surgery, Gynecology and Obstetrics* **177 (Suppl.)**, S11–S17.

Wilson, S.E. (1996) Prevention of infectious complications following abdominal trauma and tube thoracostomy. *Infectious Diseases Clinical Practice* **5 (Suppl. 1)**, S9–S14.

Wilson, S.E. (1997) Results of a randomized, multicentre trial of meropenem versus clindamycin/tobramycin for the treatment of intra-abdominal infections. *Clinical Infectious Diseases* **24 (Suppl. 2)**, S197–S206.

Wininger, D.A. and Fass, R.J. (1996) Antibiotic-impregnated cement and beads for orthopedic infections. *Antimicrobial Agents and Chemotherapy* **40**, 2675–9.

Wiseman, L.R. and Lamb, H.M. (1997) Cefpirome: a review of its antibacterial activity, pharmacokinetic properties and clinical efficacy in the treatment of severe nosocomial infections and febrile neutropenia. *Drugs* **54**, 117–40.

Wiseman, L.R., Wagstaff, A.J., Brogden, R.N. and Bryson, H.M. (1995) Meropenem: a review of its antibacterial activity, pharmacokinetic properties and clinical efficacy. *Drugs* **50**, 73–101.

Wispelwey, B. and Scheld, W.M. (1995) Brain abscess. In Mandell, G.L., Bennett, J.E. and Dolin, R. (eds), *Mandell, Douglas and Bennett's principles and practice of infectious diseases*, 4th edn. Vol. 1. New York: Churchill Livingstone, 887–900.

Withington, S., Chambers, S.T., Beard, M.E. *et al.* (1998) Invasive aspergillosis in severely neutropenic patients over 18 years: impact of intranasal amphotericin B and HEPA filtration. *Journal of Hospital Infection* **38**, 11–18.

Wittmann, D.H. and Condon, R.E. (1991) Prophylaxis of postoperative infections. *Infection* **19 (Suppl. 6)**, S337–S344.

Wittmann, D.H., Aprahamian, C. and Bergstein, J.M. (1990) Etappenlavage: advanced diffuse peritonitis managed by planned multiple laparotomies utilizing zippers, slide fasteners, and velcro analogue for temporary abdominal closure. *World Journal of Surgery* **14**, 218–26.

Wittmann, D.H., Bergstein, J.M. and Frantzides, C. (1991) Calculated empirical antimicrobial therapy for mixed surgical infections. *Infection* **19 (Suppl. 6)**, S345–S350.

Wittmann, D.H., Schein, M. and Condon, R.E. (1996) Management of secondary peritonitis. *Annals of Surgery* **224**, 10–18.

Wong, S.W., Fernando, D. and Grant, P. (1997) Leg wound infections associated with coronary revascularization. *Australian and New Zealand Journal of Surgery* **67**, 689–91.

Wong-Beringer, A., Jacobs, R.A. and Guglielmo, B.J. (1998) Lipid formulations of amphotericin B: clinical efficacy and toxicities. *Clinical Infectious Diseases* **27**, 603–18.

Wood, C.A. and Wisniewski, R.M. (1994) β-Lactams versus glycopeptides in treatment of subcutaneous abscesses infected with *Staphylococcus aureus*. *Antimicrobial Agents and Chemotherapy* **38**, 1023–6.

Wynd, M.A. and Paladino, J.A. (1996) Cefepime: a fourth-generation parenteral cephalosporin. *Annals of Pharmacotherapy* **30**, 1414–24.

Yellin, A.E., Heseltine, P.N.R., Berne, T.V. *et al.* (1985) The role of *Pseudomonas* species in patients treated with ampicillin and sulbactam for gangrenous and perforated appendicitis. *Surgery, Gynecology and Obstetrics* **161**, 303–7.

Zhou, H., Goldman, M., Wu, J. *et al.* (1998) A pharmacokinetic study of intravenous itraconazole followed by oral administration of itraconazole capsules in patients with advanced human immunodeficiency virus infection. *Journal of Clinical Pharmacology* **38**, 593–602.

INDEX

Note: *italic page references* refer to tables. **Bold page references** refer to figures.

β-lactam ring 68–9, 70, **70**
β-lactamase inhibitors 87–8
β-lactamases
   see also extended-spectrum β-lactamases
   antibacterial resistance 34–6, **35**, 70, 114
   tests for 163–4
β-lactams 68–88
   see also cephalosporins; penicillins
   β-lactamase inhibitors 87–8
   antibacterial resistance 36–7, 38, 42
   antibacterial spectrum *86*
   carbapenems 84–5
   combined therapy 65
   excretion 54
   general information 68–70
   intra-abdominal infections *194*, 196
   investigational 88
   mechanisms of action 29
   monobactams 87
   PAE 65–6
   pregnancy *56*
   side-effects 61, 63
ABCD (Amphotec) 128, **129**, *129*, 130
abdominal infections 139–40, 242
abdominal trauma 187, **188**, 189
ABLC 129–30, *129*
abortion 209
abscess 9–10, 11–13, 197
   adnexal 210
   brain 169–72, *170*
   epidural 172–3
   haematogenous osteomyelitis 227
   intra-abdominal 13
   intramedullary of the spinal cord 173–4
   liver 203–6
   metastatic 12–13
   paraspinal 13
   pulmonary cavity 183–5, *184*
   renal 208–9
7-ACA *see* 7-aminocephalosporanic acid
acute physiology and chronic health evaluation-II (APACHE-II) score 5, 9, *188*
6-APA *see* 6-aminopenicillanic acid
acyclovir 148–50, *150*
adaptive resistance 90
administration of antibiotics
   local 116–17
   oral 50, 116, 122
   parenteral 50–51, 117, 122

   route 49–51, **51**, 115–17, 122
   timing 115–17, 122
adnexal abscess 210
aerobic bacteria 111, 112
   see also staphylococci; streptococci
   bite wounds *223*
   diabetic foot infection *220*
   Enterobacteriaceae 8, 35–6, *36*, 39–40, *41*, 197
   growth kinetics 16–18
   intra-abdominal infections *192*, 196
   Pasteurella 222
   Pseudomonas aeruginosa 7, 38, 74, 218, 221
Al555G mutation 91
D-alanyl-D-alanine 37–8
albendazole 157, 158
AmBi (AmBisome) 128, 130
American Society of Anaesthesiologists (ASA) 6, 113
7-aminocephalosporanic acid (7-ACA) 26, 69, 77, **80**
6-aminopenicillanic acid (6-APA) 68, **70**
aminoglycosides 88–93
   administration 91–3, *93*
   antibacterial spectrum *86*, 89
   bacterial resistance 36, 44, 90
   clinical use 90–93
   combined therapy 65
   dosing schedules 91–3, *93*
   drug interactions 50–51, 88–9
   general properties 88–9
   gentamicin 44, 88–9, 92, 164–5
   intra-abdominal infections *195*, 198–9
   mechanisms of action 30–31
   PAE 65, 66
   pharmacokinetics 89–90
   pregnancy *56*
   renal failure *52*
   side-effects 63, 90–93
   tobramycin 92
aminopenicillins 71, 72–3, 74
amoxycillin 71, 72–3, 74, 87–8
amoxycillin-clavulanate 87–8
Amphotec *see* ABCD
amphotericin B 126–31, *126–9*
   ABCD (Amphotec) 128, **129**, *129*, 130
   ABLC *129–30*, *129*
   AmBi 128, 130
   fungal resistance *146*, 147
   immunocompromised patients 244

invasive aspergillosis 138, 244
systemic candidosis 141, 244
ampicillin 44, 71, *72–3*, 74
ampicillin-sulbactam 88
amputation 120–21
anaerobic bacteria
  see also facultative anaerobes; obligate anaerobes
  anti-anaerobe agents 98–101, 199
  antimicrobial resistance 45–6
  Bacteroides 41, 45–6, 190, *191–5*, 193, 196, 197, *210*, 213, 215, 220, 223
  Clostridium 62, 100, *191–3*, 206–8, *210*, 215, 217, 218, 220, 223
  growth kinetics 16–18
  intra-abdominal infections 190, *192*, 196
  Peptostreptococcus 191–2, *210*, 213, 215, 220, 223
  Porphyromonas 191, 222, 223
  Prevotella 191, 193, *210*, 213, 215, 222, 223
aneurysm 233
antagonistic drugs 65
anti-anaerobe agents 98–101
  antibacterial spectrum 99
  chloramphenicol 31, 100–101
  clindamycin 99–100
  general properties 98
  metronidazole 9, 46, 65, 98–9
  peritonitis 199
antibacterial agents 25–123
  β-lactams 68–88, *72–3*, *75*, *76*, *78–9*, *80*, *82*, *86*
  adverse reactions *60–61*
  aminoglycosides 86, 88–93, *93*
  anti-anaerobe agents 98–101, *99*
  bacterial resistance 31–48, *35*, *41*
  cotrimoxazole 105–6
  development 67–8, **67**
  fusidic acid *103*, 104
  general information 25–66, **28**
  glycopeptides 101–4, *103*
  history 25–7
  incompatibilities 59
  investigational 107–8
  macrolides 106
  mechanisms of action 27–31, **28**
  pregnancy *55–7*
  quinolones 94–8
  renal failure *52–3*
  rifamycins *103*, 104–5
  tetracyclines 107
  topical 108–10, *109*
  treatment strategies and principles 49–66
  trimethoprim 105–6
  varieties 67–110

antibacterial prophylaxis in surgery 111–23
  administration of antibiotics 115–17
  appendectomy 114
  biliary surgery 114–15
  choices 117–21, *119–20*, 122–3
  clean procedures 115
  conclusions 121–2
  general principals 111–14
  invasive investigational procedures 115
  organ transplants 123
  regimens 117–21, *119–20*, 122–3
  SDD 122–3
antibacterial resistance 31–48
  β-lactams 36–7, 38, 42, 70
  acquired 32–4
  adaptive 90
  aminoglycosides 36, 44, 90
  biochemical mechanisms 34–9, **35**
  cephalosporins 113–14
  clinical problems 39–46, *41*
  drug modification 34–6, *35*
  glycopeptides 37–8, 44–5, 101
  intracellular drug accumulation 38
  metabolic bypass 38–9
  natural (intrinsic) 31–2
  overuse of antibacterials 27, 39
  penicillins 10, 34, 35, 37, 44, 45
  prevention strategies 46–8
  quinolones 38, 94
  single dose therapy 113–14
  target site alteration 36–8
  tolerance 34
  topical antibacterial agents 108
antibiotic tolerance 34
antibiotic-associated colitis 206–8
antifungal agents 124–47
  amphotericin B 126–31, *126–9*
  azoles 126, *126*, 131–6, *132*, *135*, 145–7, *146*
  disease treatment 136–46, *137*
  flucytosine *126*, 136, 146, *146*
  intra-abdominal infections *195*, 202–3
  modes of action 124–6, **125**
  polyenes 126–31, *126–9*, 138, 141, *146*, 147, 244
  pregnancy *128*
  resistance 142, 146–7, *146*, **147**
  side-effects *127*, 130–31, *132–3*, 136
  terbinafine 136
  varieties 126–36, *126*
antimicrobial agents 23–165
  antibacterial 25–123
    general information 25–66
    prophylaxis in surgery 111–23
    varieties 67–110
  antifungal 124–47

antiparasitic 156–8
antiviral 148–56
laboratory aspects of use 160–65
antimicrobial resistance 10
see also antibacterial resistance
amphotericin B *146*, 147
fungal 142, 146–7, *146*, **147**
gene transfer 18, 19–20
intra-abdominal infections 196–7
sensitivity tests 162
viral 152, 154
antiparasitic agents
hydatid disease 156–8, **157**, *159*
miscellaneous 158, *159*
pregnancy *128*
Taenia infection 158, *159*
antpseudomonal penicillins 74
antiviral agents 148–56, **149**
HBV/HCV 155–6
herpes 148–51, *150*
HIV 151–5, *153*
pregnancy *128*
APACHE-II *see* acute physiology and chronic health evaluation-II
appendectomy 114, 118, 121
appendicitis *192*
area under the concentration-time curve (AUC) for free drug 54
arthritis, septic 236, *237*
arthroplasty 116, 120
ASA *see* American Society of Anaesthesiologists
aseptic techniques 3–4, 6
aspergillomas 139
aspergillosis, invasive 137–9, *137*, 244
aspiration 183–4, 203
AUC *see* area under the concentration-time curve
avoparcin 38
azithromycin 106
azoles 126, 131–6
black moulds 145
drug interactions 135–6, *135*
fluconazole *132*, 133–4, *135*, 141–2, 146–7, *146*
imidazoles *126*, 131–3, *132*
intraconazole *132*, 134, *135*, 138
ketoconazole 132–3, *132*, *135*
resistance 146–7, *146*
triazoles *126*, *132*, 133–4
aztreonam 87

bacteraemia 12
bacteria *see* antibacterial agents; antibacterial prophylaxis in surgery; antibacterial resistance; biology of bacteria

bacterial synergistic gangrene 216
bacteriophages 18
Bacteroides
antimicrobial resistance *41*, 45–6
B. fragilis 41, 45–6
intra-abdominal infections 190, *191–5*, 193, 196, 197
other infections *210, 213, 215, 220, 223*
Basidiobolus 145
bathing, post-operative 7
benzimidazoles 157
benzylpenicillin 26, 70–71, **72**
Bernard, H.R. 112
Betadine *see* iodine PVP complex
bile 54, 114–15
biliary surgery 114–15, 118, 121
biliary tract infection 203–6
biofilms 123
biology of bacteria
cell wall 15–16, *15*, 17
gene transfer 18–20
general properties 14–15, *14*, 15
growth kinetics 16–18
normal flora 20–22, *21–2*
bite wounds 221–4, *223*
black moulds 145
blood
monitoring 160, 164–5
supply 6–7
bone-marrow
toxicity 64
transplants 137–9
Bottone, E.J. 143
brain abscess
general information 169–70
microbial aetiology 170–71, *170*
treatment strategies 171–2
break-point techniques 161
breast prosthesis 144
Brook, I. 212
Brotzu, G. 26
Burke, J.F. 111, 117
burn wounds 108–10, 144

Candida
antimicrobial resistance 146–7, *146*
C. albicans 139–40, 146–7, **162**
catheter-related infections 182
fluconazole therapy 133–4
HIV 155
systemic candidosis 139–42
candidosis *137*, 139–42, 244
CAPD *see* continuous ambulatory peritoneal dialysis
carbapenems 84

cardiac surgery 117, 118
cardiovascular infections
    catheter-related 181–2
    vascular grafts 179–81
catheter-related infections
    central 181–2
    intravenous 10, 11, 181–2
cell membrane 27, 124, **125**
cell wall
    bacterial 15–16, **15**, **17**, 27–9, **28**, 32, 38
    fungal 124, *125*
cellular defences 238, 239
Celsus 11
Center for Disease Control, Atlanta 144
central nervous system, transplant patient infections 242
cephalosporins 77–84, *78–9*, **80**, *82*
    antibacterial spectrum *82*
    bacterial resistance 113–14
    cephalosporin C 26, 77
    cephalosporin N 26
    cephamycins 81
    choosing 117–18, 120, 121
    development **80**
    ESBLs 163
    first-generation 77, *78*, **80**, 81, 118
    fourth-generation *79*, **80**, 83
    intra-abdominal infections 199
    loracarbef 83
    moxalactam 83
    pharmacokinetics 83–4
    pregnancy 56
    quinolone linkage 98
    renal failure 52
    second-generation *78*, **80**, 81, 120
    semi-synthetic 68–9, 77
    side-effects 61, 64, 74, 77, 84
    third-generation
        choosing 118, 121
        development **80**
        ESBL emergence 35–6, 39–40, 163
        general information *78–9*, 81–3
        pneumococcal resistance 45
    true 81
cephamycins 81
cephems *see* cephalosporins
cerebrospinal fluid (CSF), drug penetration 54
Chain, E. 25
chest infections, pulmonary cavity abscess 183–5, *184*
childhood diseases, haematogenous osteomyelitis 225–30, **228**, *230*
chlamydiae 14, *14*
chloramphenicol 31, *57*, 100–101

cilastatin 85
ciprofloxacin 65, *86*, 95–6, 97, *195*
clarithromycin 106
Classen, D.C. 117
clean procedures 115
clinafloxacin *86*, *95*, 97, *195*
clindamycin *53*, *57*, 99–100
Clonorchis sinensis 158, *159*
clostridial myonecrosis 217–18
Clostridium
    C. difficile 62, 206–8
    C. perfringens 100
    C. septicum 217, 218
    infections *see* anaerobic bacteria
cMRSA *see* community methicillin-resistant *Staphylococcus aureus*
CMV *see* cytomegalovirus
co-amoxyclav *see* amoxycillin-clavulanate
Cole, W.R. 112
colitis, antibiotic-associated 206–8
combination therapy 64–5, 66
community methicillin-resistant *Staphylococcus aureus* (cMRSA) 42, 43, 48
comorbidity 113
Conidiobolus 145
conjugation 19–20, 33
contamination *see* microbial contamination
contiguous osteomyelitis 230–31
continuous ambulatory peritoneal dialysis (CAPD) 189
contraception
    IUD 209
    oral 58, 105
cotrimoxazole 105–6
Cruse, P. 5–6
cryptococcosis 137, 145–6, 244
Cryptococcus neoformans 145–6, 244
CSA *see* cyclosporin
CSF *see* cerebrospinal fluid
Cullen 211
Culver, D.H. 6, 113
cutaneous zygomycosis 144
cyclosporin (CSA) 240
cytochrome P_{450} 124, **125**, 135–6, *135*
cytomegalovirus (CMV) 150, 151, 155, 242–3

Dennis, F. 25
dermatological surgery 117
diabetic foot infections
    general points 218–19
    microbiology *220*
    treatment strategies 219–21
dialysis patients, dosing schedules 52–3, 54–5, 92–3

diarrhoea, antibiotic associated 62
digestive tract, selective decontamination 122–3
dilution techniques 160–61
disc tests 160
distribution volume 51–3
DNA
   gyrase 38
   synthesis inhibition 31
   transfer 18–19, 33–4
Domagk, G. 25
dosing schedules 51–3, *52–3*
drainage 13
dressings 7
drug addiction 231
drug interactions *59–61*
   aminoglycosides 50–51, 88–9
   azoles 135–6, *135*
   gastrointestinal adsorption 58
   hepatic metabolism 58–9
   penicillin 50–51, 88–9
   synergistic 65, 88–9
drug modification 34–6, *35*
duration of surgery 112, 113, 122

E-test 161, **162**
ears
   ototoxicity 63, 91
   surgery 121
Echinococcus 156, 158, *159*
echo-guided percutaneous cyst puncture 157–8
Edwin Smith Surgical Papyrus 3, 169
Ehrlich, P. 25
elective procedures 111, 116, 122
endogenous sources of infection 7–8
Entamoeba histolytica 158, *159*
Enterobacteriaceae
   antimicrobial resistance 35–6, *36*, 39–40, *41*
   Escherichia coli 8, 197
Enterobacter 40
   E. cancerogenus 40
enterococci
   antimicrobial resistance 44
   combined therapy 65
   intra-abdominal infections 201
   regimens for 118
   VRE 37–8, *41*, 43–5, 48
Enterococcus faecalis 44
Enterococcus faecium 44
entomophthoramycosis 145
epidural abscess
   intracranial 172
   spinal 172–3
episiotomy 209

Epstein-Barr virus 243
ergosterol 124
erythromycin 106
ESBLs *see* extended-spectrum β-lactamases
Escherichia coli 8, 197
excretion, antibiotics 54
exogenous sources of infection 7
Exophiala 145
extended-spectrum β-lactamases (ESBLs)
   antimicrobial resistance 35–6, *36*, 39–40, *41*, 48
   sensitivity tests 163
   treatment options *41*

F+/F− cells 19
F factor 19
facial infections 175–6
facultative anaerobes 111, 112
   see also staphylococci; streptococci
   bite wounds 223
   diabetic foot infection 220
   growth kinetics 18
   intra-abdominal infections 190, 191, *191*, *192*, 196
   necrotizing fasciitis 212, *213*
   Pasteurella 222
   Pseudomonas aeruginosa 7, 38, 74, 218, 221
famciclovir 149
Fasciola hepatica 158, *159*
Fedden 211
female genital tract, intra-abdominal infections 191, *193*, 209–10, *210*
fertility (F) factor 19
fevers 64
Finegold, S. 9
Fleming, A. 25, 26
Fling 3
Florey, H. 25, 26
flucloxacillin 63, 71, 72, 86, 120–21
fluconazole *132*, 133–4, *135*, 141–2, 146–7, *146*
flucytosine *126*, 136, 146, *146*
flukes 158
5-fluorocytosine *see* flucytosine
fluoroquinolones 38, 44, 94–8, *95–6*
Foord, R. 5–6
foot infections, diabetic 218–221, *220*
Foothills Hospital project 5, 112
foscarnet 150–51, *150*
Fournier's gangrene 211, 216
Frazier, E.H. 212
fungal infections *see* antifungal agents; mycoses
furazolium hydrochloride 110
Fusarium 145

fusidic acid 53, 57, 65, *103*, 104

Galen 11
ganciclovir 150, *150*, 151
gangrene *see* necrotizing fasciitis
gastrointestinal
    adsorption 58
    side-effects 62
    surgery 118
    zygomycosis 143–4
gatifloxacin 97
gene transfer
    acquired resistance 33–4
    conjugation 19–20, 33
    mutation 18
    transduction 18–19, 33
    transformation 18, 33
genitalia *see* female genital tract
gentamicin 44, *86*, 88–9, 92, 164–5
Gilbert, D.N. 92
glycopeptide-resistant enterococci 37–8, 41, 43–5, 48
glycopeptides 101–4, *103*
    bacterial resistance 37–8, 44–5, 101
    general properties 101, *103*
    mechanisms of action 29
    teicoplanin 37–8, 44–5, 101, 102–4, *103*
    vancomycin 59, *61*, 101–2, *103*
Gram-negative bacteria 15–16, **17**
    β-lactams 70, 81
    antibiotic considerations 69
    antimicrobial resistance 31–2, 34–5, 38, 48, 70
    arthroplasty 120
    Bacteroides 41, 45–6, 190, *191–5*, 193, 196, 197, 210, 213, 215, 220, 223
    brain abscess 169–70, *170*, 171, 172
    cell wall 28–9
    cephalosporins 83
    Enterobacteriaceae 8, 35–6, *36*, 39–40, *41*, 197
    epidural abscess 172–3
    ESBL production 163
    intra-abdominal infections *191*, 201
    monobactams 87
    necrotizing fasciitis 212, *213*, 215
    penicillin 71, 74
    Pseudomonas aeruginosa 7, 38, 74, 218, 221
    pulmonary cavity abscess 183–5, *184*
    salmonellae 231, 233
    vascular graft infection 179–80
Gram-positive bacteria 15–16, **17**
    see also staphylococci; streptococci; enterococci

β-lactams 70, 77
    antibacterial resistance 48
    antibiotic considerations 69
    cell wall 28
    intra-abdominal infections *191*
    necrotizing fasciitis *213*
    resistance 70
Gregory, J.E. 142
grepafloxacin 95–6, 97, 98
Groves, A. 3
growth kinetics, bacterial 16–18
growth promoters 38, 46
gynaecological infections 209–10

haematogenous osteomyelitis 225–30, **228**, *230*
hair removal 6
Haley, R.W. 6, 112
half-life 53–4
handwashing 3, 6
HBV *see* hepatitis B virus
HCV *see* hepatitis C virus
head infections
    facial 175–6
    oral cavity 176–8, *177*
    sinusitis 175, *176*
HEPA (high-efficiency particulate air) filtration 138
hepatic metabolism, drug interactions 58–9
hepatitis, antimicrobial side-effects 63
hepatitis B virus (HBV) 243, 244
    treatment 155–6
hepatitis C virus (HCV) 243
    treatment 155–6
hepatitis G virus (HGC) 243, 244
herpes
    antiviral agents 148–51, *150*
    CMV 150, 151, 155, 242–3
    HIV infected patients 155
herpes simplex 243
herpes zoster 243
HGV *see* hepatitis G virus
high-efficiency particulate air (HEPA) filtration 138
HIV antiviral agents 151–5, *153*
    opportunistic infections 155
    protease inhibitors 148, 152–4
    reverse transcriptase inhibitors 151–2
    side-effects *153*, 154
    treatment strategies 152–5
HIV-1 151, 152
HIV-2 151
host defences 238–40, *239*
humoural defences 238
hydatid disease 156–8, *157*, 159
hygiene 3–4, 6

hypersensitivity 61, 74, 84
hysterectomy 210

IASC *see* intramedullary abscess of the spinal cord
idiopathic gangrene of the scrotum *see* Fournier's gangrene
imidazoles *126*, 131–3, *132*
imipenem *53*, *54*, *56*, 84–5, *86*
immunocompromised patient infections 123, 238–45
  bacterial 240–42
  fungal 244
  host barriers 238–40, *239*
  opportunistic 155, 217, 238
  toxoplasmosis 245
  Vibrio infection 216
  viral 240, *241*, 242–4
indigenous microbiota 20–22, *21–2*, 62
indinavir 152, *153*, 154–5
infection control committees 47
infection risk factors 112–14
inflammation 11
integrons 33, 34
intensive care units 10–11
interactions, drugs *see* drug interactions
intra-abdominal infections 186–210, *204–5*
  antibiotic-associated colitis 206–8
  biliary tract infection 203–6
  gynaecological infections 209–10
  liver abscess 203–6
  peritonitis 186–203
  renal abscess 208–9
  therapeutic considerations 193–203, *194–5*, *202*, *204–5*
intracellular drug accumulation 38
intramedullary abscess of the spinal cord (IASC) 173–4
intrauterine contraceptive device (IUD) 209, *210*
intravenous administration 50–51, *51*
intravenous drug addiction 231
invasive aspergillosis 137–9, *137*, 244
investigational antibacterial agents 88, 107–8
investigational procedures, invasive 115
iodine PVP complex 110
itraconazole *132*, 134, *135*, 138
IUD *see* intrauterine contraceptive device

Jevons 36
joint replacement therapy 233–6, *235*
Jones, J. 211
jumping genes 33

Kaposi's lesions 155

ketoconazole 132–3, *132*, *135*
ketolide antimicrobials 106
kidneys
  abscess 208–9
  antibiotic side-effects 63, 90–91
  failure 52–3, 54–5, 85
Kirschner, M. 187

laboratory aspects of antimicrobial use 160–65
  antimicrobial sensitivity tests 160–64, **162, 163**
  patient serum assays 160, 164–5
laboratory reports, interpretation 161–4
lactation 56–7
large bowel surgery 118
latent interval 7
Lesher 94
levofloxacin 95–6, 97
Lister, J. 4, 111
liver
  abscess 203–6
  antibiotic side-effects 58–9, 63
local administration 116–17
locomotor infections
  osteomyelitis 225–33
  prosthetic 233–6, *235*
  septic arthritis 236, *237*
loracarbef 83
low-affinity penicillin-binding proteins 37
lung conditions
  pulmonary aspergillosis 137–9
  pulmonary cavity abscess 183–5, *184*
  pulmonary zygomycosis 143
  transplant patients 242
lysogenic bacteria 18

MacEwen 169
macrolides 31, *57*, 106
Mader 229
mafenide cream 109
Mayo Clinic 8–9
*mec*A gene 37, 163
Medical Research Council, UK 236
Meleney 186, 211
meropenem *53*, *56*, 85
metabolic bypass 38–9
methicillin-resistant *Staphylococcus aureus* (MRSA) 8
  antibacterial resistance 36–7, *41*, *42*, *43*, 46–8
  control guidelines 46–8, *47*
  sensitivity tests 162–3
methicillin-susceptible *Staphylococcus aureus* (MSSA) 42, 46
metronidazole 9, 46, *53*, *57*, 65, 98–9, *195*

MIC *see* minimum inhibitory concentration
microbial contamination 5–6, 111–13, 114
microbiology *see* biology of bacteria
minimum inhibitory concentration (MIC) 161
MODS *see* multiple organ dysfunction syndrome
MOF *see* multiple organ failure
monobactams 87
moxalactam 83
moxifloxacin 95, 97
MRSA *see* methicillin-resistant *Staphylococcus aureus*
MSSA *see* methicillin-susceptible *Staphylococcus aureus*
mucocutaneous defences 238, 239
mucormycosis *see* zygomycosis
multiple organ dysfunction syndrome (MODS) 12
multiple organ failure (MOF) 12
mupirocin 108, 110
mupirocin-resistant *Staphylococcus aureus* 43
mutation
   AI555G 91
   spontaneous 18, 32–3, 35
mycobacteria 16
Mycobacterium tuberculosis 189, 233
mycoplasms 14, *14*
mycoses 136–46
   see also antifungal agents
   black moulds 145
   brain abscess 170–71, *170*
   cryptococcosis 145–6, 244
   entomophthoramycosis 145
   Fusarium 145
   immunocompromised patients 240, 244
   invasive aspergillosis 137–9, *137*, 244
   necrotizing fasciitis 217
   osteomyelitis 233
   Scedosporium 145
   systemic candidosis *137*, 139–42
   zygomycosis *137*, 142–4
myositis 215–16, 217

nail puncture wounds 221
nalidixic acid 94
National Research Council *ad hoc* Committee on Trauma 5, 112
NCCLS guidelines, *Candida* 142
neck infections 175–8
necrotizing fasciitis 209, 211–17
   clinical features *214*
   microbiology 212–13, *213*, 214
   mortality rates 214
   treatment strategies 213–16, *215*
   variants 216–17
necrotizing myositis 215–16, 217
necrotizing pneumonia 183–4
necrotizing soft-tissue infections 211–18
   clostridial myonecrosis 217–18
   necrotizing fasciitis 211–17
   necrotizing myositis 215–16, 217
   non-clostridal anaerobic cellulitis 218
needle-stick injury
   hepatitis B virus 156
   HIV 154–5
nelfinavir *153*, 154
neonates, *Clostridium difficile* infection 207
neural side-effects 63
neurocysticercosis *see Taenia* infection
neurosurgical infections 169–74
   brain abscess 169–72, *170*
   epidural abscess 172–3
   IASC 173–4
   spinal infusion 174
neurotrophic ulcer 219
nitroimidazoles 31, *195*
   metronidazole 9, 46, *53*, *57*, 65, 98–9, *195*
NNRTIs *see* non-nucleoside reverse transcriptase inhibitors
non-clostridal anaerobic cellulitis 218
non-nucleoside reverse transcriptase inhibitors (NNRTIs) 152, *153*, 154
nose surgery 121
nucleic acids
   see also DNA
   antifungal agents 125, **125**
   functional inhibition 31
   synthesis inhibition 31
nucleoside analogues 148, 151–4, *153*

obligate anaerobes 111, 112
   antibiotic considerations 69
   Bacteroides see anaerobic bacteria
   bite wounds 222, *223*
   brain abscess 170, *170*
   Clostridium see anaerobic bacteria
   diabetic foot infection *220*
   epidural abscess 172–3
   growth kinetics 16–18
   intra-abdominal infections 191, *191*, *193*, 210
   necrotizing fasciitis 212, *213*, 214, *215*
   Peptostreptococcus see anaerobic bacteria
   Porphyromonas see anaerobic bacteria
   Prevotella see anaerobic bacteria
   pulmonary cavity abscess 183–4, *184*
   significance 9, 20
Ochroconis gallopavum 145

odontogenic infections 176–8, *177*
open-fractures 120–21
opportunistic infections
    definition 238
    HIV 155
    necrotizing fasciitis 217
oral cavity infections 176–8, *177*
oral contraceptives, antibiotic interactions 58, 105
oral therapy 50, 116, 122
organ transplants *see* transplant patient infections
orthopaedic infections 116, 120
    osteomyelitis 225–33
    prosthetic 233–6, *235*
    septic arthritis 236, *237*
osteomyelitis 225–33
    contiguous 230–31
    fungal 233
    haematogenous 225–30, **228**, *230*
    intravenous drug addiction 231
    microbiology 226, *227*
    mycobacterial 233
    post-traumatic 230–31, **232**
    salmonellae 231, 233
ototoxicity 63, 91
overuse of antibacterials 27, 39
oxazolidinones 107–8

PAE *see* post-antibiotic effect
parasites *see* antiparasitic agents
parenteral administration 50–51, 117, 122
Pasteurella 222
patient factors, surgical infection 5
PBP2a *see* penicillin-binding protein 2a
PBPs *see* penicillin-binding proteins
pelvic cellulitis 210
pelvic surgery 118, 120
penciclovir 149
penicillin-binding protein 2a (PBP2a) 37
penicillin-binding proteins (PBPs) 29, 37
penicillinases *see* β-lactamases
penicillins 70–7, 72–3, **75**
    aminoglycoside synergism 88–9
    aminopenicillins 71, 72–3, 74
    amoxycillin 71, 72–3, 74, 87–8
    ampicillin 44, 71, 72–3, 74, 88
    antibacterial spectrum 76
    antipseudomonal 74
    antistaphylococcal 71
    bacterial resistance 10, 34, **35**, 37, 44, 45
    benzylpenicillin 26, 70–71, 72
    broad-spectrum 71, 74
    combined therapy 65
    development 75
    drug interactions 50–51, 88–9
    extended-spectrum 74
    flucloxacillin 63, 71, 72, 120–21
    history 25–6
    natural 70–71
    penicillin V 71, 72
    penicillin G 26, 70–71, 72
    penicillinase-stable 34, 71
    piperacillin 72–3, 74, 88
    pregnancy 56, *56*
    renal failure 52
    semi-synthetic 68
    side-effects 61, 63, 64, 74, 77
    ticarcillin 72–3, 74, 88
penicilloyl derivative 74
peptidoglycan 15, 28–9, **30**
Peptostreptococcus see anaerobic bacteria
perineal phlegmon *see* Fournier's gangrene
periplasmic space 16
peritonitis 186–203
    abdominal trauma 187, **188**, 189
    antimicrobial recommendations 201–3, *202*
    clinical results 197–200
    drug delivery 200
    general points 186–9
    microbiology 189–3, *191*, *192*, *193*
    model basis for therapy 197
    mortality rates 187, **188**, 200
    primary 186, *187*, 189
    secondary 186, *187*, 189, *204*
    spontaneous 186, *187*, 189
    tertiary 186, *187*
    therapy 193–203, *194–5*, *202*
Pfanner 211
pharmacokinetic factors 49–58
    administration routes 49–51, *50*, *51*, 115–17, 122
    breakdown 54
    cerebrospinal fluid penetration 54
    cidal vs. static action 57–8
    distribution 51–3
    dosing schedules 51–3, *52–3*
    excretion 54
    half-life 53–4
    lactation 56–7
    pregnancy 55, 56–7, *56–7*
    protein binding 51–4
    renal failure 52–3, 54–5
phenoxymethylpenicillin 71, **72**
piperacillin 72–3, 74
piperacillin-tazobactam 86, 88
plasmids 19–20, 33
pneumococci 37, *41*, 45, 183–4
Pneumocystis carinii 158, *159*
pneumonia, necrotizing 183–4

polyenes
    amphotericin B 126–31, *126–9*, 138, 141, *146*, 147, 244
    fungal resistance *146*, 147
porin channels 38
Porphyromonas see anaerobic bacteria
post-antibiotic effect (PAE) 65–6, 90
post-traumatic osteomyelitis 230–31, **232**
pregnancy, pharmacokinetic factors 55, 56–7, *56–7*, 128
Prevotella see anaerobic bacteria
prokaryotic cells 14–15, *14*, 27
prontosil red 25
prophylaxis *see* antibacterial prophylaxis in surgery
prosthesis 10, 11
    breast implants 144
    joint infections 233–6, *235*
protease inhibitors 148, 152–4, *153*
protein
    binding 51–4
    synthesis inhibition 28, 29–31, 125–6, *125*
Pseudomonas aeruginosa 7, 38, 74, 218, 221
Public Health Laboratory Service, UK 236
pulmonary aspergillosis 137–9
pulmonary cavity abscess 183–5, *184*
pulmonary zygomycosis 143
puncture wounds 221–4, *223*

Q/D *see* quinupristin/dalfopristin
quinolones 31, 94–8
    administration 96
    antibacterial spectrum 94, 95–6, *96*
    bacterial resistance 38, 94
    ciprofloxacin 65, 86, 95–6, 97, *195*
    clinafloxacin 86, 95, 97, *195*
    examples 65, 95–6, 97
    gatifloxacin 97
    general properties 94
    grepafloxacin 95–6, 97, 98
    intra-abdominal infections *195*
    levofloxacin 95–6, 97
    moxifloxacin 95, 97
    nalidixic acid 94
    PAE 65
    pregnancy 57
    renal failure 53
    side-effects 97–8
    sparfloxacin 95–6, 97
    trovafloxacin 95–6, 97, 98
    VRE 44
quinupristin/dalfopristin (Q/D) 65, 107

R determinants *see* resistance determinants

R factors *see* resistance factors
RACS *see* Royal Australasian College of Surgeons
rashes 61, 62
regimens, antibiotic 117–21, *119–20*, 122–3
renal abscess 208–9
renal failure, therapeutic considerations 52–3, 54–5, 85
renal side-effects 63, 90–91
resistance *see* antibacterial resistance; antimicrobial resistance
resistance (R) determinants 19, 20
resistance (R) factors 19
resistant transfer factor (RFT) 19
reverse transcriptase (RT) inhibitors 151–4, 153
RFT *see* resistant transfer factor
rhinocerebral infections 143
Rhizopus microsporus var. *rhizopodiformis* 142–4
Ribaud, P. 138
ribosomes 27, 29–30
rickettsiae 14, *14*
rifampicin 31, *53*, *57*, 65, *103*, 104–5
rifamycins 31, 65, *103*, 104–5
risk factors for surgical infection 5, 112–14
    antimicrobial resistance 10
    aseptic technique 4
    blood supply 6–7
    changing microbial causation 8–9
    endogenous influences 7–8
    exogenous sources 7
    latent interval 7
    microbial contamination 5–6
    patient factors 5
    post-operative dressings 7
    protected microbial environments 9–10
    skin shedding 6
    technology 10–11
    wound closure 9
    wound size 6
ritonavir *153*, 154
roxithromycin 106
Royal Australasian College of Surgeons (RACS), Basic Clinical Sciences Examination ix
RT inhibitors *see* reverse transcriptase inhibitors

salmonellae 231, 233
sanquinavir 152, *153*, 154
Scedosporium 145
Scheld, W.M. 170
scrotal gangrene 216

selective decontamination of the digestive tract (SDD) 122–3
Seley, G. 112
Semmelweiss 3
SENIC (Study on the Efficacy of Nosocomial Infection Control) risk index 6, 112–13
sensitivity tests 160–4, **163**
   automated techniques 161, 163
   break-point techniques 161
   dilution techniques 160–61
   disc tests 160
   E-test 161, **162**
   interpreting reports 161–4
septic arthritis 236, *237*
septicaemia 12
serum assays 160, 164–5
sexually transmitted diseases 120
side-effects
   β-lactams 61, 63
   aminoglycosides 63, 90–93
   anti-herpes agents 150–51
   antifungal agents *127*, 130–31, 132–3, 136
   bone-marrow 64
   cephalosporins 61, 64, 74, 77, 84
   chloramphenicol 100–101
   cotrimoxazole 105–6
   fever 64
   gastrointestinal 62
   hepatitis 63
   HIV antiviral agents *153*, 154
   hypersensitivity 61
   imipenem 85
   indigenous flora changes 62
   meropenem 85
   neural 63
   penicillins 61, 63, 64, 74, 77
   quinolones 97–8
   rashes 61, 62
   renal 63, 90–91
   rifampicin 106
   vancomycin 102
sidofovir 149
silver nitrate 109–10
silver sulphadiazine 110
sinks, contaminated 7
sinusitis 123, 175, *176*
Sir William Dunn School of Pathology, Oxford 25
SIRS *see* systemic inflammatory response syndrome
skin infections 211–24
   bite/puncture wounds 221–4, *223*
   diabetic foot infections 218–221, *220*
   necrotizing 211–18

skin shedding 6
soft-tissue infections 211–24
   bite/puncture wounds 221–4, *223*
   diabetic foot infections 218–221, *220*
   necrotizing 211–18
Solomkin, J.S. 200
sparfloxacin 95–6, 97
spinal cord, abscess 172–4
spinal infusion 174
staged abdominal repair (STAR) 9, 187
staphylococci
   see also Staphylococcus aureus; *Staphylococcus epidermidis*
   antimicrobial resistance 10
   catheter-related infections 181–2
   necrotizing fasciitis *215*
   prosthetic infections 234–5, *235*
   vascular graft infection 179–81
Staphylococcus aureus
   see also methicillin-resistant *Staphylococcus aureus*
   antibiotic resistance 114
   arthroplasty 120
   brain abscess 169–70, *170*, 171, 172
   catheter-related infections 182
   cMRSA 42, 43, 48
   endogenous sources 7–8
   epidural abscess 172–3
   haematogenous osteomyelitis 225–6, 229, *230*
   MSSA 42, 46
   mupirocin-resistant 43
   penicillin 71
   pulmonary cavity abscess 183–5, *184*
   vascular graft infection 179–80
Staphylococcus epidermidis, arthroplasty 120
STAR *see* staged abdominal repair
streptococcal scrotal gangrene *see* Fournier's gangrene
streptococci
   see also Streptococcus pneumoniae
   brain abscess 169–70, *170*, 171
   epidural abscess 172–3
   necrotizing fasciitis *215*
   prontosil red 25
Streptococcus pneumoniae 37, 41
Streptococcus pyogenes 215
streptogramins 65, 107
Study on the Efficacy of Nosocomial Infection Control (SENIC) risk index 6, 112–13
sulphamethoxazole 105–6
sulphamylon cream 109
sulphanilamide 25
sulphonamides 25, 31, 38–9

surgical infection 3–13, 167–245
  see also abscess
  antibacterial prophylaxis 111–23
  cardiovascular 179–82
  chest 183–5, *184*
  definition 3, 4, 113
  head 175–8
  history 3–4
  host response 11–13
  immunocompromised patients 238–45
  intra-abdominal 186–210
  locomotor 225–37
  neck 175–8
  neurosurgical 169–74
  orthopaedic 225–37
  principal factors 4–11, *5*
  skin 211–24
  soft-tissue 211–24
  transplant patients 238, 239–45
Surgical Infectious Disease Society, North America ix
Surgical Wound Infection Task Force 4
synergy 65, 88–9
systemic candidosis *137*, 139–42
systemic inflammatory response syndrome (SIRS) 12

Taenia infection 158, *159*
tapeworms 156–8, *159*
target site alteration 36–8
targeted screening 68
technology 10–11
teicoplanin 53, 57, 101, 102–4, *103*
  bacterial resistance 37–8, 44–5
terbinafine 136
tetracyclines 30, 107
thymidine kinase 149–50
ticarcillin *72–3*, 74
ticarcillin-clavulanate 88
tobramycin 92
tolerance, antibiotic 34
topical antibacterial agents 108–10, *109*
Toxoplasma gondii 158, *159*
toxoplasmosis 245
transduction 18–19, 33
transformation 18, 33
transplant patient infections 123, 238, 239–45
  bacterial 240–42, **241**
  fungal 244, **241**
  host barriers 238, 239–40, *239*

  toxoplasmosis 245
  viral 151, 240, **241**, 242–4
transposons 19, 33
treatment strategies and principles 49–66
  combination therapy 64–5, 66
  drug interactions 58–9, *59–61*
  pharmacokinetic factors 49–58
  post-antibiotic effect 65–6
  side-effects 61–4
triazoles *126*, *132*, 133–4
trimethoprim 39, 105–6
trovafloxacin 96, 97, 98
tuberculosis
  osteomyelitis 233
  peritonitis 189

ulcers, neurotrophic 219
urinary tract infections 121, 242

vaginal flora 191, *193*, 209–10
vancomycin 53, 57, 59, 61, 101–2, *103*
vancomycin-resistant enterococci (VRE) 37–8, *41*, 43–5, 48
vascular graft infection 179–81
ventilator contamination 123
Vibrio infection 216–17
viral infections
  see also antiviral agents
  immunocompromised patients 240, *241*, 242–4
VRE *see* vancomycin-resistant enterococci

Waksman, S. 26
Waldvogel 225
water, contaminated 7
Weinberg 186
Whittman, D.H. 9
WHO *see* World Health Organization
Wilson, S.E. 211
Wispelwey, B. 170
Wittmann, D.H. 187
Working Party Report 1998 43
World Health Organization (WHO) 39
wounds
  classification systems 5–6, 112
  closure 9
  infection 4
  size 6

zidovudine 151–2, *153*, 154–5
zygomycosis *137*, 142–4